Catalysis of Organic Reactions

CHEMICAL INDUSTRIES

A Series of Reference Books and Textbooks

Consulting Editor
HEINZ HEINEMANN
Heinz Heinemann, Inc.,
Berkeley, California

Volume 1: Fluid Catalytic Cracking with Zeolite Catalysts, *Paul B. Venuto and E. Thomas Habib, Jr.*

Volume 2: Ethylene: Keystone to the Petrochemical Industry, *Ludwig Kniel, Olaf Winter, and Karl Stork*

Volume 3: The Chemistry and Technology of Petroleum, *James G. Speight*

Volume 4: The Desulfurization of Heavy Oils and Residua, *James G. Speight*

Volume 5: Catalysis of Organic Reactions, *edited by William R. Moser*

Volume 6: Acetylene-Based Chemicals from Coal and Other Natural Resources, *Robert J. Tedeschi*

Volume 7: Chemically Resistant Masonry, *Walter Lee Sheppard, Jr.*

Volume 8: Compressors and Expanders: Selection and Application for the Process Industry, *Heinz P. Bloch, Joseph A. Cameron, Frank M. Danowski, Jr., Ralph James, Jr., Judson S. Swearingen, and Marilyn E. Weightman*

Volume 9: Metering Pumps: Selection and Application, *James P. Poynton*

Volume 10: Hydrocarbons from Methanol, *Clarence D. Chang*

Volume 11: Foam Flotation: Theory and Applications, *Ann N. Clarke and David J. Wilson*

Volume 12: The Chemistry and Technology of Coal, *James G. Speight*

Volume 13: Pneumatic and Hydraulic Conveying of Solids, *O. A. Williams*

Volume 14: Catalyst Manufacture: Laboratory and Commercial Preparations, *Alvin B. Stiles*

Volume 15: Characterization of Heterogeneous Catalysts, *edited by Francis Delannay*

Volume 16: BASIC Programs for Chemical Engineering Design, *James H. Weber*

Volume 17: Catalyst Poisoning, *L. Louis Hegedus and Robert W. McCabe*

Volume 18: Catalysis of Organic Reactions, *edited by John R. Kosak*

Volume 19: Adsorption Technology: A Step-by-Step Approach to Process Evaluation and Application, *edited by Frank L. Slejko*

Volume 20: Deactivation and Poisoning of Catalysts, *edited by Jacques Oudar and Henry Wise*

Volume 21: Catalysis and Surface Science: Developments in Chemicals from Methanol, Hydrotreating of Hydrocarbons, Catalyst Preparation, Monomers and Polymers, Photocatalysts and Photovoltaics, *edited by Heinz Heinemann and Gabor A. Somorjai*

Volume 22: Catalysis of Organic Reactions, *edited by Robert L. Augustine*

Volume 23: Modern Control Techniques for the Processing Industries, *T. H. Tsai, J. W. Lane, and C. S. Lin*

Volume 24: Temperature-Programmed Reduction for Solid Materials Characterization, *Alan Jones and Brian McNicol*

Volume 25: Catalytic Cracking: Catalysts, Chemistry, and Kinetics, *Bohdan W. Wojciechowski and Avelino Corma*

Volume 26: Chemical Reaction and Reactor Engineering, *edited by J. J. Carberry and A. Varma*

Volume 27: Filtration: Principles and Practices, second edition, *edited by Michael J. Matteson and Clyde Orr*

Volume 28: Corrosion Mechanisms, *edited by Florian Mansfeld*

Volume 29: Catalysis and Surface Properties of Liquid Metals and Alloys, *Yoshisada Ogino*

Volume 30: Catalyst Deactivation, *edited by Eugene E. Petersen and Alexis T. Bell*

Volume 31: Hydrogen Effects in Catalysis: Fundamentals and Practical Applications, *edited by Zoltán Paál and P. G. Menon*

Volume 32: Flow Management for Engineers and Scientists, *Nicholas P. Cheremisinoff and Paul N. Cheremisinoff*

Volume 33: Catalysis of Organic Reactions, *edited by Paul N. Rylander, Harold Greenfield, and Robert L. Augustine*

Volume 34: Powder and Bulk Solids Handling Processes: Instrumentation and Control, *Koichi Iinoya, Hiroaki Masuda, and Kinnosuke Watanabe*

Additional Volumes in Preparation

Reverse Osmosis Technology: Applications for High-Purity-Water Production, *edited by Bipin S. Parekh*

Catalysis of Organic Reactions

edited by

Paul N. Rylander

Special Chemicals Division
Engelhard Corporation
Edison, New Jersey

Harold Greenfield

First Chemical Corp.
Pascagoula, Mississippi

Robert L. Augustine

Department of Chemistry
Seton Hall University
South Orange, New Jersey

MARCEL DEKKER, INC. New York and Basel

Library of Congress Cataloging-in-Publication Data

Catalysis of organic reactions / edited by Paul N. Rylander, Harold Greenfield, Robert L. Augustine.
p. cm. -- (Chemical industries ; v. 33)
"Papers presented at the Eleventh Conference on Catalysis of Organic Reactions in Savannah, Georgia, in April 1986 ... sponsored by the Organic Reactions Catalysis Society"--Pref.
Includes bibliographies.
ISBN 0-8247-7927-4
1. Chemistry, Organic--Synthesis--Congresses. 2. Catalysis--Congresses. I. Rylander, Paul Nels. II. Greenfield, Harold. III. Augustine, Robert L.
IV. Conference on the Catalysis of Organic Reactions (11th : 1986 : Savannah, Ga.) V. Organic Reactions Catalysis Society. VI. Series.
QD262.C354 1988
547.1'395--dc19 87-36855
CIP

MARCEL DEKKER, INC.
270 Madison Avenue, New York, New York 10016

Current printing (last digit):
10 9 8 7 6 5 4 3 2 1

PRINTED IN THE UNITED STATES OF AMERICA

Preface

This volume of *Catalysis of Organic Reactions* is a collection of the papers presented at the Eleventh Conference on Catalysis of Organic Reactions in Savannah, Georgia, in April, 1986. The conference was sponsored by the Organic Reactions Catalysis Society (ORCS), an affiliate of the Catalysis Society of North America.

The major purpose of the ORCS has been to provide a forum for the presentation and discussion of examples of the use of catalysis in organic synthesis. The range of attention has been and continues to be very broad, as exemplified by the papers in this book. It encompasses theoretical and practical aspects of both homogeneous and heterogeneous catalysis in a variety of areas of academic and commercial interest and importance.

An excellent summary of the history of these conferences has been provided by Robert L. Augustine in the preface to *Catalysis of Organic Reactions* (Marcel Dekker, Inc., 1985).

It is a pleasure to acknowledge gratefully the support of the Eleventh Conference by generous contributions from Air Products and Chemicals, Ayerst Laboratories Research Inc., Degussa Corp., E. I. Du Pont deNemours and Company, Engelhard Corp., Johnson Matthey, Merck & Co., Inc., Parr Instrument Co., Procter & Gamble Co., G. D. Searle Research and Development, Tennessee Eastman Co., Uniroyal Chemical Co., and Warner-

Lambert/Parke-Davis. The support given by these companies did much to make the conference a success.

Our thanks go to Jean DeRosa for her skillful efforts in preparing all of the manuscripts for publication.

Paul N. Rylander
Harold Greenfield
Robert L. Augustine

Contents

PREPARATION OF AMINES AND ANILINES

HETEROGENEOUS CATALYSTS

SELECTIVE REACTIONS

SELECTED TOPICS

Contributors

Kuzunari Akiyoshi, Purdue University, West Lafayette, Indiana

Gary F. Allen, Mobay Chemical Corp., New Martinsville, West Virginia

R. L. Augustine, Seton Hall University, South Orange, New Jersey

Mihaly Bartok, Jozsef Attila University, Szeged, Hungary

Apurba Bhattacharya, Merck & Co., Inc., Rahway, New Jersey

James W. Brockington, Air Products and Chemicals, Inc., Allentown, Pennsylvania

Nancy L. Brungard, Engelhard Corp., Menlo Park, Edison, New Jersey

Alan J. Chalk, Givauden Corp., Clifton, New Jersey

Charles G. Coe, Air Products and Chemicals, Inc., Allentown, Pennsylvania

Ulf-H Dolling, Merck & Co., Inc., Rahway, New Jersey

Hugh L. Dryden, Jr., G. D. Searle, Skokie, Illinois

Robert J. Farrauto, Engelhard Corp., Menlo Park, Edison, New Jersey

Anne M. Gaffney, ARCO Chemical Co., Newtown Square, Pennsylvania

Owen J. Goodmonson, G. D. Searle, Skokie, Illinois

Edward J. J. Grabowski, Merck & Co., Inc., Rahway, New Jersey

Harold Greenfield, Uniroyal Chemical Co., Naugatuck, Connecticut

V. F. Grenda, Merck & Co., Inc., Rahway, New Jersey

Roger A. Grey, ARCO Chemical Co., Newtown Square, Pennsylvania

Richard G. Herman, Lehigh University, Bethlehem, Pennsylvania

K. High, Seton Hall University, South Orange, New Jersey

Melvin C. Hobson, Engelhard Corp., Menlo Park, Edison, New Jersey

John T. Hortenstine, Jr., McNeil Pharmaceutical, Spring House, Pennsylvania

Chao-Yang Hsu, Sun Refining and Marketing Co., Marcus Hook, Pennsylvania

David L. Hughes, Merck & Co., Inc., Rahway, New Jersey

Sandor Karady, Merck & Co., Inc., Rahway, New Jersey

Lawrence J. Karas, ARCO Chemical Co., Newtown Square, Pennsylvania

K. P. Kelly, Seton Hall University, South Orange, New Jersey

Kamil Klier, Lehigh University, Bethlehem, Pennsylvania

J. F. Knifton, Texaco Chemical Co., Austin, Texas

T. Konoike, University of Pittsburgh, Pittsburgh, Pennsylvania

John R. Kosak, E. I. Du Pont deNemours and Co., Deepwater, New Jersey

Alan P. Kozikowski, University of Pittsburgh, Pittsburgh, Pennsylvania

J. M. Larkin, Texaco Chemical Co., Austin, Texas

John J. Leonard, ARCO Chemical Co., Newton Square, Pennsylvania

J. J. Lin, Texaco Chemical Co., Austin, Texas

J. E. Lyons, Sun Refining and Marketing Co., Marcus Hook, Pennsylvania

Wilhelm F. Maier, University of California, Berkeley, California

Russell E. Malz, Jr., Uniroyal Chemical Co., Naugatuck, Connecticut

Cynthia A. Maryanoff, McNeil Pharmaceutical, Spring House, Pennsylvania

John E. Mills, McNeil Pharmaceutical, Spring House, Pennsylvania

Arpad Molnar, Jozsef Attila University, Szeged, Hungary

Ei-ichi Negishi, Purdue University, West Lafayette, Indiana

Ferenc Notheisz, Jozsef Attila University, Szeged, Hungary

P. O'Hagan, Seton Hall University, South Orange, New Jersey

Dan Ostgard, University of Southern Illinois, Carbondale, Illinois

William E. Pascoe, Eastman Kodak Company, Rochester, New York

James R. Pilney, G. D. Searle, Skokie, Illinois

A. Ritter, University of Pittsburgh, Pittsburgh, Pennsylvania

Kenneth M. Ryan, Merck & Co., Inc., Rahway, New Jersey

Mike Scaros, G. D. Searle, Skokie, Illinois

Geza Sirokman, Jozsef Attila University, Szeged, Hungary

Gerard V. Smith, University of Southern Illinois, Carbondale, Illinois

John A. Sofranko, ARCO Chemical Co., Newtown Square, Pennsylvania

Robin C. Stanzione, McNeil Pharmaceutical, Spring House, Pennsylvania

George Suld, Sun Refining and Marketing Co., Marcus Hook, Pennsylvania

Hsiang-Ning Sun, ARCO Chemical Co., Newtown Square, Pennsylvania

Tomotsu Takahashi, Purdue University, West Lafayette, Indiana

M. M. Thompson, Seton Hall University, South Orange, New Jersey

Gamini A. Vedage, Air Products and Chemicals, Inc., Allentown, Pennsylvania

R. W. Warner, Seton Hall University, South Orange, New Jersey

Leonard M. Weinstock, Merck & Co., Inc., Rahway, New Jersey

John P. Westrich, G. D. Searle, Skokie, Illinois

Catalysis of Organic Reactions

Homogeneous Catalysis

1

Techniques for Directing the Course of Selective Palladium-Catalyzed Oxidations of Olefins and Aromatics Using Homogeneous and Heterogeneous Catalyst Systems

James E. Lyons, George Suld, and Chao-Yang Hsu

Applied Research and Development Department
Sun Refining and Marketing Company
Marcus Hook, PA

ABSTRACT

Palladium holds an important position in olefin oxidation. Vinylic oxidation products or their stable isomers, including acetaldehyde, acetone, and vinyl acetate, are manufactured by related reactions involving nucleophilic attack on a Pd(II) olefin π-allyl complexes which may undergo nucleophilic attack to give allylic rather than vinylic products. In this way, a family of industrially important α, β-unsaturated alcohols, esters and acids can be produced in catalytic liquid-phase oxidations. Oxidative addition of olefinic substrates to supported palladium catalysts may give surface π-allylic species which undergo nucleophilic attack. A catalytic application of this type is the mild palladium-catalyzed oxidation of propylene to acrylic acid in water. Oxidation of aromatic substrates by both soluble and supported palladium catalysts gives phenols, aryl acetates or products of oxidative dimerization. Electrophilic attack on the ring by Pd(II) is the likely first step in these reactions. Thus, the use of palladium catalysts gives new and potentially useful routes to a wide spectrum of chemical oxygenates from both olefinic and aromatic precursors.

I. INTRODUCTION

One of the most synthetically useful and industrially important classes of catalytic oxidation reactions involves nucleophilic attack on palladium-olefin π-complexes to give vinylic oxidation products and reduced palladium species which can be re-oxidized in situ by a number of oxidants including Cu(II).[1,2] In this manner, vinylic oxidation products or their stable isomers, including acetaldehyde, acetone and vinyl acetate, are manufactured by a series of related catalytic reactions. These reactions occur either in solutions of palladium complexes or over heterogeneous supported palladium catalysts. It is interesting to note that the palladium catalyzed oxidation of ethylene to acetaldehyde was initially carried out over a supported palladium catalyst, but is now practiced in solution, whereas the oxidation of ethylene to vinyl acetate was first practiced in the liquid phase and now industrial production of vinyl acetate employs supported palladium catalysts.[1]

Other synthetically important pathways are available to palladium-olefin complexes which produce not vinylic, but allylic, oxidation products. Palladium π-allyl intermediates can form by oxidative addition to the metal center. Nucleophilic attack at the termini of the π-allyl ligand gives rise to allylic oxidation products and reduced palladium species. As with Wacker chemistry, these reactions can be made to occur catalytically in the presence of palladium either in homogeneous solution or over heterogeneous supported catalysts. The development of this chemistry for the production of a large and important family of allylic oxidation products, including α, β-unsaturated alcohols, esters, ketones, and carboxylic acids is possible; yet until recently, only limited application of these reactions to catalytic syntheses has been made.

Another instance in which palladium catalysts, either supported or in solution, can be used to advantage is in the selective oxidation of the aromatic ring. To date, most of these reactions have required strong oxidants and high oxidation state palladium aryl intermediates have been implicated. The selective air-oxidation of aromatics to phenolic derivatives is possible, however, and could be of considerable utility.

In this paper, the role of supported palladium catalysts in the allylic oxidation of olefins will be discussed and those factors which favor allylic oxidation over vinylic oxidation will be considered. Relationships between homogeneous and heterogeneous catalysts will be developed and analogies between solution and surface species will be drawn. Evidence will be presented which is consistent with the formation of π-allyl intermediates in both homogeneous and heterogeneous catalytic reactions. In addition, the results of oxidations of aromatic rings using molecular oxygen in the presence of palladium catalysts will be presented. The data suggest the involvement of high oxidation state palladium aryl intermediates. Methods for the control of reaction selectivity will be given.

II. RESULTS AND DISCUSSION

Oxidation of Olefins. There are several pathways available for the oxidation of α-olefins catalyzed by Pd(II) in solution, (eq. 1). The pathway which

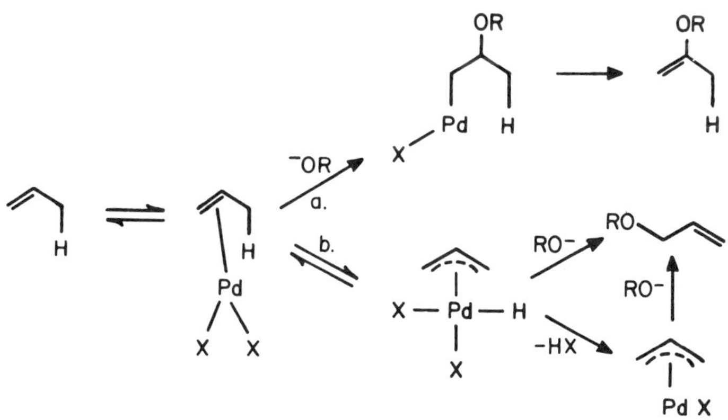

(1)

predominates can be controlled by the nature of the Pd(II) complex which is used as the catalyst. For example, palladium acetate or palladium chloride can catalyze predominant vinylic oxidation, (eq. 1a) whereas palladium trifluoroacetate of π-allyl palladium trifluoroacetate gives predominant allylic oxidation (eq. 1b). It has been suggested that Pd(II) complexes with strongly electron withdrawing ligands are highly electrophilic and oxidatively add an allylic C-H bond, resulting in Pd(IV) π-allyl intermediates.[3]

A catalytic application of the use of palladium trifluoroacetate for the formation of an allylic oxidation product is the high yield production of allyl acetate from propylene (eq. 2).[4]

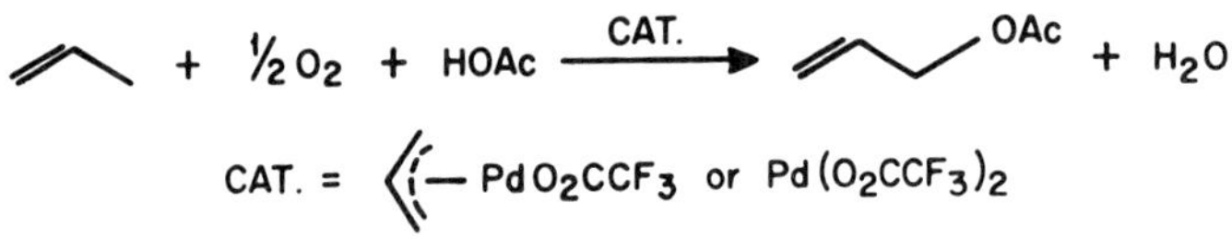

(2)

There appears to be another means of encouraging allylic over vinylic oxidation of α-olefins. This involves the use of relatively weak inorganic bases to assist in proton removal. Thus the addition of sodium acetate, carbonate or bicarbonate promoted formation of palladium(II) π-allyl complexes from either $PdCl_2$ and propylene or di-μ-chloro-dichlorobispropenedipalladium.[5] It is interesting that the products of the stoichiometric oxidation of propylene by palladium acetate shift from predominantly vinylic to predominantly allylic as sodium acetate is progressively added to the solution (Table I).[6] Furthermore, it has been found that addition of sodium acetate to solutions of palladium(II) gives fair yields of allyl acetate during the catalytic oxidation of propylene in acetic acid.[7,8] Low oxidation

TABLE I

The Effect of [NaOAc] on Propylene Oxidation Products at 25°C[a,b]

NaOAc M	3-Acetoxy-Propene, %	2-Acetoxy-Propene, %	Cis- + Trans-1-Acetoxypropene, %
Nil	2	96	2
0.098	13	86.5	0.5
0.49	71	26	3
0.90	94	6	<0.5

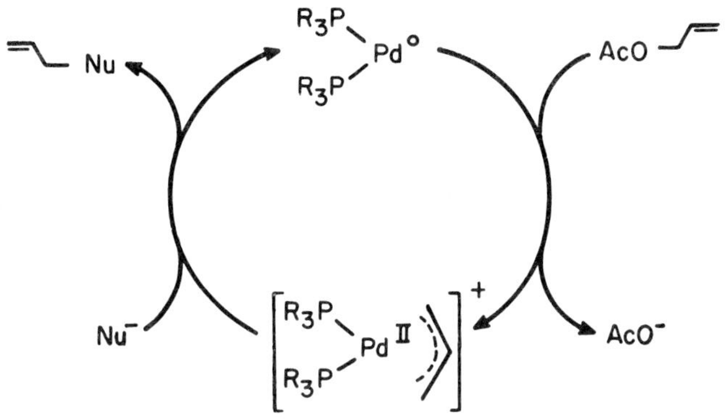

Scheme 1. Catalytic Cycle for Palladium Catalyzed Allylation

state palladium has been implicated in these reactions[8] and heterogeneous catalysts such as Pd/SiO_2 treated with sodium or potassium acetate are very effective for catalyzing the oxidation of propylene to allyl acetate at elevated temperature in the vapor phase.[9]

Finally, another route to catalytically active π-allyl intermediates proceeds via oxidative addition to a coordinatively unsaturated palladium(O) center. This usually requires a good leaving group in the allylic position. A cationic π-allyl complex can be isolated and shown to be a reaction intermediate (Scheme 1, eq. 3).[10] When the leaving group is H, as it is in an α-olefin, the reaction is thermodynamically disfavored unless the hydrogen which would be produced can be oxidatively converted to water.

 (3)

Heterogeneous catalysts are known to activate the C-H bond of α-olefins and hence, it became of interest to determine whether an efficient catalytic olefin oxidation could be accomplished in a manner analogous to that shown in Scheme 1 over a supported palladium(O) center (Scheme 2).

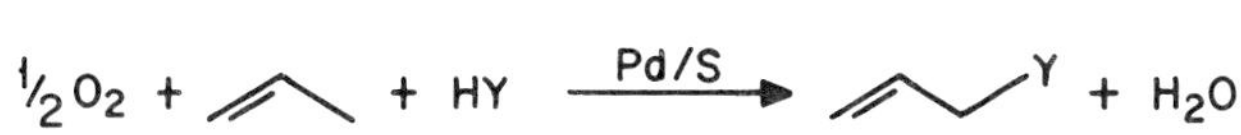 (4)

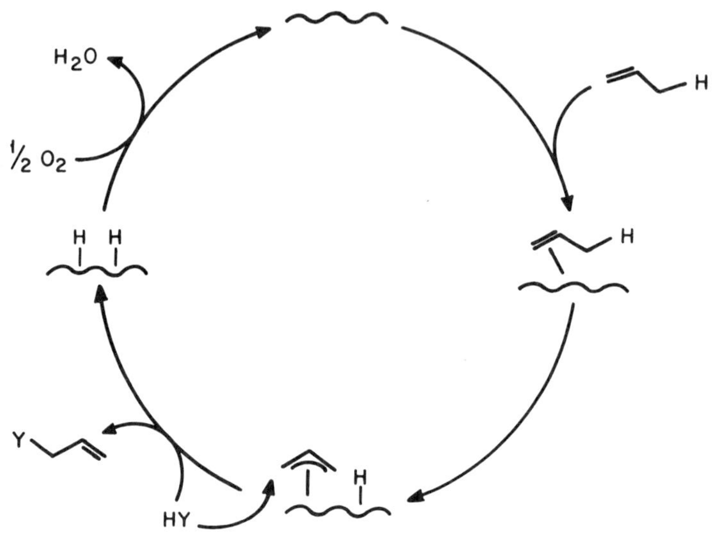

Scheme 2. Oxidative Addition of Propylene to a Pd Surface

A route of this kind was proposed (HY=HOH) for the conversion of propylene to acrylic acid via allyl alcohol.[11-14] Low conversions (turnover numbers <1) were observed under mild conditions[11] and poor selectivities ($<68\%$, T.O.N. ~ 10) were observed under forcing conditions.[13]

Propylene Oxidations Over Supported Palladium Catalysts. We have examined Reaction 4 (HY=HOAc, HOH) and find that it occurs over heterogeneous palladium catalysts under mild conditions in the liquid phase to give high yields of allylic oxidation products. An efficient catalyst for these reactions is 10% palladium on carbon (Table II). When HY is acetic acid, allyl acetate is the major product (Table III) and when HY is water, acrylic acid is produced in high yields (Table IV). Presumably, in the latter case, allyl alcohol is the initial product which is rapidly converted to acrylic acid on the surface of the catalyst (eq. 5)

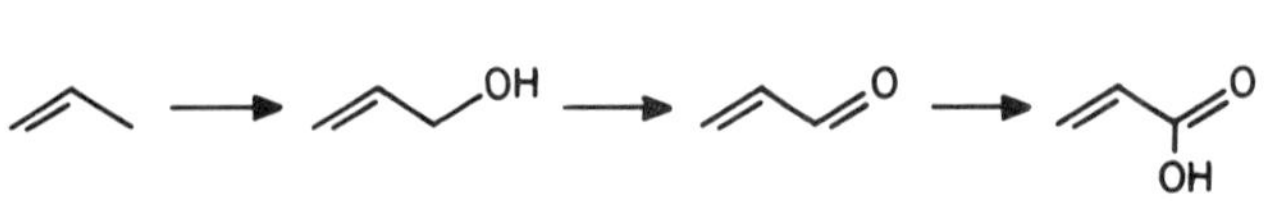

(5)

TABLE II

Propylene Oxidations Over Supported Palladium Catalysts in Water[a]

Catalyst	Gas Uptake Mol/g. atom Pd/hr	Sel to Acrylic Acid Wt. % in Liquid
1% Pd-C	12.4	31
5% Pd-C	10.8	72
10% Pd-C	10.1	88
20% Pd-C	7.5	91
5% Pd-Al_2O_3	4.2	78
5% Pd-SiO_2	0.8	--

<u>Activation of the Catalyst.</u> These reactions have several interesting characteristics which must be taken into account if high rates and selectivities are to be achieved. First, it is very beneficial to pre-treat the catalyst with propylene in the absence of oxygen at a temperature of at least about 65°C prior to conducting the oxidation. If this is done, higher rates and selectivities are observed and the catalyst is made active for low temperature oxidations. Catalysts which were inactive at 40°C or 50°C become active for allylic oxidations at these temperatures (Table IV).

During the period when the catalyst is treated with propylene, some allylic oxidation product is formed, but a considerable amount of vinylic oxidation product is also formed. For example, over 70% of the by-product acetone (0.67 moles per g. atom Pd present) produced in the oxidation of propylene in water over 10% Pd-C arose during the propylene pre-treatment period (Table IV). Very little acetone was formed during the rest of the run.

A possible explanation for these observations is that during the propylene pre-treatment period, surface oxygen (peroxide or oxide) is removed and oxidized palladium sites are reduced to Pd(O) (eq. 6). The low

(6)

TABLE III

Palladium Catalyzed Oxidation of Propylene to Allyl Acetate[a]

Catalyst[b)]	NaOAc moles/l.	Activation Time, Min.	Gas Uptake Mmoles	Wt. % in Product: Allyl Acetate	Wt. % in Product: 1-Propenyl Acetate	Selectivity to Allyl Acetate, %
$Pd(OAc)_2/CuCl_2$	0.167	0	47	>0.4	0.75	<8.0
$Pd(OAc)_2/CuCl_2$	0.667	0	65	2.33	3.36	9.1
10% Pd-C	0.0	0	15	1.69	1.70	29.9
10% Pd-C	0.167	0	20	1.89	1.75	39.1
10% Pd-C	0.333	0	38	5.79	1.65	69.7
10% Pd-C	0.0	15	32	4.94	0.83	85.6
10% Pd-C	0.167	30	41	16.04	1.71	90.6
10% Pd-C	0.333	30	50	17.57	0.98	94.7
10% Pd-C	0.500	30	49	17.40	0.38	97.9
10% Pd-C	0.667	30	53	18.57	TR	>99.0

TABLE IV

Effects of Additives and Activation Temperature on Rate and Selectivity of Palladium-Catalyzed Oxidations of Propylene in Water[a]

Activation T, C	Reaction T, C	Additive mmoles	Reaction Products[b], mmol CO_2	Acetaldehyde	Acetic Acid	Acetone	Acrolein	Acrylic Acid	Turnover No. mmoles Acrylic/g.atom Pd/hr
80	65	--	2.3	0.5	0.8	0.8	0.2	12.9	3.2
80	50	--	0.7	0.2	0.5	0.8	0.3	7.7	1.9
80	40	--	0.4	0.1	0.4	0.4	0.1	8.3	1.2
40	40	--	--	--	--	--	--	--	--
50	50	--	--	--	0.3	0.3	--	--	--
65	65	--	2.3	0.4	0.7	1.0	0.3	14.4	3.5
65	65	BHT[c]	2.4	0.4	1.0	0.6	0.2	17.3	4.3
--[d]	65	--	2.5	0.4	1.1	1.1	0.3	9.7	2.4
65	65	LiCl[e]	na	--	--	3.4	--	0.1	--
65	--[f]	--	0.25	--	0.3	0.7	0.1	0.9	--

oxidation state metal centers may then oxidatively add the olefin to form surface π-allyl intermediates (eq. 7) which go on to produce allylic oxidation

(7)

products (Scheme 2). Whether or not this is actually occurring, the propylene pre-treatment serves to form a more active catalyst for selectively producing allylic oxidation products.

Several things are particularly detrimental to both the activity and selectivity in allylic oxidations over supported palladium catalysts. The presence of a strong oxidant such as MnO_4^-, $S_2O_8^=$, AgO, Ag + O_2, NO_3^-, etc. slows the rate of allylic oxidation and causes vinylic oxidation to either compete or predominate. Pre-treatment of the palladium catalyst with oxygen in the absence of propylene at elevated temperatures prior to reaction also has a deleterious effect on allylic oxidation. The presence of halide ion is extremely detrimental to the reaction, promoting instead a Wacker reaction which forms acetone to the virtual exclusion of allylic oxidation (Table IV).

These observations suggest that the active catalyst for these allylic oxidations could be low oxidation state palladium, and that conditions or reagents that cause higher oxidation state palladium to be formed on the catalyst surface lead to vinylic oxidations.

Radical Inhibitors. Addition of a free radical inhibitor such as 2,6-di-*tert*-butyl-*p*-cresol, BHT, has been found to have a highly beneficial effect. Higher yields of acrylic acid result from reactions conducted in the presence of BHT (Table IV). This implies that the allylic oxidation is not free radical in nature. However, deleterious free radical reactions, such as further oxidation or polymerization of acrylic acid, might be inhibited.

By properly activating the catalyst in water and conducting the reaction at 65°C in the presence of BHT, it is possible to smoothly oxidize propylene to acrylic acid in greater than 90% yield (eq. 8). Turnover numbers exceed 4 moles of acrylic acid produced per gram atom of

$$CH_2{=}CH{-}CH_3 + \tfrac{3}{2}\,O_2 \xrightarrow[65^{\circ}C,\ 5\,BARS]{10\%\ Pd/C,\ BHT} \underset{>90\%}{CH_2{=}CH{-}CO_2H} \tag{8}$$

palladium per hour under these conditions. As the acrylic acid concentration builds up, however, the reaction rate slows slightly. Nonetheless, it was possible to continuously produce acrylic acid in aqueous solution at an average rate of about 3 moles acrylic acid/gram atom Pd/hour for a period of 50 hours in this manner (Fig. 1). At any point in the operation, it was possible to restore initial activity by filtering the catalyst, rinsing off adsorbed acrylic acid with acetic acid and water and reactivating with propylene at <65°C. No leaching of the catalyst into the aqueous phase was detected and a clear water-white solution was obtained on filtration which exhibited no activity toward oxidation under reaction conditions in the absence of added catalyst. Thus, palladium on carbon is an efficient, long-lived catalyst for the selective generation of acrylic acid from propylene in aqueous solution under very mild conditions.

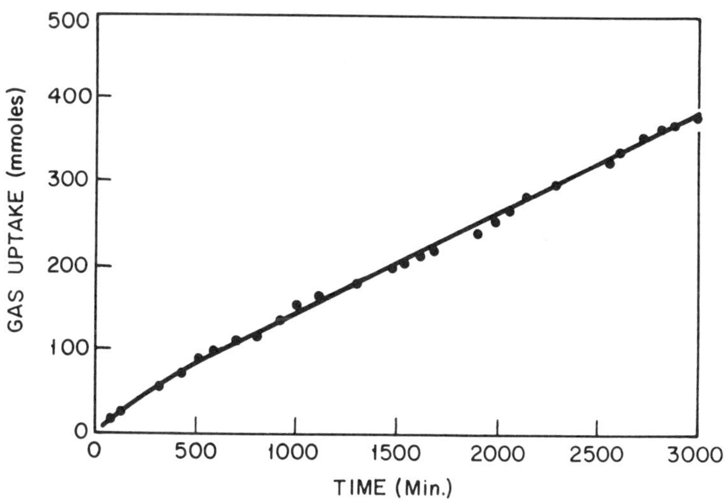

Fig. 1. Oxidation of propylene to acrylic acid. A 60/40 $O_2/C_3^{=}$ gas mixture (4 BARS) reacted in 30 ml of water containing 1 g of 10% Pd/C

Allyl Acetate Formation. Propylene pre-treatment is useful in preparing an active catalyst for allylic oxidation in acetic acid to form allyl acetate as well (Table III). Care to prevent over-oxidation of the catalyst is even more crucial for reactions in acetic acid than for those carried out in water. Addition of acetate ion is beneficial for allylic oxidations over supported palladium catalysts in acetic acid just as was found with soluble catalysts[7,8] and with vapor phase oxidations over Pd/SiO_2.[9] Palladium slowly leaches from the catalyst, imparting low-level catalytic activity to the solution. As we have seen, however, acetate ion will direct even Pd(II)-catalyzed oxidations toward allylic products.[6-8]

Butene Oxidations. Butene-1, *cis*- and *trans*-butene-2 and isobutylene were oxidized over supported palladium catalysts, both in water and in acetic acid. Butenes gave crotonic acid in water but, as had been noted earlier,[11] the α,β-unsaturated ketone and aldehyde were also major products (eq. 9).

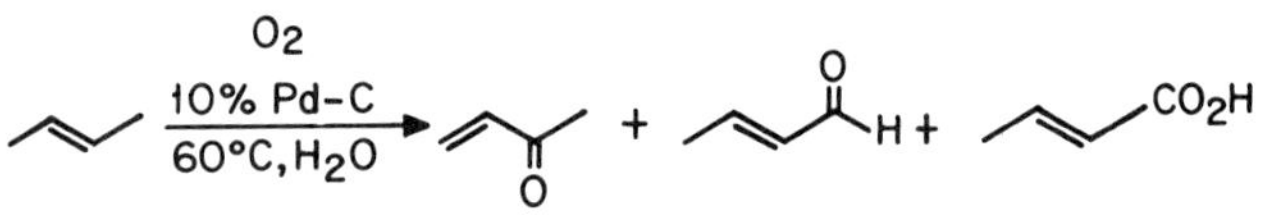

(9)

Isobutylene gave considerably more methacrolein than methacrylic acid (eq. 10). Perhaps the 2-methyl group causes desorption of the intermediate from the surface before it oxidizes further to the acid.

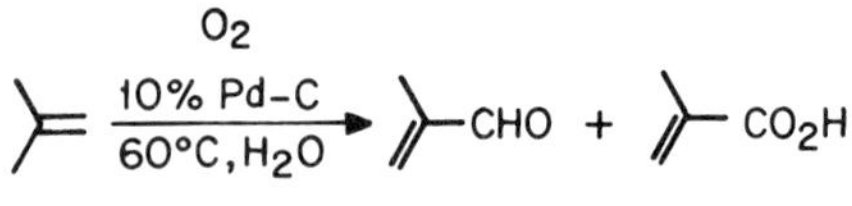

(10)

Oxidation of either *cis*- or *trans*-2-butene over palladium in acetic acid gave three times as much branched product as linear allylic acetate (eq. 11, Table V). Although *cis*-2-butene-1-acetate was initially produced from *cis*-2-butene, and *trans*-2-butene-1-acetate was formed from *trans*-2-butene, geometrical isomerization occurred with time. It was found that supported palladium catalysts catalyze the isomerization of the products

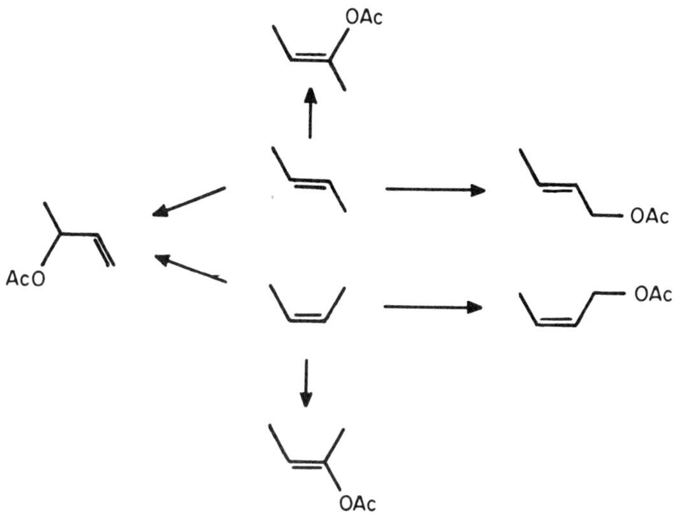

(11)

TABLE V

The Oxidation of Cis- and Trans-2-Butene Over Palladium in Acetic Acid

2-Butene	HOAc, mls	Catalyst	Grams Acetate Produced	Selectivity, % Monoacetates Linear Allylic	Branched Allylic	Vinylic	Diacetates
Cis-	30	10% Pd-C	2.9	15.4	45.1	9.0	30.5
Trans-	30	10% Pd-C	4.0	17.9	43.2	7.2	31.3
Cis-	30	5% Pd/Al_2O_3	2.4	17.0	57.3	5.3	19.2
Trans-	30	5% Pd/Al_2O_3	3.6	15.1	48.5	7.4	26.9

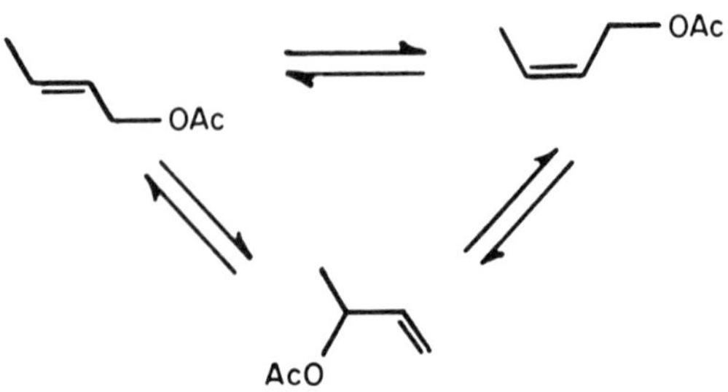

(12)

(eq. 12) preventing highly regio- and stereoselective reactions. The 3/1 ratio of branched to linear acetates formed during butene oxidations over palladium in acetic acid is similar to the product mixture which is obtained by oxidizing palladium π-crotyl complexes with either $Tl(OAc)_3$ or $Pb(OAc)_4$ (eq. 13).[15]

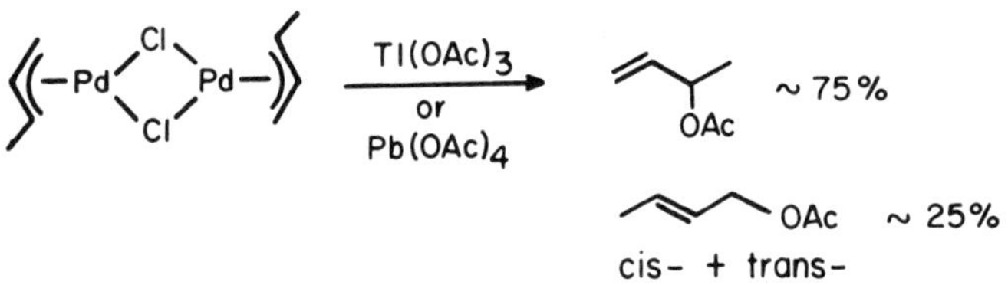

(13)

Oxidation of Aromatics Using Strong Oxidants. Palladium acetate catalyzes the oxidation of aromatics by strong oxidants in acetic acid to give aryl acetates and coupling products under mild conditions (Eq. 14).[16] Unusual

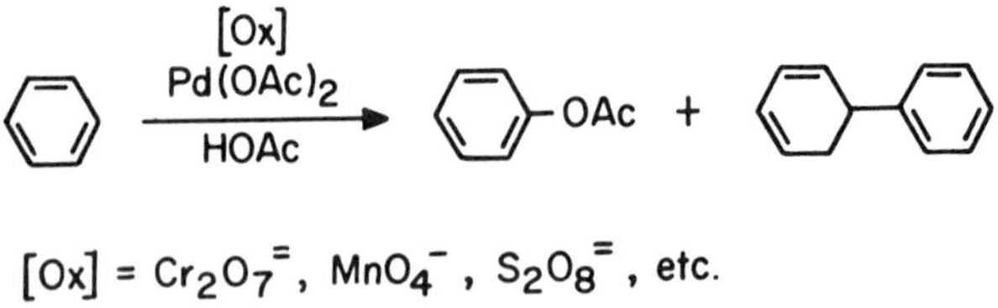

(14)

regioselectivity has been reported for this reaction.[17] It has been suggested that this reaction proceeds vis electrophilic attack by $Pd(OAc)_2$ on the aromatic ring to give a palladium(II) aryl intermediate (Eq. 15) which can

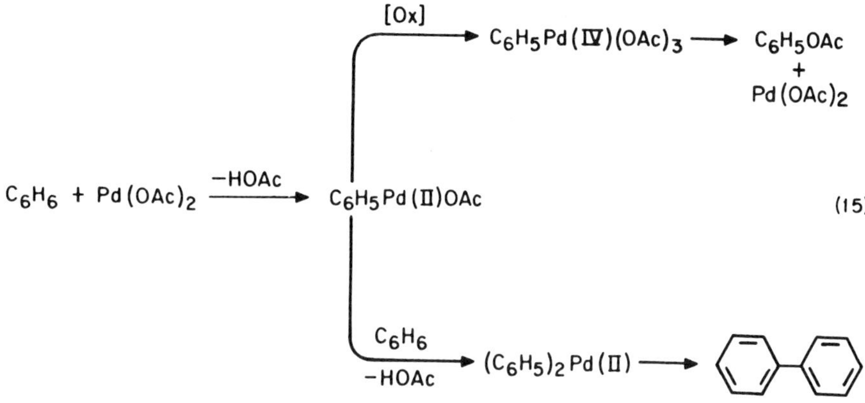

(15)

either oxidize to give phenyl acetate or couple to give biphenyl.[18] This mechanism is supported by the observations that: (a) partial rate factors for oxidation of substituted aromatics are consistent with electrophilic palladation of the ring, (b) isotope effects for forming both phenyl acetate and biphenyl are the same, suggesting a common intermediate, and (c) the phenyl acetate/biphenyl ratio is proportional to the concentration of strong oxidant used.

A number of investigations have shown that catalytic oxidative dimerization can be made to predominate over acetoxylation in solvents other than pure acetic acid.[19] Although a comprehensive study of the factors affecting this reaction has not been done to date, reaction conditions and solvents play a major role.

Aromatic Oxidations Using O_2. It would be of practical synthetic interest to carry out these reactions using air or oxygen as the oxidant, rather than requiring stoichiometric consumption of expensive strong oxidants. However, attempts to use molecular oxygen - even in the presence of Cu(II) in hopes of reoxidizing palladium as in the Wacker reaction - were unsuccessful under mild conditions. Oxygen was successfully used as the oxidant for Reaction 14 when high temperatures were employed.[20] In this case, both $Pd(OAc)_2$ and supported palladium catalysts were effective.

Little work has been done on the acetoxylation of substituted aromatics using molecular oxygen as the oxidant. We have explored the palladium-catalyzed oxidation of phenyl acetate as a route to derivatives of the industrially important dihydroxyaromatics: catechol, resorcinol and hydroquinone. We were interested in finding conditions under which O_2 could be used as the oxidant and in determining those factors which controlled acetoxylation and those which controlled dimerization. Furthermore, we were interested in elucidating and understanding reaction regioselectivity.

Palladium acetate was found to catalyze the oxidation of phenyl acetate at 145°C using 800 psig of an oxygen-containing gas (Table VI, eq. 16). Phenylene diacetates and diacetoxy biphenyls were formed as major

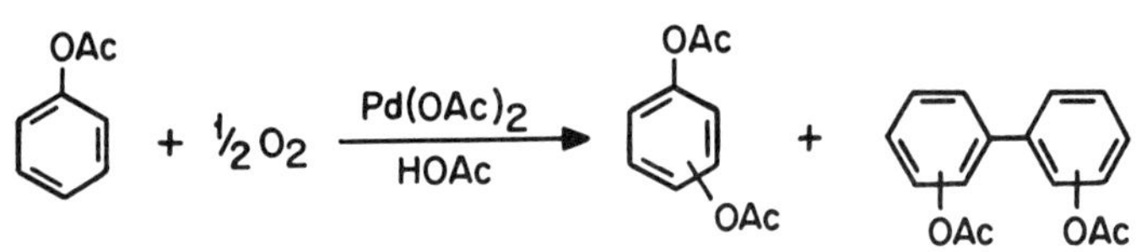

(16)

reaction products. In all cases, the three possible phenylene diacetates were formed. Only two of the six possible diacetoxybiphenyls were produced in significant quantities: o,m'-diacetoxybiphenyl and o,p'-diacetoxybiphenyl. The yields of major reaction products vary considerably with temperature, oxygen partial pressure, phenyl acetate concentration and added acetate.

Effect of Temperature and Pressure. Although $Pd(OAc)_2$ catalyzes the oxidative dimerization of aromatics under mild conditions, acetoxylation of the ring becomes important at higher temperatures (Fig. 2). Acetoxylation selectivity increases with increasing oxygen partial pressure as well (Fig. 3). It is interesting to compare this effect with the increase in ring acetoxylation on increasing oxidant concentration in the $Pd(OAc)_2$ catalyzed oxidation of benzene in acetic acid (Table VII). The fact that higher temperatures and greater oxygen partial pressures favor acetoxylation relative to oxidative dimerization is consistent with the requirement that palladium be in a higher oxidation state for acetoxylation than for oxidative dimerization in cases where molecular oxygen is the oxidant.

TABLE VI

Oxidation of Phenyl Acetate Catalyzed by $Pd(OAc)_2$ in Acetic Acid[a]

% O_2 In Gas	Added Substance, mmoles	Product Analyses, mmoles Produced: $C_6H_4(OAc)_2$ [b] o-	m-	p-	Total	$(AcOC_6H_4)_2$ o-,m′	o-,p′	Total	AcC_6H_4OAc [b] o-	m-	p-	Total
4	--	1.9	2.3	2.5	6.7	37.4	21.2	58.6	0.8	Tr	Tr	0.8
21	--	6.8	7.2	7.8	21.8	33.3	19.2	52.5	0.5	Tr	Tr	0.5
40	--	7.0	9.1	11.7	27.8	33.4	19.0	52.4	0.8	Tr	Tr	0.8
4	KOAc, 15.6	1.9	2.5	3.4	7.8	20.5	10.5	31.0	0.9	0.5	Tr	1.4
4	CsOAc, 15.6	1.9	2.0	3.1	7.0	8.1	4.1	12.2	1.0	Tr	Tr	1.0
40	CsOAc, 15.8	3.4	5.1	6.5	15.0	9.3	4.8	14.1	Tr	1.1	0.9	2.0
4	Ac_2O, 159	4.8	2.6	3.7	11.1	35.1	18.8	53.9	26.5	1.3	0.7	28.5
4	Ac_2O, 53	2.5	2.3	3.2	8.0	37.9	20.6	58.5	6.7	0.7	0.3	7.7

TABLE VII

Product Distributions for the ½Pd(OAc)$_2$¼-Catalyzed Oxidation of Benzene by Dichromate[a] and Phenyl Acetate by Oxygen[b]

Benzene Oxidation			Phenyl Acetate Oxidation		
$Na_2Cr_2O_7$, mmole	% Yield[c)] C_6H_5OAc	$(C_6H_5)_2$	O_2 Partial Pressure, psig	% Yield[c)] $C_6H_4(OAc)_2$	$(C_6H_4OAc)_2$
0	<1	22			
1	99	23	32	223	1953
3	190	33	168	728	1750
7.5	353	19	320	927	1747

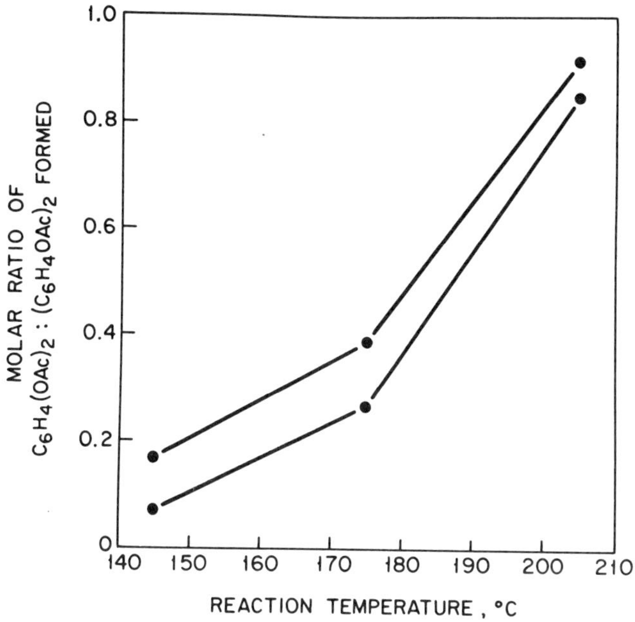

Fig. 2. Effect of reaction temperature on the acetoxylation/dimerization ratio

Effect of Metal Acetates. The presence of alkali metal acetate has little effect on the rate of acetoxylation at low partial pressure of oxygen (Fig. 4), but has a pronounced inhibiting effect on oxidative dimerization. Metal acetate may complex readily with both $[Pd(OAc)_2]$, which is trimeric in acetic acid and [ArPdOAc], reducing their effectiveness in coordinating phenyl acetate (eq. 17, 18, 19).[21] Acetoxylation may be less affected by added acetate than is oxidative dimerization because complexation of

$$2[Pd(OAc)_2]_3 + 6MOAc \rightleftharpoons 3M_2Pd_2(OAc)_6 \quad (17)$$

$$M_2Pd_2(OAc)_6 + 2MOAc \rightleftharpoons 2M_2Pd(OAc)_4 \quad (18)$$

$$(AcOC_6H_4)PdOAc + 2MOAc \rightleftharpoons M_2Pd(OAc)_3(C_6H_4OAc) \quad (19)$$

phenyl acetate with $[M_2Pd(OAc)_3(C_6H_4OAc)]$ may be more difficult than oxidizing $[M_2Pd(OAc)_3(C_6H_4OAc)]$ to a Pd(IV) intermediate (eq. 15). It is

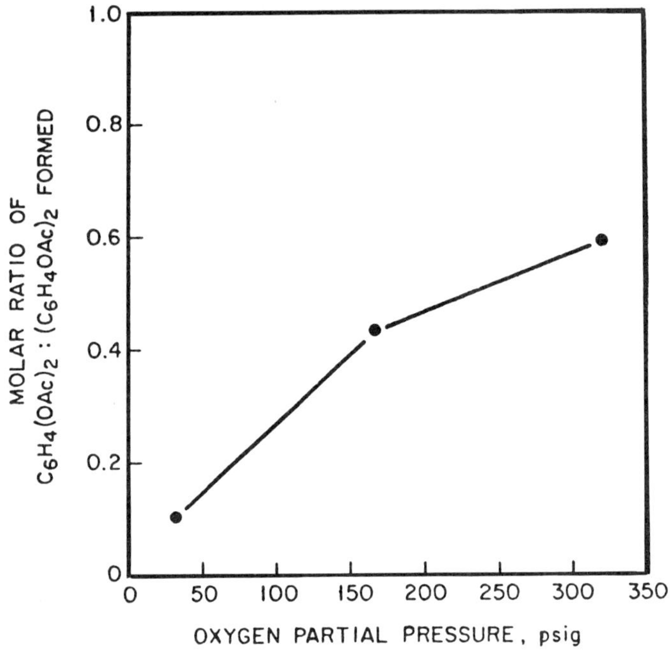

Fig. 3. Effect of oxygen partial pressure on the acetoxylation/dimerization ratio.

well known, for example, that halogens readily oxidize M_2PdX_4 to M_2PdX_6.[22] It is seen (Table VI) in oxidations carried out at high oxygen partial pressure in the presence of a large excess of CsOAc, that although the acetoxylation yield is reduced less than two-fold, the yield of oxidative dimerization product is lowered by a factor of about 4.

Effects of Phenyl Acetate Concentration-Predominant Acetoxylation. As would be expected, decrease in phenyl acetate concentration disfavors oxidative dimerization more than acetoxylation. Fig. 5 shows the effect of lowering the phenyl acetate concentration on the ratio of acetoxylation to dimerization.

In summary, those conditions which favor acetoxylation over oxidative dimerization are high temperature, high oxygen partial pressure, high

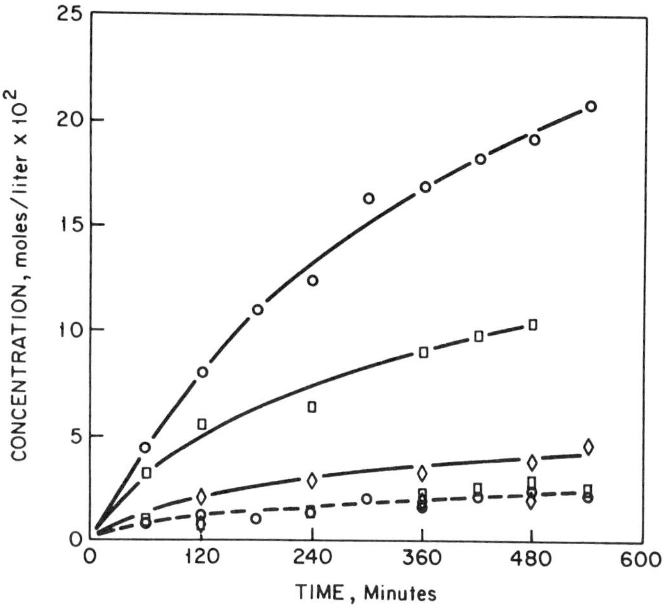

Fig. 4. Effect of alkali metal acetate concentration on acetoxylation rate.

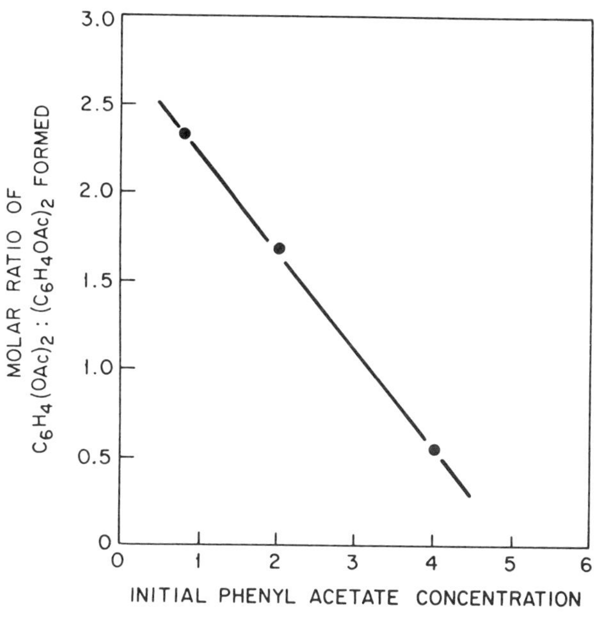

Fig. 5. Effect of phenylacetate concentration on the acetoxylation: dimerization ratio.

acetate ion concentration, and low phenyl acetate concentration. Under these conditions, (175°C, 320 psi O_2, 0.78M PhOAc, excess CsOAc) selective acetoxylation occurs. The molar ratio of the phenylene diacetates to biphenyl diacetates formed under these conditions was 94.5/1. Unfortunately, the low phenyl acetate concentrtion, coupled with high CsOAc concentration, depressed the overall rate to a rather low level.

Regioselectivity and Mechanism. The relative amounts of the three isomeric phenylene diacetates produced during acetoxylation of phenyl acetate catalyzed by $Pd(OAc)_2$ using O_2 as oxidant are in the order *o*-<*m*-<*p*-; however, the differences in yields are slight. This pattern is similar to that found for $Pd(OAc)_2$-catalyzed acetoxylation using strong oxidants which are believed to involve unselective electrophilic palladation of the aromatic ring.[18]

The oxidative dimerization reaction which occurs during $Pd(OAc)_2$-catalyzed oxidation using O_2 is both unique and selective: *o,m'*- and *o,p'*-biphenyl diacetates are formed in approximately a 2:1 ratio. Neither *m,m'*- nor *m,p'* were detected and only traces of *o,o'*- and p,*p'*- were formed.

A mechanism which is consistent with the observed product profile is given in Scheme 3. Electrophilic attack on the aromatic ring could produce the three palladium aryl complexes of which IIIa would be the most stable due to chelation with the *o*-acetate group. The *m*- and *p*-palladium aryls, (IIIa and IIIb) might then rapidly oxidize to the respective phenylene diacetates as would IIIc to some extent. The more stable palladium(II)aryl, IIIc, however, may be the precursor through which the dimerization reactions largely proceed. In this case, steric crowding is critical and rings orient so as to position the acetate group as far from nearby ligands as possible, VIa and VIb. Examination of space-filling molecular models of the proposed intermediates, VIa and VIb, indicates a clear preference for the acetate group in the incoming phenyl acetate to be as remote as possible from the palladium center. This positions the ring for palladation either *meta*- or *para*-. Since there are two *meta* and a single *para* position for attack, statistical probability predicts the 2:1 ratio for the *o,m'*-: *o,p'*-isomers that we observe.

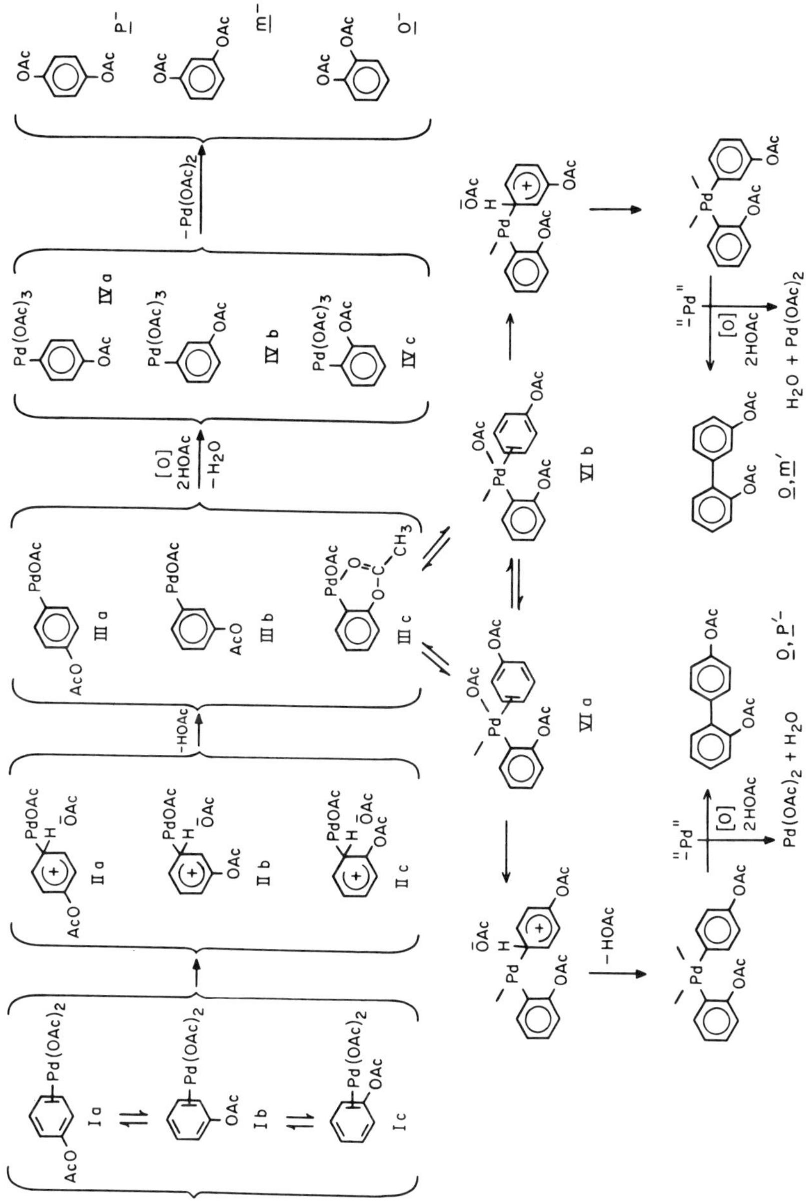

Scheme 3

<u>Supported Palladium Catalysts for Aromatic Oxidation</u>. Since benzene can be oxidized in acetic acid using both homogeneous and heterogeneous palladium catalysts,[20] we compared supported palladium catalysts to $Pd(OAc)_2$ for oxidation of phenyl acetate. Supported catalysts required temperatures of $>175^oC$ for significant activity using O_2 as the oxidant. At these temperatures carbon was not a suitable support. Palladium (5%) on alumina catalyzed oxidation at high temperature (175^oC) and high oxygen partial pressure (320 psig) to give predominant ring acetoxylation. Even at high temperatures over three-hour reaction periods, phenyl acetate conversion was low (5%); however, regioselectivity was considerably different from the homogeneous system. The major reaction product over 5% Pd/Al_2O_3 was *p*-phenylene diacetate (eq. 20) with very little *o*- and only traces of the *m*-isomer.

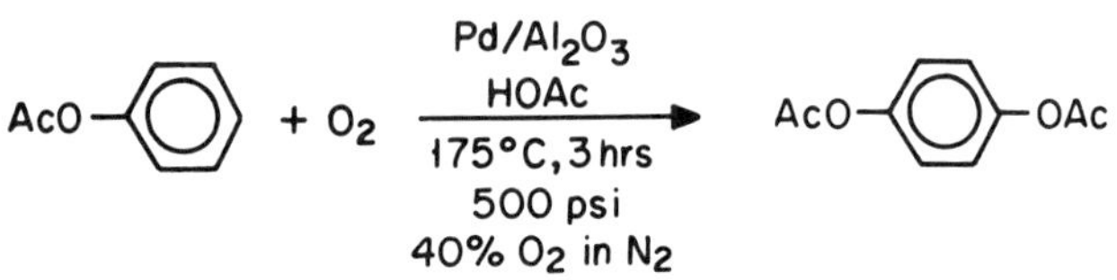

(20)

It has long been known that $Pd(OAc)_2$ catalyzes the oxidation of toluene in acetic acid to give benzylic oxidation products.[23] The catalyst appears to be more active when reactions are run in the presence of activated carbon.[23] Low oxidation state palladium has been implicated in benzylic oxidations.[24] We have found that palladium on carbon is effective for benzylic oxidations under mild conditions (eq. 21) under which little or

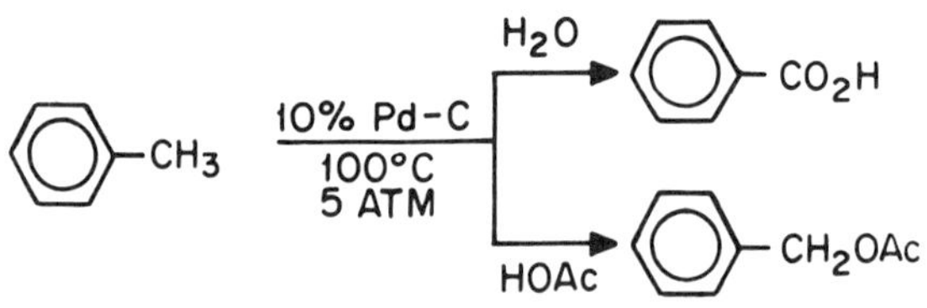

(21)

no ring oxidation occurs. This encourages one to speculate, in close analogy with olefinic systems, that over supported metal catalysts, higher oxidation state palladium species lead to ring oxidation, whereas lower oxidation state species are responsible for benzylic oxidation.

SUMMARY

Nucleophilic attack on a Pd(II) π-complex gives an intermediate which can lead to a vinylic oxidation product or its stable isomer. If carried out in the presence of a suitable oxidant which is regenerable in air, the reaction is catalytic. Chloride ion greatly favors vinylic oxidation. Such reactions have proven to be of enormous industrial utility for oxidizing olefins to vinyl esters, ketones and aldehydes over the past three decades. Supported palladium catalysts have been developed which produce these products as well.

Another class of industrially and synthetically important oxidation products - α,β-unsaturated alcohols, esters and carboxylic acids - also arise <u>via</u> palladium catalyzed oxidation of olefins, but these catalytic reactions have not been extensively investigated. Several factors favor the allylic oxidation of olefins over vinylic oxidation. In homogenous solutions of Pd(II) π-complexs, weak inorganic bases aid formation of π-allyl complexes which give rise to allylic oxidation products. Strongly electron withdrawing anionic ligands promote oxidative addition of C-H bonds of coordinated olefin to give π-allyl intermediates which can undergo nucleophilic attack to produce allylic, rather than vinylic oxidation products. Supported palladium catalysts which can be maintained in a low oxidation state appear to oxidatively add α-olefins to produce π-allyl surface species and give allylic oxidation products, whereas the addition of strong oxidants, chloride ion or use of forcing conditions promotes formation of vinylic oxidation products over these catalysts.

Palladium catalyzed aromatic oxidations proceed in two distinct directions - ring oxidation or oxidative coupling. In solution, oxidative coupling seems to be favored by low oxidation state ($Pd^{II} \rightleftharpoons Pd^{O}$) catalysis, whereas ring oxidation is favored by high oxidation state ($Pd^{IV} \rightleftharpoons Pd^{II}$) systems. Similar trends are apparent when heterogeneous palladium catalysts are used.

Alkyl aromatic compounds may undergo competitive oxidation at the alkyl side chain. Benzylic oxidation is favored by low oxidation state palladium, whereas more electrophilic high oxidation state species attack the ring.

By applying these considerations, it becomes possible to direct olefin oxidations along new and interesting allylic oxidation pathways. We are also learning how to promote aromatic oxidations using palladium catalysts. Ring oxidation, oxidative coupling or side chain oxidation can be accomplished selectively by the proper choice of catalyst conditions.

REFERENCES

1. P.M. Henry, **Palladium Catalyzed Oxidation of Hydrocarbons,** D. Reidel Publishing Company, Boston, 1980.
2. P.M. Maitlis, **The Organic Chemistry of Palladium, Vol. II Catalytic Reactions,** Academic Press, New York, 1971.
3. B.M. Trost and P.J. Metzner, J. Am. Chem. Soc., **102**, 3572 (1980).
4. L. Saussine, J.-P. Laloz and H. Mimoun, French Patent 2,450,802 (1980).
5. A.D. Ketley and J. Braatz, Chem. Commun., 169 (1968).
6. S. Winstein, J. McCaskie, H.-B. Lee and P.M. Henry, J. Am. Chem. Soc., **98**, 6913 (1978).
7. British Patents 1,101,055; 1,101,056; 1,027, 396.
8. I.P. Stolyarov, M.N. Vargaftik, O.M. Nefedov and I.I. Moiseev, Kinetika i Kataliz, **23**, 376 (1982).
9. W. Swodenk and G. Scharfe, U.S. Patent 3,925,452 (1975).
10. P.R. Auburn, P.B. Mackenzie and B. Bosnich, J. Am. Chem. Soc., **107**, 2033 (1985).
11. T. Seiyama, N. Yamazoe, J. Hojo and M. Hayakawa, J. Catal., 173 (1972).
12. K. Fujimoto and T. Kunugi, J. Jpn. Petrol. Inst., **17**, 739 (1974).
13. R. David and J. Estienne, U.S. Patent 3,624,147 (1971).
14. J.A. Hinnenkamp, U.S. Patent 4,435,598 (1984).
15. W. Kitching, T. Sakakiyama, Z. Rappoport, P.D. Sleezer, S. Winstein, and W.G. Young, J. Am. Chem. Soc., **94**, 2329 (1972).

16. P.M. Henry, J. Org. Chem., **36**, 1886 (1971).

17. (a) L. Eberson and L. Gomez-Gonzalez, Acta Chem. Scand., **27**, 1162 (1973); **27** 1249 (1973); **27**, 1253 (1973). (b) L. Eberson and L. Jonsson, Acta Chem. Scand., **28**, 597 (1974); **28**, 771 (1974); **30**, 361 (1976). (c) L. Eberson and L. Jonsson, Acta Chem. Scand., **28**, 771 (1974); Liebigs Ann. Chem., 233 (1977).

18. L.M. Stock, K. Tse, L.J. Vorvick and S.A. Walstrum, J. Org. Chem., **46**, 1759 (1981).

19. (a) R. Van Helden and G. Verberg, Rec. Trav. Chim., **84**, 1263 (1965). (b) M.O. Unger and R.A. Fouty, J. Org. Chem., **34**, 18 (1969). (c) F.R.S. Clark, R.O.C. Norman, C.B. Thomas and J.S. Willson, J.C. S. Perkin I, 1289 (1974). (d) H. Iataaki and H. Yoshimoto, J. Org. Chem., **38**, 76 (1973). (e) H. Itatani and H. Yoshimoto, Chem. Ind., 674 (1971). (f) A.I. Rudenkov, G.U. Mennenga, L.N. Rachkovskaya, K.I. Matveev and I.V. Kozhevnikov, Kinetika i Kataliz, **18**, 915 (1977). (g) Y. Itahara, Chem. Ind. 330 (1982).

20. V.H.-J. Arpe and L. Hornig, Erdol und Kohle, **23**, 79 (1970).

21. R.J. Pandey and P.M. Henry, Can. J. Chem., **52**, 1241 (1974).

22. Reviewed in F.A. Cotton and G. Wilkinson, **Advanced Inorganic Chemistry**, 2nd Edition, John Wiley & Sons, NY (1966) pp. 1035-38.

23. D.R. Bryant, J.E. McKeon and B.C. Ream, Tetrahedron Lett., 3371 (1968); J. Org. Chem., **33**, 4123 (1968).

24. M.K. Starchevskii, M.N. Vargaftik and I.I. Moiseev, Kinetika i Kataliz, **20**, 1163 (1979).

2

Syngas Reactions Involving C-3 Substances

J.J. Lin, J.M. Larkin and J.F. Knifton

Texaco Chemical Co.
Austin, TX

ABSTRACT

Reaction of synthesis gas with a variety of allylic substrates has been examined using homogeneous cobalt, rhodium and ruthenium catalysis. 2-Hydroxytetrahydrofuran, γ-butyrolactone, C_4-aliphatic carboxylic acids, pyrroline, 2-pyrrolidinone and C_4-acetates have each been selectively prepared through modification of the catalyst structure, syngas composition and operating conditions. The desired product is generated in each case according to whether the predominant chemoselective or regioselective reaction path involves carbonylation, hydroformylation or homologation. A subsequent cyclization step may be used to markedly improve overall selectivity.

I. INTRODUCTION

The mainstream of synthesis gas research during the last decade has emphasized the direct production of large volume, commodity chemicals, such as methanol, ethanol, ethylene glycol, acetic acid, acetic anhydride and light olefins.[1] The 1984-86 drop in oil prices has in many cases, however, delayed the commercialization of this technology. The scope of our research has now shifted, therefore, to the potential role of CO/H_2 as a building block for the preparation of higher value, and smaller volume, "specialty" chemicals. In this paper, we focus on our efforts to use available, propylene-based, C-3 substrates for the synthesis of C-4 specialty chemicals. The reactions under study are summarized in Scheme 1. They

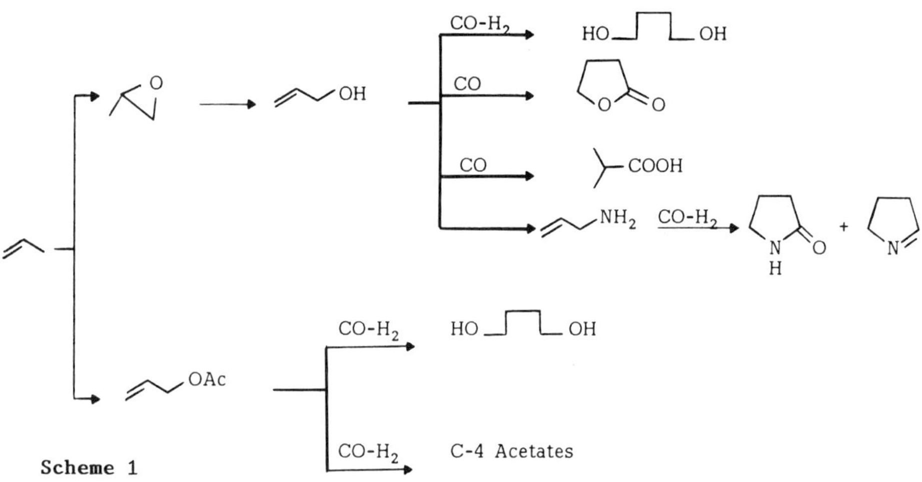

Scheme 1

include the generation of a variety of C_4-aliphatic oxygenates and amines, and their derivatives, including the preparation of 1,4-butanediol, γ-butyrolactone, C_4-carboxylic acids, 2-pyrrolidinone, pyrroline and C_4-acetates. The selective production of these materials requires syngas as the basic building block, be it through carbonylation, hydroformylation or homologation chemistry, in conjunction with intramolecular lactonization, lactamization, amination and acetalization. The predominant chemoselective or regioselective reaction path is controlled very often by the choice of catalyst - in this work we have used cobalt, rhodium-phosphine, rhodium-iodide and cobalt-ruthenium based systems. Some of the more significant preparative results include:

(a) The generation of 2-hydroxytetrahydrofuran from allyl alcohol in up to 96% selectivity.

(b) γ-Butyrolactone formation from allyl alcohol in 90% selectivity.

(c) The selective homologation of allyl acetate to C_4-acetates.

(d) The selective carbonylation of allyl alcohol to C_4-carboxylic acids.

(e) Pyrrolidinone and pyrroline preparation from allyl amine.

II. RESULTS AND DISCUSSION

A. 1,4-Butanediol from Allyl Acetate and Allyl Alcohol

1,4-Butanediol (U.S. market 250M ton/yr) is currently manufactured by the Reppe process involving acetylene and paraformaldehyde, or by butane oxidation.[2] Research into alternative syngas routes has been pioneered by General Electric's allyl acetate hydroformylation process (Scheme 2).[3,4]

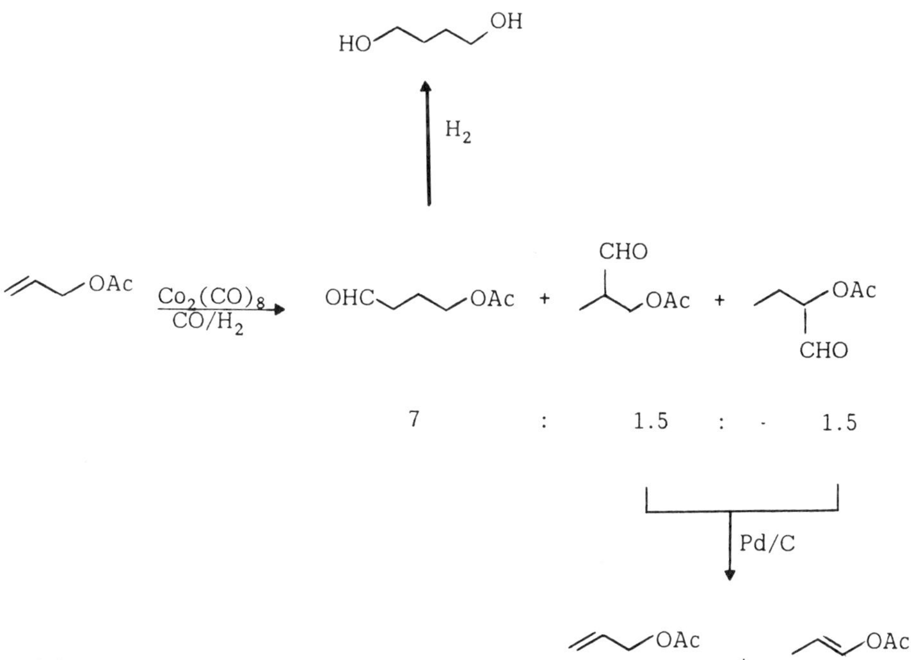

Scheme 2

Under typical operating conditions (125°C, 3000 psi), the linearity of the 4-acetoxybutyraldehyde intermediate is ca. 74%.[4] An additional step is required to recycle the branched aldehyde by-products.

During our earlier studies on methanol homologation catalyzed by homogeneous cobalt carbonyl-based catalysts, it was discovered that the addition of weakly-coordinating ligands, such as triphenylgermane, diphenylsulphide, and succinonitrile[5] enhances the cobalt catalyst stability and its reactivity. When these and related catalyst systems were evaluated for allyl acetate hydroformylation, it was found that the desired synthesis could be conducted under very mild conditions (typically 90°C, 800 psi), although the linearity of the oxo product was only in the 48-66% range. Results are summarized in Table I.

Other C-3 substrates have also been extensively investigated as suitable precursors for 1,4-butanediol production. While the du Pont- and Celanese-patented processes utilize acrolein,[6,7] Pittman[8] and others[9] have demonstrated that allyl alcohol may be selectively hydroformylated to

TABLE I

Hydroformylation of Allyl Acetate[a] (by low-pressure cobalt catalysis)

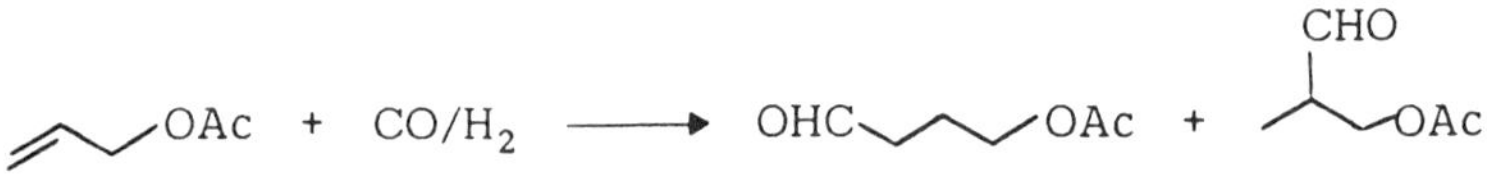

(I) (II)

Catalyst-Promoter	Selectivities (%) (I)	(II)
$Co_2(CO)_8$-Ph_3GeH	48	36
$Co_2(CO)_8$-Ph_2S	57	11
$Co_2(CO)_8$-2,2'-dipyridyl	66	14
$Co_2(CO)_8$-succinonitrile	61	19

[a] Reaction conditions: 90°C, 800 psi, CO/H_2 (1:1)

Allyl acetate, 100 mmole; $Co_2(CO)_8$, 1.0 mmole, promoter, 0.25→2.0 mmole; solvent, 35 g.

produce 4-hydroxybutyraldehyde in ca. 80% selectivity. In our preliminary work on allyl alcohol hydroformylation (Eq. 1) catalyzed by $HRh(CO)(PPh_3)_3$, we found that the major side reactions - allyl alcohol isomerization to propionaldehyde, and isobutyraldehyde formation via dehydration of 3-hydroxy-2-methylpropanal followed by hydrogenation (Eq. 2 and 3) - are closely tied to the presence of excess triphenylphosphine and

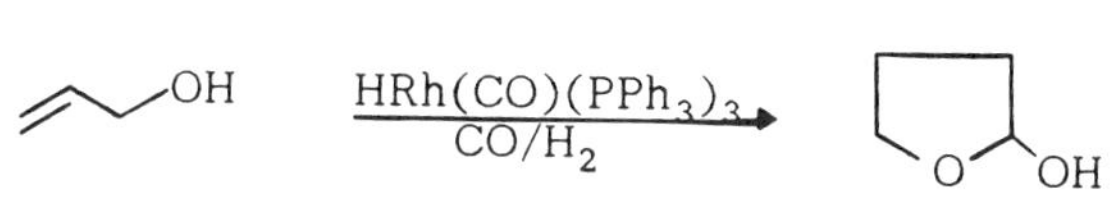

(1)

(2)

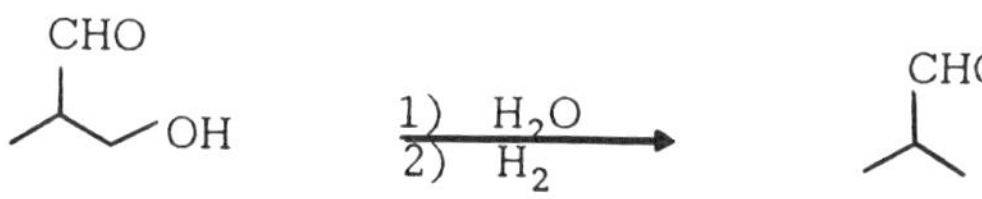

(3)

the addition of inert solvent. The results in Table II indicate a significant decrease in 2-hydroxytetrahydrofuran selectivity if either triphenylphosphine or p-dioxane are absent.

Although these side reactions can be suppressed by the use of protecting groups, such as t-butyl[10] and formate,[11] a preferable solution to the problem of improved, 1,4-butanediol selectivity would be to generate the 2-hydroxytetrahydrofuran intermediate (Eq. 4) at temperatures mild enough that the competing reactions 2 and 3 are extremely slow.

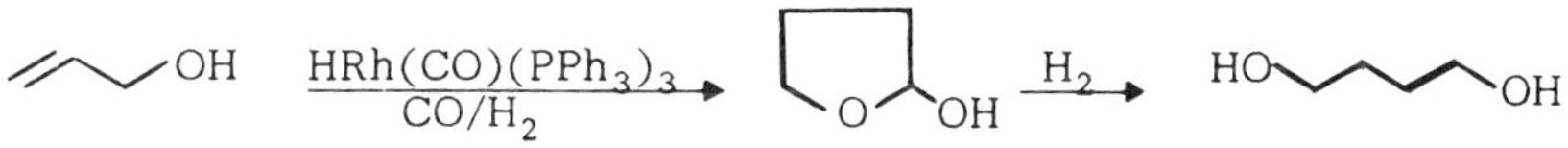

(4)

We now find that the rhodium-catalyzed hydroformylation of allyl alcohol to 2-hydroxytetrahydrofuran may be conducted in 96% selectivity at 100% allyl alcohol conversion under extremely mild conditions in the presence of ketonic solvents such as acetophenone and 2-undecanone.[12] The system is very productive (the rate of hydroformylation is typically 0.2 g

TABLE II

Hydroformylation of Allyl Alcohol[a]

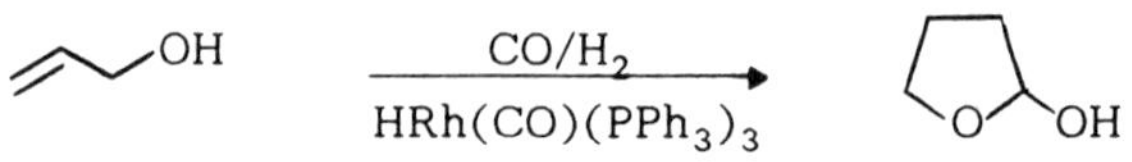

(I)

Added Ligand	Solvent	Conditions	Conversion	Selectivity (%) (I)	Selectivity (%) C_3-byproducts
Ph_3P	None	CO/H_2=1:2 ca. 85°C	85	31	34
Ph_3P	p-Dioxane	CO/H_2=1:2 ca. 85°C	>95	68	32
None	p-Dioxane	CO/H_2=1:2 ca. 100°C	>95	29	29

[a] Reaction conditions: 800 psi
Allyl alcohol, 10.0 g; $HRh(CO)(PPh_3)_3$, 0.05 mmole; PPh_3, 50 mmole; p-dioxane, 10.0 g.

TABLE III

2-Hydroxytetrahydrofuran Synthesis from Allyl Alcohol[a]

Catalyst	Solvent	Allyl Alcohol Conv. (%)	2-Hydroxytetra-hydrofuran Selectivity (%)
$HRh(CO)(PPh_3)_2$-PPh_3	Acetophenone	~100	96
" "	2-Undecanone	86	92

[a]Reaction conditions: 60°C; 800 psi, CO/H_2 (1:1).

Allyl alcohol 10.0 g; $HRh(CO)(PPh_3)_3$, 0.05 mmole; PPh_3, 50 mmole;

solvent, 10.0 g

product/cc/ hr) and the stable rhodium catalyst system allows multiple catalyst cycling without deactivation. The intermediate 2-hydroxytetrahydrofuran is readily isolated by fractional distillation and hydrogenated to 1,4-butanediol in excellent yield.

B. γ-Butyrolactone from Allyl Alcohol

Presently γ-butyrolactone is manufactured from maleic anhydride or via acetylene-based Reppe chemistry.[2] It finds extensive usage as a chemical intermediate in the generation of lactam derivatives. Allyl alcohol can also be employed as a feedstock for γ-butyrolactone synthesis; two reports illustrate allyl alcohol carbonylation to vinylacetic acid and γ-butyrolactone using rhodium- and nickel-based catalysis.[13] A non-iodide promoted cobalt catalyst allows ca. 60% yield of γ-butyrolactone.[14]

In our work we have achieved 90% selectivity to γ-butyrolactone by use of the homogenous dicobalt octacarbonyl-N,N,N',N'-tetramethylethylenediamine catalyst couple.[15] Results are summarized in Table IV. It may be noted that:

(a) Competing isomerization of allyl alcohol to propionaldehyde can be completely eliminated by the use of lower operating temperatures (ca. 60°C).

(b) A carbon monoxide-rich syngas favors the desired carbonylation step.

(c) N,N,N',N'-Tetramethylethylenediamine (TMEDA) addition allows 90% selectivity to γ-butyrolactone at 100% allyl alcohol conversion per pass.

(d) The presence of benzyl cyanide solvent ensures improved selectivity to the desired lactone.

C. Butyric Acids from Allyl Alcohol

The Monsanto process for acetic acid manufacture features an extremely active, homogeneous, iodide-promoted rhodium catalyst for methanol carbonylation.[16] The same catalyst is also effective for olefin hydroxycarbonylation to carboxylic acids, by CO/H_2O.[17] Under syngas

TABLE IV

γ-Butyrolactone from Allyl Alcohol and Carbon Monoxide

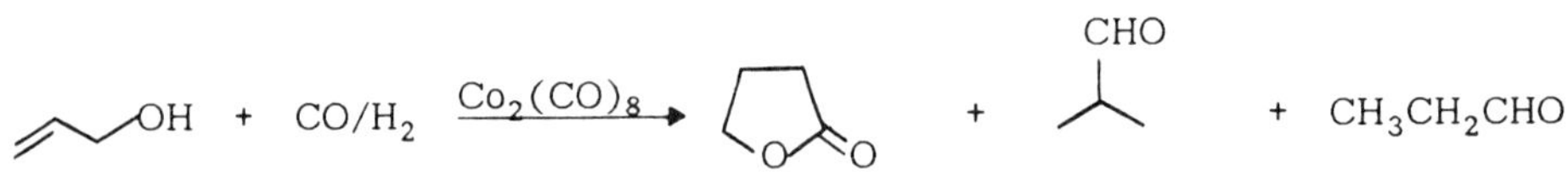

(I) (II) (III)

Promoter	Syngas ratio	Conditions	Allyl alcohol (conv. %)	Selectivities, % I	II	III
2-Hydroxypyridine[a]	CO/H_2 1:2	1150 psi 80°C, 18 hrs	100	62	23	15
" [a]	"	1000 psi 60°C, 18 hrs	50	80	20	0
" [a]	2:1	3500 psi 100°C, 2 hrs	100	79	21	0
" [b]	"	3000 psi 100°C, 2 hrs	100	74	13	13
TMEDA[c] [a]	1:2	1000 psi 84°C, 18 hr	100	90	5	5

[a] Allyl alcohol 4.0 g, $Co_2(CO)_8$ 1.0 mmole, promoter 2.0 mmoles, benzyl cyanide solvent, 5.0 g

[b] Benzyl cyanide 1.0 g

[c] TMEDA 0.5 mmole

conditions, we find the reaction of allyl alcohol, catalyzed by Rh(acac)$_3$-I_2, provides a new route to C_4-aliphatic carboxylic acids.[15] Chemoselective carbonylation conditions are typically 180°C, 1000-2000 psi, CO/H_2 (1:2). The normal and branched butyric acids are generated in 47% and 34% yields, respectively (Eq. 5).

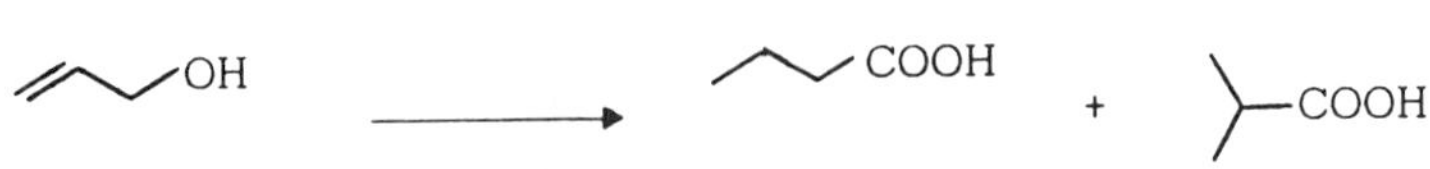

(5)

34% 47%

It is speculated that the formation of C_4-carboxylic acids occurs via an intermediate allylic rhodium complex, followed by migratory CO insertion and olefin reduction (see Scheme 3).

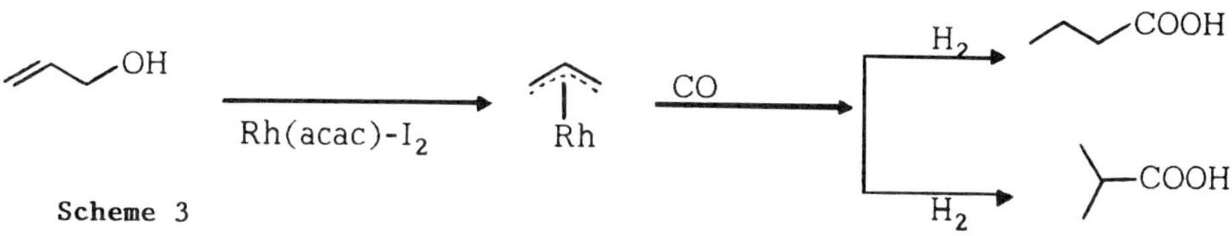

Scheme 3

D. Allyl Acetate Homologation

We had earlier disclosed that ruthenium - possibly in combination with dicobalt octacarbonyl and certain quaternary phosphonium salts, such as tetra-n-butylphosphonium bromide - is a very effective catalyst for alcohol homologation,[18] as well as the generation of acetic acid homologues.[19] We now report that analogous catalyst formulations (i.e., RuO_2-$Co_2(CO)_8$,Bu_4PBr, 1:0.25:10 molar ratio) afford the homologation of allyl and isopropenyl acetates to n-butyl and 2-methylpropyl acetates as the major products (Eqs. 6 and 7). Typical synthesis conditions are 200°C, 2000-4000 psi, CO/H_2 (1:1).

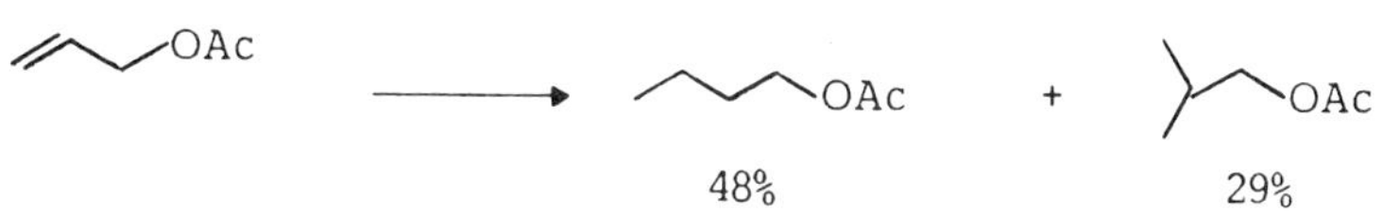

(6)

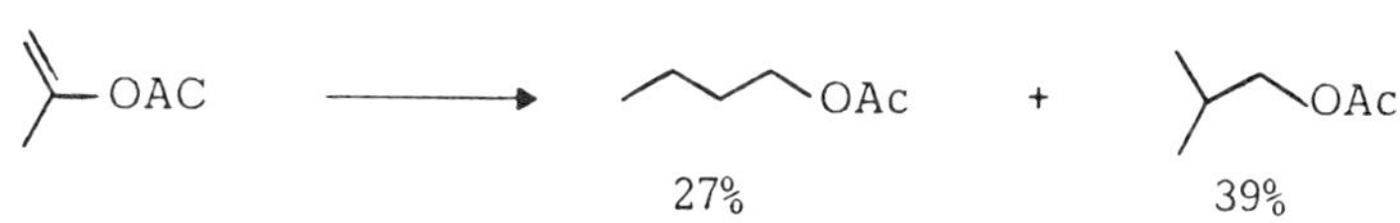

(7)

E. Pyrroline/2-Pyrolidinone Synthesis from Allyl Amine

In an extension of our allyl alcohol syngas technology to cyclic products (see Section II, A and B) we have also demonstrated the preparation of two heterocyclic amines, 2-pyrrolidinone and pyrroline (an intermediate in pyrrolidine synthesis) from allyl amine. These syngas/cyclization sequences are both catalyzed by $HRh(CO)(PPh_3)_3$ plus excess phosphine; experimental conditions are 75-80°C, 800 psi, CO/H_2 (1:1). Typical product compositions are illustrated below (Eq. 8). Pyrroline selectivity of 50% with toluene as

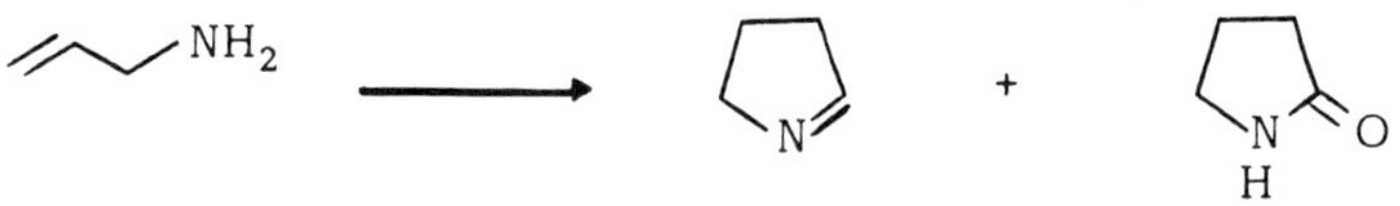

(8)

solvent:			
	toluene	50%	50%
	dimethyl phthalate	92%	0%

solvent may be contrasted with the 92% selectivity and yield in dimethyl phthalate.

Speculative synthetic routes to these two products are depicted in Scheme 4. Carbonylation of the allyl amine, via the initial formation of an acyl-rhodium species, followed by cyclization would yield 2-pyrrolidinone. Hydroformylation, on the other hand, gives pyrroline.

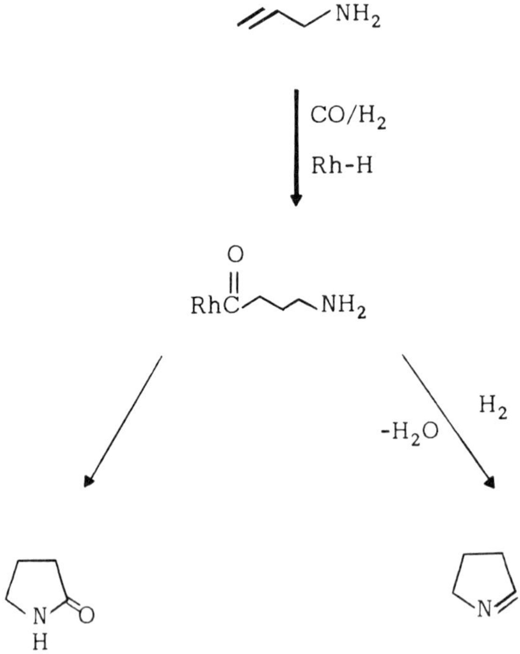

Scheme 4

III. CONCLUSIONS

Syngas reactions with various allylic precursors, particularly allyl alcohol, allyl acetate and allyl amine, have been examined in this work as a function of catalyst and synthesis gas composition, choice of added solvent and operating conditions. While 2-hydroxytetrahydrofuran, γ-butyrolactone, C_4-acetates, C_4-carboxylic acids, pyrroline and 2-pyrrolidinone have each been prepared selectively in this work, the more general conclusions that may be drawn from the data are as follows:

(1) The chemoselectivity observed for each of the C_3-substrates considered here is very much controlled by the nature of the allylic functional group - be it hydroxyl, acetate or amine. Regioselectivity at the unsymmetrical double bond is enhanced in the case of allyl alcohol and allyl amine by a subsequent cyclization step.

(2) The choice of catalyst determines whether the predominant reaction of the allylic precursor is (a) carbonylation - as is the case of cobalt, rhodium-phosphine and rhodium-iodide based catalysts, (b) hydroformylation - by certain cobalt carbonyl-ligand and rhodium-phosphine combinations and (c) methylene homologation - by ruthenium modified catalysis.

(3) Cyclization can be used to markedly improve the overall selectivity of the synthesis and in many cases it is the key to the unusually high product selectivity. Four types of cyclization sequence have been recognized in this work:

carbonylation-lactonization - leading to γ-butyrolactone
carbonylation-lactamization - leading to pyrrolidinone
hydroformylation-amination - giving pyrroline
hydroformylation-acetalization - to 2-hydroxytetrahydrofuran.

IV. ACKNOWLEDGEMENT

The authors wish to thank Texaco Inc. for permission to publish this paper, and Messrs. M.R. Swenson, R. Gonzales and P. Lopez for experimental assistance.

REFERENCES

1. See for example: J. Haggin, Chem. Eng. News, p. 23 (June 8, 1981); P.H. Spitz, Chemtech, 295 (May 1977); R.A. Sheldon, **Chemicals from Synthesis Gas**, D. Reidel, Dordrecht, 1983; G.W. Parshall, **Homogeneous Catalysis**, Wiley-Interscience, New York, 1980; R.L. Pruett, Annals N.Y. Acad. Sci., **295**, 239 (1977); J.F. Knifton, Platinum Metals Rev. **29**, 63 (1985); W. Keim, **Catalysis in C_1 Chemistry**, D. Reidel, Dordrecht, 1983).
2. A.M. Brownstein and H.L. List, Hydrocarbon Process., 159 (Sept. 1977); Chem Systems Report #83-2 (Dec. 1983); European Chem. News, p. 31 (Nov. 26, 1984); Chem Week, p. 13 (Nov. 21, 1984); Europ. Chem. News, p. 22 (Jan. 6, 1986); Chem Week, p. 9 (June 12, 1985).
3. W.E. Smith and J. Gerhart, U.S. Patent 4,039,592 (1977) to General Electric.
4. Chem Systems PERP Report 1980/81, p. 192 (Dec. 1980).
5. J.J. Lin and J.F. Knifton, U.S. Patents 4,476,326 (1984) and 4,409, 405 (1983) to Texaco.
6. C.C. Cumbe and K.K. Bhatia, U.S. Patent 3,929,915 (1975) to duPont.
7. P.D. Taylor, T.H. Vanderspurt, and A.B. Conciatori, U.S. Patent 4,161, 616 (1979) to Celanese.
8. C.U. Pittman and W.D. Honnick, J. Org. Chem., **45**, 2132 (1980).
9. M. Tamura and S. Kumano, Chem. Econ. Eng. Rev., **12**, 32 (1980); Europ. Chem. News, p. 28 (March 10, 1980); N.A. DeMunck, J.P.A. Notenboom, J.E. DeLeur, and J.J.F. Scholten, J. Mol. Catal., **11**, 233 (1981).
10. N. Harris, A.J. Dennis, and G.E. Harrison, U.S. Patent 4,503,275 (1985) to Davy McKee.
11. J.J. Lin and J.M. Larkin, unpublished results.
12. J.J. Lin and J.F. Knifton, U.S. Patents 4,529,808 (1985) and 4,533,742 (1985) to Texaco.
13. H. Fernholz and D. Freudenberger, U.S. Patents 3,980,671 (1976) and 4,140,865 (1979) to Hoechst.
14. A. Matsuda, Bull. Chem. Soc. Jpn., **41**, 1876 (1968).
15. J.J. Lin, unpublished data.

16. J.F. Roth, J.H. Craddock, A. Hershman, and F.E. Paulik, Chemtech, 600 (1971).

17. D.E. Morris and G.V. Johnson, Symp. Rhodium Homogeneous Catal., Proceedings, 1978, p. 113.

18. J.F. Knifton, R.A. Grigsby and J.J. Lin, Organometallics, **3**, 62 (1984).

19. J.F. Knifton, J. Mol. Catal., **11**, 91 (1981).

3

Hydroformulation of Terpenes and Related Molecules

A.J. Chalk

Givaudan Corporation
Delawana Avenue
Clifton, New Jersey

ABSTRACT

The use of hydroformylation for the synthesis of perfumery aldehydes is reviewed. Suitable olefins, including a variety of terpenes, were hydroformylated under mild conditions (500 psig, 60-140°C) using rhodium catalysts. The selective monohydroformylation of various polyolefins was examined and compared to expectations based on the rate constants found in the literature for specific double bonds. The effect of neighboring alcohol and ester groups on the reaction was also investigated.

I. INTRODUCTION

Aldehydes are one of the most widely used groups of compounds in perfumery. The aldehyde functional group, when it occurs alone as in the fatty aldehydes makes a powerful and valuable olfactive contribution which offsets the disadvantage of its instability to oxidizing conditions.

The hydroformylation of olefins is potentially one of the least expensive syntheses for many aldehydes but companies with facilities and expertise in this technology have shown little interest in perfumery aldehydes because of the small volumes involved. Ruhrchemie has long supplied the perfumery industry with "Oxo" products from styrene, diisobutylene and dicyclopentadiene but only because these were intermediates for products required by other industries (pharmaceuticals, plastics, etc.). Shell produces linear alcohols by hydroformylation but only as mixtures (the Neodols) for the manufacture of detergents. Celanese attempted to commercialize the C_7 and C_9 aldehydes since these are intermediates for the corresponding acids but they have shown no interest in hydroformylating pure olefins higher than 1-octene.

Many patents describe the synthesis by hydroformylation of the more important perfumery aldehydes such as citronellal, hydroxycitronellal, 10-undecanal, undecanal and 2-methylundecanal.[1a-e] Processes for the last two products are apparently economically viable since these products are now made by Ruhrchemie in Europe and Chemfleur in the United States. Other commercial perfumery aldehydes which Chemfleur has apparently made by hydroformylation are those obtained from styrene, α-methylstyrene and limonene.

Naturally occurring terpenes are a useful source of olefins which are cheap enough to produce aldehydes and alcohols of interest to the perfumery industry. Some products on the market are shown in Fig. 1. While we do not know whether they are produced by hydroformylation of the corresponding terpenes as shown, it would seem to be the logical choice. Rhodium would be the catalyst of choice, since hydroformylation with dicobalt octacarbonyl is reported to result in structural rearrangements in terpenes, such as α-pinene, due to the acidic character of cobalt tetracarbonylhydride.[2] Specificity and yields were also found to be poor for the cobalt catalyst with a range of terpenes.[3] Although it has been claimed that camphene undergoes hydroformylation without rearrangement when $Co_2(CO)_8$ is used as catalyst,[4] it would seem prudent to use rhodium or possibly cobalt in combination with phosphines to avoid formation of the highly acidic catalyst $HCo(CO)_4$. A number of patents claim the hydroformylation of terpenes and related molecules or claim the products of hydroformylations as new perfumery chemicals. Some of these are listed in Table 1. Where available, reaction conditions used in examples are given for rhodium catalysts. In this report we investigate the possibility of hydroformylating similar olefins at lower pressures ($\sim$30 atm.).

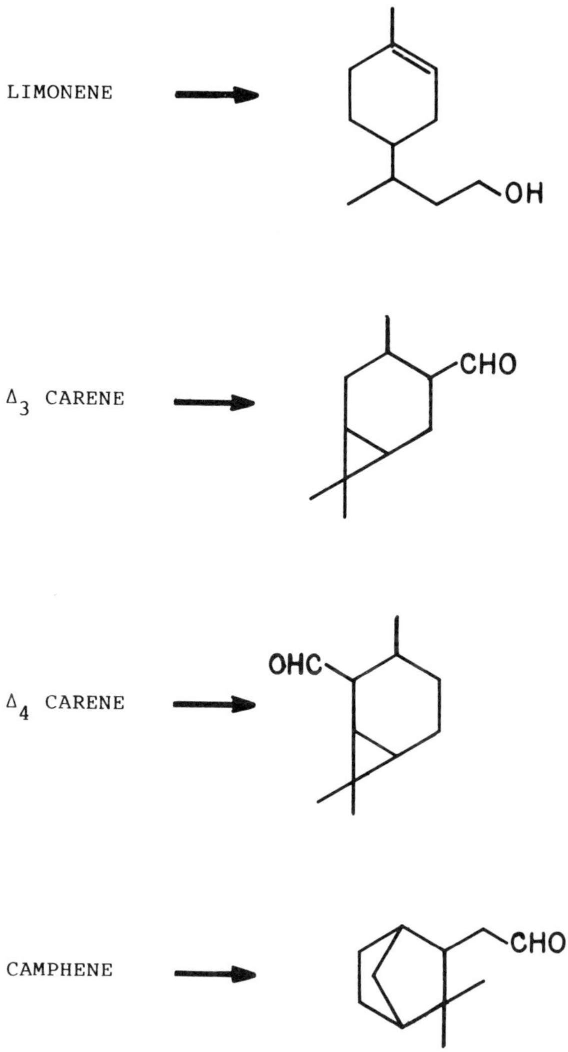

Fig. 1. Hydroformylation of Terpenes

TABLE 1

U.S. Patents

Olefin	Conditions: Pressure atmosphere	Conditions: Temperature °C	L/Rh[a] m/m	Number	Year	Company
α-Terpinene	200	130	50	4283561	1978	Henkel
Caryophyllene	120	120	50	4270007	1978	Henkel
Limonene	280	150	100	4337363	1979	Ruhrchemie
Limonene, Camphene, β-Pinene, α-Terpinene	200	125	100	4334100	1982	Henkel
β-Terpineol, Limonene epoxide	50	100	H[b]	4352937	1982	Shell

[a]Ratio of ligand (PPh_3) to rhodium in moles/mole

[b]Heterogeneous rhodium catalyst

II. COBALT VERSUS RHODIUM AS CATALYST

Table 2 compares preliminary results for a cobalt/trialkylphosphine catalyst versus a rhodium/triphenylphosphine catalyst. Both reactions can be carried out conveniently with a total pressure of only 500 psig, but the former requires a molar ratio of $H_2/CO = 2$ since alcohols are formed, while the latter requires $H_2/CO = 1$ since aldehydes are produced. In both cases the temperature was that required to give a convenient rate of gas uptake. Yields are based on GC analysis of distilled products. For each substrate a higher temperature was required for cobalt and a less selective reaction was obtained than was the case for rhodium. For limonene the product mixture for cobalt was too complex for separation while linalool gave no distillable products. These results clearly established rhodium as the catalyst of choice.

Cobalt was an alternative for olefins such as cyclododecene which gave only one isomer as product and this catalyst could be advantageous if the desired product was the alcohol rather than the aldehyde. However, in view of the more selective reactions catalyzed by rhodium, it was used in virtually all subsequent exploratory reactions.

III. OLEFIN STRUCTURE

Table 3 gives results for some monosubstituted olefins. Yields are based on the olefin charged and would be higher if the olefin was recovered and recycled since olefin hydrogeneration and isomerization were negligible in most cases. The most reactive double bond was that of styrene which reacts so rapidly at 80° that it is difficult to prevent a rise in temperature due to the exothermic nature of the reaction. Other monosubstituted double bonds react slowly at 80 to 100°. The high reactivity is for the position α to the phenyl group however. When this position is virtually blocked as in α-methylstyrene, a very unreactive double bond results, less reactive than that of limonene, β-pinene or 2,4,4-trimethylpentene. For α-methylstyrene, the reactivity of the position α to the phenyl group allows the formation of about 5% of an aldehyde next to a quaternary carbon. No such product was found in the case of the other 1,1-disubstituted olefins. A number of

TABLE 2

Rhodium Versus Cobalt for Hydroformylation

Substrate	Catalyst[a]	Temperature °C	Time hours	Conversion[c] %	Alkane[d] %	Yield[e] %	Linearity[f] %	Residue[g] %
Cyclododecene	Co	180	16	44	0	32	N/A	0
	Rh[b]	95	16	65	0	59	N/A	5
Limonene	Co	180	20	87	27	h	N/A	23
	Rh	120	19	98	0	69	100	8
3,7-Dimethyl-	Co	180	17	80	13	35	80	27
1,6-octadiene	Rh	80	4	97	0	93	85	0
Linalool	Co	180	16	100	0	h	N/A	100
	Rh	100	3	97	0	86	N/A	0

[a]Co Cat. - Molar ratio Substrate: $P(C_8H_{17})_3:Co(OAc)_2 = 2:4 \times 10^{-2}:10^{-2}$

H_2 = 333 psi   CO = 166 psi

Rh cat. - Molar ratio Substrate: $PPh_3:RhHCO(PPh_3)_3 = 2:760 \times 10^{-4}:10^{-4}$

H_2 = 250 psi   CO = 250 psi

[b]No PPh_3 used

[c]Conversion of substrate to all products

[d]Hydrogenation of substrate

[e]Yield for Co refers to alcohol products based on substrate charged

Yield for Rh refers to aldehyde or equivalent products based on substrate charged

[f]Where applicable refers to % linear product compared to all isomeric products

[g]Weight of residue after flash distillation from catalyst/weight of substrate charged

[h]See text

TABLE 3

Hydroformylation with $RhHCO(PPh_3)_3$[a]

Olefin	Temperature °C	Time hours	Conversion to aldehydes[c] %	Aldehyde yield[c] %	Linearity[c] %
Styrene	80	4	100	92	14
p-t-Butylstyrene	70	48	98	95	17
α-Methylstyrene	140	24	83	64	95
3,3-Dimethyl-1-butene	80	21	94	84	93
β-Pinene	110	40	70	52	100
2,4,4-Trimethylpentene[b]	130	30	40	27	100
Cyclooctene	100	20	91	80	N/A
Cyclododecene	120	10	65	42	N/A
2,6,10-Trimethyl-1,5,9-Undecatriene	110	48	72	84	100

[a]Conditions as Table 2 (for rhodium catalyst) except as noted.

[b]PPh_3 reduced by a factor of 4.

[c]As Table 2

trisubstituted olefins were found to be unreactive under the conditions described in Table III. Even some hindered 1,1-disubstituted olefins such as caryophyllene and longifolene did not give any significant amount of products after 20 hours at 140°C and 500 psi total pressure.

A. β-Pinene

Earlier work has shown that two isomers (cis & trans) result from hydroformylation of β-pinene. The cis/trans ratio increasing as the temperature is lowered or the pressure raised.[2] This is consistent with the expectation that the cis isomer is kinetically preferred since it results from coordination of the rhodium to the less hindered side of the olefin while the trans isomer should be more stable thermodynamically.

Although we employed constant pressure and similar temperatures we were surprised to find that the ratio of the two products varied with the duration of the reaction and with the catalyst (Table 4). The two isomers were monitored by capillary GC (20 M) and the ratio found to agree with the ratio of the areas of the proton NMR peaks at δ = 1.05 and 0.90 ppm. These peaks for methyl protons identify the two isomers for both the aldehydes and corresponding alcohols. Identification was made on the alcohols. Using Europium $(FOD)_3d_{30}$ the 1.05 ppm peak moved furthest downfield and was assigned to the cis isomer which corresponds to the aldehyde with longer retention time.

The conditions of the second experiment were identical to the first except that a different batch of catalyst was used. The lower cis/trans ratio and rapid formation of α-pinene lead us to compare catalyst batches and the second was found to have significantly more chloride present. Use of $RhClCO(PPh_3)_2$ as sole catalyst was found to give a similar effect in the third experiment. Recycling this catalyst resulted in a higher cis/trans ratio, presumably due to loss of chloride as HCl. Reducing the amount of excess triphenyl phosphine was also found to decrease the cis/trans ratio of the product aldehydes (catalyst 4). When excess triphenylphosphine was eliminated, isomerization of β- to α-pinene became the dominant reaction. When triphenylphosphine was replaced by a bulky phosphite ligand (catalyst 5), the catalytic activity for hydroformylation was also greatly reduced and no reaction was observed until the temperature was raised to 140°C. The

TABLE 4

β-Pinene Hydroformylation[a]

Catalyst[b]	Temperature °C	Time hours	Conversion to aldehydes %	Aldehyde % Cis	Trans	α-Pinene %
1	100	25	34	90	10	0
	110	18	68	86	14	1
	110	39	75	83	17	3
1'	100	40	80	31	69	15
2	100	15	52	39	61	9
		27	86	37	63	13
3	120	41	39	74	26	2
4	110	20	15	48	52	2
		40	53	54	46	7
5	140	19	10	8	92	60

[a]1.7 mole β-pinene, 0.1 m mole catalyst, $[H_2] = [CO] = 250$ psi

[b]1 = $RhHCO(PPh_3)_3$ + 76 mm PPh_3; 1' = as 1 but different batch of catalyst (see text); 2 = $RhClCO(PPh_3)_2$ + 76 mm PPh_3; 3 = 2 recycled; 4 = $RhHCO(PPh_3)_3$ + 8 mm PPh_3; 5 = $Rh(Acac)_3$ + 18 mm tri(2-t-butylphenyl)phosphite

principle reaction was then isomerization to α-pinene but the small amount of aldehyde produced had an extremely low cis/trans ratio.

The formation of trans aldehyde requires that the catalyst coordinate to the more sterically hindered face of the olefin. This would be favored by a less bulky catalyst and therefore by a reduction in the ligand concentration or the use of a more weakly coordinating ligand. These considerations rationalize the results with catalysts 4 and 5. An explanation for the effect of chloride is less obvious. The isomerization of β- to α-pinene which accompanies the formation of trans product suggests an intermediate having a bond from rhodium to a quaternary carbon. Conceivably fission of this bond, either homolytically, or with HCl heterolytically, could cause an isomerization which would lead to rhodium coordinated to the more hindered face of the olefin and thence to trans product. It should be mentioned that the use of the bulky phosphite ligand (catalyst 5) was examined in view of a report[5] that it results in a more active catalyst for disubstituted olefins similar to β-pinene. Unfortunately this was not found to be the case in this instance.

B. Polyolefins

Hydroformylation of a molecule containing two or more double bonds should proceed by consecutive reactions. For the simplest case of a diolefin we have:

$$\underset{\text{I}}{\text{Diolefin}} \xrightarrow[1]{k_1} \underset{\text{II}}{\text{Monoaldehyde}} \xrightarrow[2]{k_2} \underset{\text{III}}{\text{Dialdehyde}}$$

The changes in the concentrations of I, II and III are shown in Fig. 2 for the case where $k_1 = 2k_2$.[6] The concentration of monoaldehyde rises to a maximum which was calculated for various values of k_1/k_2 and plotted together with the corresponding conversion and selectivity in Fig. 3. From Fig. 3 it can be seen that in order to prepare the monoaldehyde in reasonable yield, e.g. 77% we require that $k_1 \simeq 10\ k_2$. In this case we have a selectivity of 84% at 92% conversion. Rate constants (10^3 k. min.^1) for various double bonds are given in the literature[7] for hydroformylation with rhodium at 110°C as follows: styrene (124)> monoolefins (40-55)> 1,2-disub-

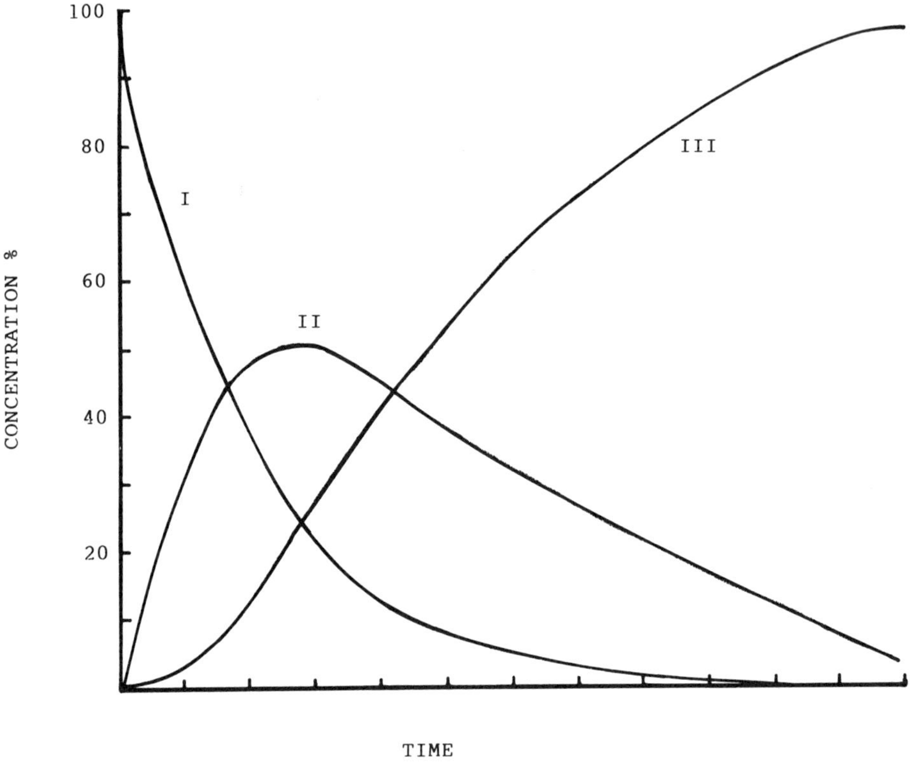

$$[II]_{Max} = A_o e^{-k_2 t} = A_o e \ \ln k_2/k_1 \Big/ k_1/k_2 - 1$$

Fig. 2. Concentration Changes for I → II → III

stituted olefins (35-40) > 1,1-disubstituted olefins (∼26) > cyclohexene (6) > trisubstituted olefins (∼3) > tetrasubstituted olefins (∼0.7).

From these values we expect to be able to selectively hydroformylate a mono- or disubstituted double bond in the presence of a trisubstituted double bond. Table 5 contains results for 3,7-dimethyl-1,6-octadiene, limonene and 1,5-dimethyl-1,5,9-cyclododecatriene which confirm this. No selectivity is expected however for the hydroformylation of a mono- in the presence of a disubstituted double bond. An exception to this is the hydroformylation of a monosubstituted double bond in the presence of

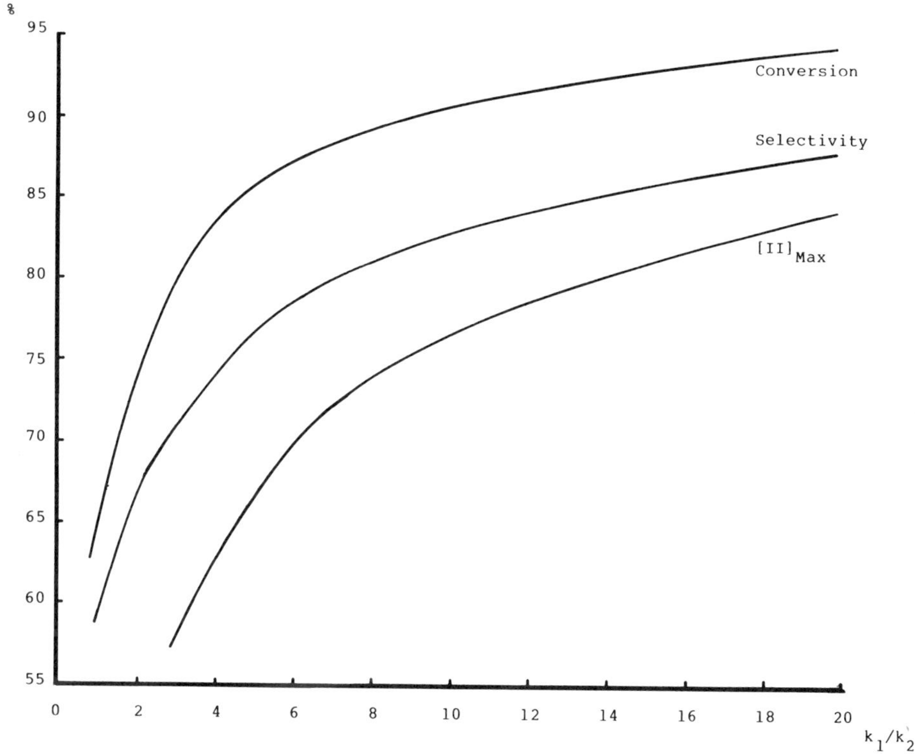

Fig. 3 Consecutive Reactions I → II → III

cyclohexene. Confirmation of this expectation was obtained for 4-vinylcyclohexene.

Other results were more surprising. Thus the strained norbornene double bond can be selectively reacted in the presence of the vinyl group (vinylnorbornene) and the (trans) disubstituted double bond is virtually untouched in the hydroformylation of 1,6-octadiene. By contrast, when both double bonds are identical as in 1,5-cyclooctadiene the selectivity is low as expected for a reaction where $k_1 \simeq k_2$. This reaction is not simple however since double bond isomerization occurs in both the starting olefin and in the monoaldehyde. Of the monoaldehydes produced, 31% was saturated (cyclooctane carboxaldehyde) while 69% were unsaturated (14% conjugated, 55% non-conjugated). The recovered diolefin was entirely 1,3-

TABLE 5

Selectivity in the Hydroformylation of Polyolefins[a]

Olefin	Temperature °C	Time hours	Conversion to aldehydes[d] %	Aldehyde yield[d] %	Selectivity to mono-aldehyde[c]
Vinylnorbornene	50-80	0.3	83	74	89
Ethylidenenorbornene	60-90	2	91	78	100
4-Vinylcyclohexene	100	34	76	60	100
1,6-Octadiene	70-80	8	80	75	96
3,7-Dimethyl-1,6-octadiene	80	4	97	93	100
1,5-Cyclooctadiene	75-100	8	87	28	40
t,t,c-1,5,9-Cyclododecatriene[b]	80	12	84	69	80
1,5-Dimethyl-1,5,9-cyclododecatriene	90	19	63	52	100
Limonene	120	19	98	69	100

[a]Conditions as Table 3 except as noted

[b]PPh_3 reduced by a factor of 4

[c]From GC areas

[d]As Table 2

cyclooctadiene. This result differs from earlier work[8] where no conjugated monoaldehydes and no dialdehydes we found. In fact we were also unable to isolate dialdehydes. However, examination of the higher molecular weight products by NMR and MS suggested hydroxyaldehydes with a molecular weight corresponding to the expected dialdehydes (168). They are therefore assumed to be the products of an intramolecular Aldol condensation (eq. 3).

IV V

The high selectivity found for trans, trans, cis-1,5,9-cyclododecatriene was unexpected. The cis double bond should hydroformylate faster than the trans, but the difference is not expected to be enough to give such a high selectivity. Examination of the products showed in fact that the selectivity could not be explained in these terms. Thus two monoaldehydes are formed in a ratio of 46/54 (C13 NMR and capillary GC) and their IR spectrum showed strong absorptions typical of both cis and trans double bonds (980 and 705 cm^{-1}). Hydroformylation of only a cis double bond should give only one product. Hydroformylation of only trans double bonds should give two. Reaction at both cis and trans double bonds should give three monoaldehydes. Double bond isomerization should of course increase these numbers but there was no sign of isomerization in the starting material or products. The evidence therefore points to a selective hydroformylation of trans double bonds. If true, this must be related to some effect which would not operate if the cis and trans double bonds were competing for coordination positions from separate molecules. Such an effect could be chelation. Thus olefin insertion by the metal hydride could be facilitated for the trans double bond by forming a chelate in which the metal alkyl is also coordinated by the cis-olefin double bond. Hydroformylation of dimethylcyclododecatriene is also more complex than it appears. The starting material was found to be a mixture of three compounds by capillary GC in the ratio 42:19:37. Only the first two reacted leaving a pure diolefin which was found by IR to be the only component having a cis disubstituted double bond. Thus, in this case also we appear to have the selective hydroformylation of the two components having trans disubstituted double bonds.

IV. UNSATURATED ALCOHOLS

A. Allylic

Hydroformylation of allylic alcohols could give rise to a number of products (eqs. 4-6). Reaction with an excess of an alcohol and a trace of acid

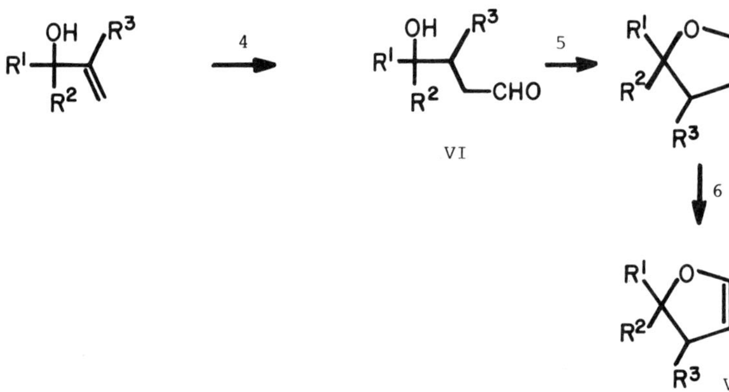

VI VII VIII

converts VI and VII to the ethers IX and X (eqs. 7 and 8). Linalool reacted

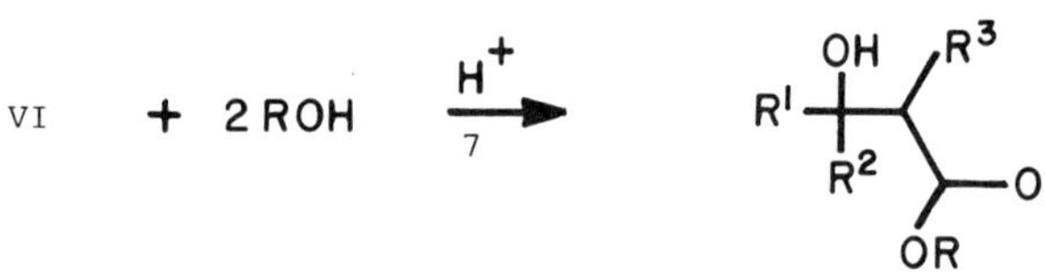

IX

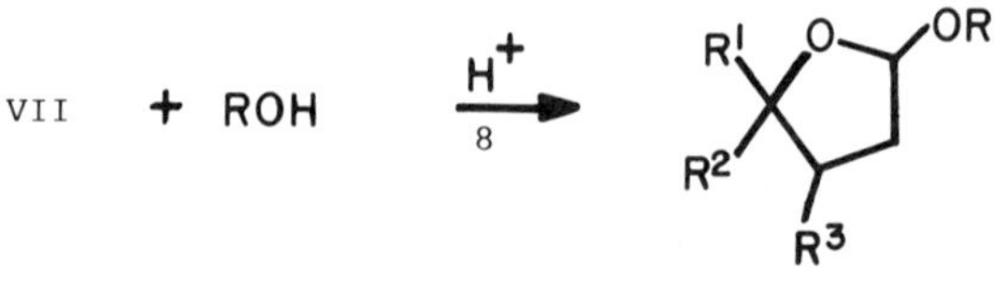

X

rapidly (Table 6) and gave a single product identified as VIIA (VII where $R' = (CH_3)_2C{=}CHCH_2CH_2-$, $R^2 = CH_3-$, $R^3 = H$). The proton NMR of VIIA had a distinctive absorption at $\delta = 5.5$ ppm due to a CH adjacent to two oxygens. Reaction with an alcohol gave the corresponding ethers eg. XA. An attempt to eliminate alcohol from XA to give the corresponding dihydrofuran was unsuccessful. Instead a cyclization occurred to give XI (eq. 9). Pinocarveol

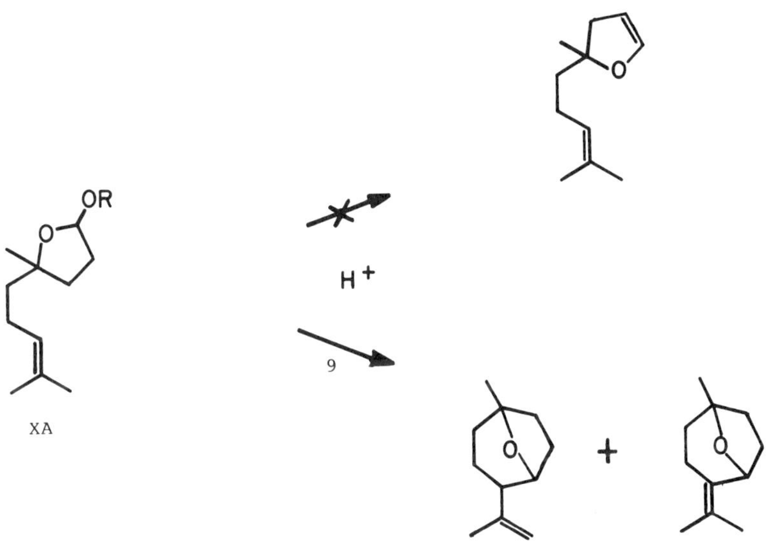

XA

XI

gave a somewhat different result. A mixture of all possible products was formed. A product of type VI (VIB) was obvious from absorptions due to an aldehyde and an alcohol. Reaction of VIB with alcohols in the presence of a trace of acid gave products of type IX. The presence of a dihydrofuran (VIIIB) was concluded from the appearance of vinyl protons at $\delta = 7.4$ and 4.1 ppm.

B. Homoallylic Alcohols, Ethers and Esters

Homoallylic alcohols could give rise to products XII-XIV (eq. 10-12). Isopulegol gave only products of type XIII. This was confirmed by reaction with alcohols which gave only products of the type XV (eq. 13).

TABLE 6

Unsaturated Alcohols and Derivatives[a]

Substrate	Temperature °C	Time hours	Conversion %	Yield %
Linalool	100	18	91	87
Pinocarveol	125	18	98	28
Isopulegol	120	15	58	50
Plinols	125	24	100	83
3,7-Dimethyl-7-hydroxyoct-1-ene	60	40	56	52
3,7-Dimethyl-7-methoxyoct-1-ene	70	20	99	83

[a]Conditions as Table 2 except as noted

[b]PPh_3 reduced by a factor of 20

XII XIII

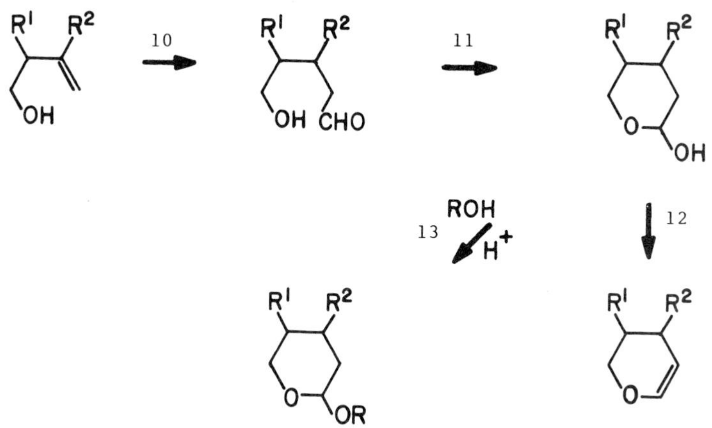

XV XIV

C. Other Alcohols

A crude mixture of the four isomeric plinols hydroformylated slowly at 125°. A mixture of the expected hydroxyaldehydes was formed (eq. 14).

3,7-Dimethyl-7-hydroxy-oct-1-ene reacted rapidly at 75° but virtually stopped after 16% conversion. A repeat reaction was similarly deactivated at 80°C. The reaction was then carried out at 60°C and after 40 hours gave 56% conversion to the expected mixture of products. The linearity (92%) was higher than that found for the parent diolefin (85%, Table 2). The corresponding methyl ether reacted without any problem at 70° to give a product with 87% linearity. Hydroformylation of unsaturated esters encountered side reactions of allylic isomerization and elimination of acetic acid. Linalyl acetate eliminates some acetic acid and isomerizes to neryl and geranyl acetates. These two reactions are connected since isomerization occurs readily in the presence of acetic acid. Hydroformylation of 2,7-octadienyl acetate suffers from similar problems. A mixture of the expected acetate aldehydes was formed. Dialdehydes were also produced however, the dominant one being XVI which was probably formed via eqs. 15-17.

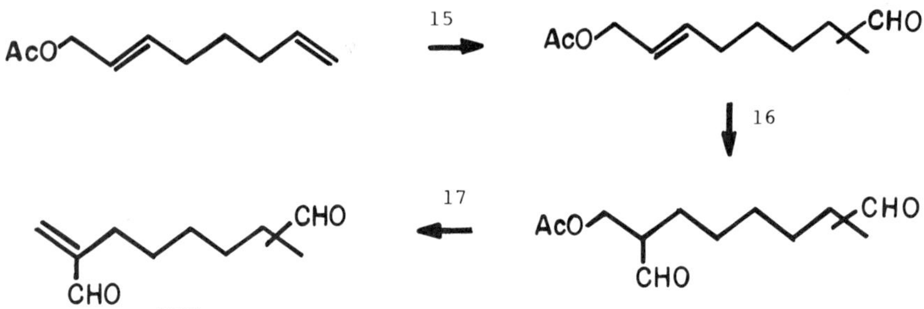

XIV

V. EXPERIMENTAL

Catalysts were prepared according to the literature: $RhClCO(PPh_3)_2$[9], $RhHCO(PPh_3)_3$[10].

Substrates were distilled prior to use.

Hydroformylations were carried out with a 500 ml low pressure, zip-open MagneDrive II packless autoclave (Autoclave Engineers).

ACKNOWLEDGEMENT

I wish to acknowledge Dr. Mingjien Chien for his assistance with the calculations.

REFERENCES

1. (a) DOS 2,050,677 (B.A.S.F.); (b) DOS 1,964,962 (B.A.S.F.); (c) British 1,555,551 (Shell); (d) U.S. 4,263,460 (Ruhrchemie); (e) U.S. 4,258,214 (Ruhrchemie).

2. W. Himmele and H. Siegel, Tetrahedron Lett., 907 (1976).
3. W.H. Clement and M. Orchin, I & EC Prod. Res. Dev., **4**, 283 (1965).
4. J.C. LoCicero and R.T. Johnson, J. Am. Chem. Soc., **74**, 2094 (1952).
5. P.W.N.M. Van Leeuwen and C.F. Roobeek, J. Organometal. Chem., **258**, 343 (1983).
6. "Physical Chemistry, W.J. Moore, 4th Ed. Prentice Hall, N.Y. 1972, p. 345.
7. "New Syntheses with Carbon Monoxide", Ed. J. Falbe, Springer-Verlag, N.Y. 1980, pp. 96-97.
8. A. Spencer, J. Organometal. Chem., **124**, 85 (1977).
9. "Inorganic Syntheses", R.E. Krieger Publishing Co., Huntingdon, N.Y. 1978, Vol. 11, pp. 99-101.
10. "Inorganic Syntheses", R.E. Krieger Publishing Co., Malabar, Fla. 1982, Vol. 15, pp. 58-59.

4

Efficient Asymmetric Alkylations via Chiral Phase-Transfer Catalysis. A Novel Dual Catalysis

U.-H. Dolling, D.L. Hughes, A. Bhattacharya, K.M. Ryan, S. Karady, L.M. Weinstock, V.J. Grenda, and E.J.J. Grabowski

Merck Sharp & Dohme Research Laboratories
Merck & Co., Inc.
Rahway, New Jersey

ABSTRACT

Catalytic asymmetric alkylations with high enantioselectivity and nearly quantitative chemical yield via chiral phase-transfer catalysis have been applied to the asymmetric synthesis of pharmaceuticals. Methylation of 6,7-dichloro-5-methoxy-2-phenyl-1-indanone with methyl chloride in 50% sodium hydroxide/toluene using N-(p-trifluoromethylbenzyl)cinchoninium bromide as catalyst produced (S)-(+)-6,7-dichloro-5-methoxy-2-methyl-2-phenyl-1-indanone in 94% ee and 95% yield. This methodology was expanded to include an asymmetric modification of the Robinson annulation employing 1,3-dichloro-2-butene (the Wichterle reagent) as a methyl vinyl ketone surrogate. Under identical conditions 6,7-dichloro-5-methoxy-2-propyl-1-indanone was alkylated to (S)-(+)-6,7-dichloro-2-(3-chloro-2-butenyl)-2,3-dihydroxy-5-methoxy-2-propyl-1-inden-1-one in 92% ee and 99% yield. Hydrolysis and condensation in sulfuric acid produced (S)-(-)-5,6-dichloro-1,2,9,9a-tetrahydro-7-methoxy-9a-propyl-3H-fluoren-3-one in 96% yield. The opposite enantiomer was prepared in 78% ee, 99% yield using N-(p-trifluoromethylbenzyl)cinchonidinium bromide as catalyst. Kinetic and mechanistic studies have provided evidence for an intermediate dimeric catalyst species and subsequent formation of a tight ion pair between catalyst and substrate. More reactive alkylating reagents alkylate tertiary amine decomposition products derived from the chiral catalyst to form new phase-transfer catalysts which in turn produce racemic product, thus lowering the overall observed asymmetric induction. The quantity of the chiral catalyst necessary was significantly reduced by the addition of achiral phase-transfer co-catalysts derived from poly(ethylene glycols). In a proposed mechanistic scheme of this novel dual catalysis the achiral co-catalyst carries out the phase-transfer part of the reaction while the high affinity of the chiral catalyst for the substrate forces the chiral alkylation exclusively through the substrate-chiral catalyst ion pair without loss of asymmetric induction.

I. INTRODUCTION

We became interested in chiral alkylations[1] in the course of the preparation of the diuretic drug candidate Indacrinone **8** which has an asymmetric quaternary carbon. The S-(+)-enantiomer has uricosuric activity while the diuretic activity resides mainly in the R-(-)-enantiomer.[2] Both activities are desirable, but the R enantiomer is much more active. To balance the therapeutic effects a 90/10 ratio of S/R enantiomers was proposed as the drug candidate. The racemic synthesis (Fig. 1) included the phase-transfer methylation of the intermediate 2-phenylindanone **5** and our attention naturally turned to incorporating chirality into the alkylation of this cyclic ketone.

Asymmetric alkylations of ketones using chiral auxiliaries have been developed into elegant three-step sequences (Fig. 2) producing excellent enantiomeric excesses (ee's).[3] The most commonly used auxiliaries are (S)-2-amino-1-methoxy-3-phenylpropane (Meyers), 1-amino-2-methoxymethyl-pyrrolidine (Enders) and (S)-valine tert-butyl ester (Koga). In general, after the derivatization of the ketone with the chiral auxiliary, the intermediate is lithiated and alkylated at -78°C. Hydrolysis generates the alkylated ketone. As an alternate approach, the use of chiral alkylating agents has met only limited success[4] (Fig. 3). Both methods employ the chiral auxiliary in stoichiometric amounts and for an economical process the auxiliary would have to be recycled. In contrast, chiral phase-transfer-mediated alkylations offer a potentially simple, one-step solution to this problem. Early reports[5] of such alkylations using ephedrinium or N-benzyl-quininium halides as catalyst claimed ee's of only 15% (Fig. 4). However, higher catalytic asymmetric inductions achieved in epoxidations[6] and Michael additions using chiral Crown ethers[7] or quaternized cinchona alkaloids[8] encouraged us to pursue the catalytic route (Fig. 5).

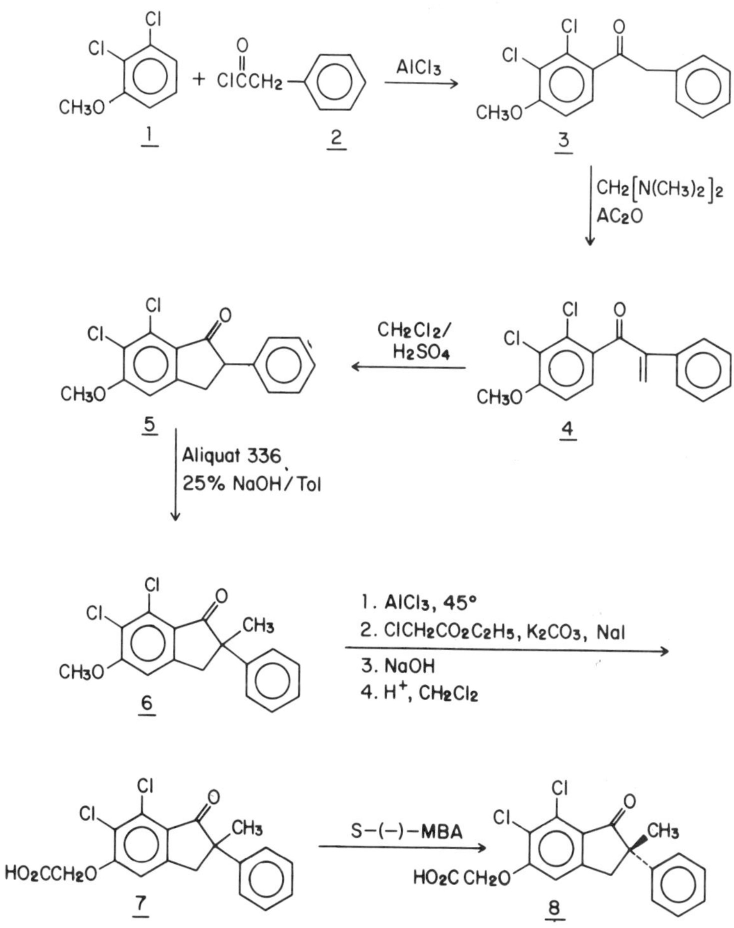

Fig. 1. Racemic synthesis and resolution of Indacrinone.

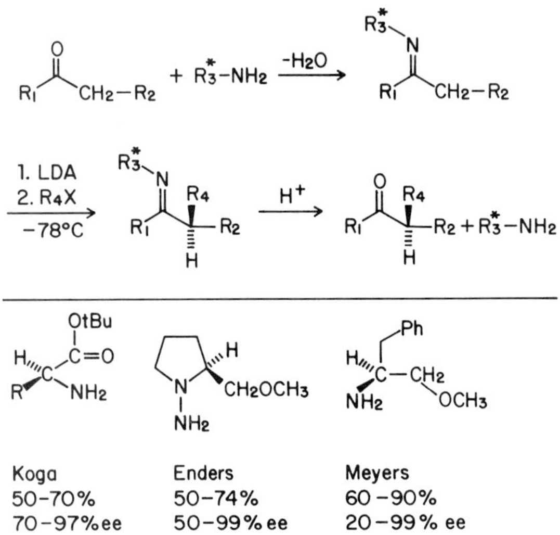

Fig. 2. Asymmetric alkylations with chiral auxiliaries.

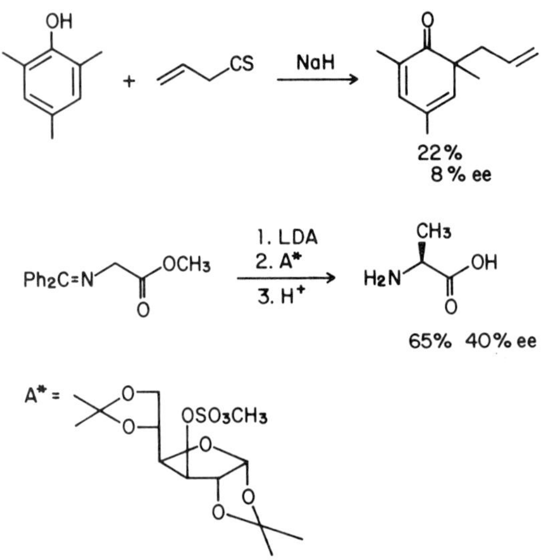

Fig. 3. Asymmetric alkylations with chiral alkylating reagents.

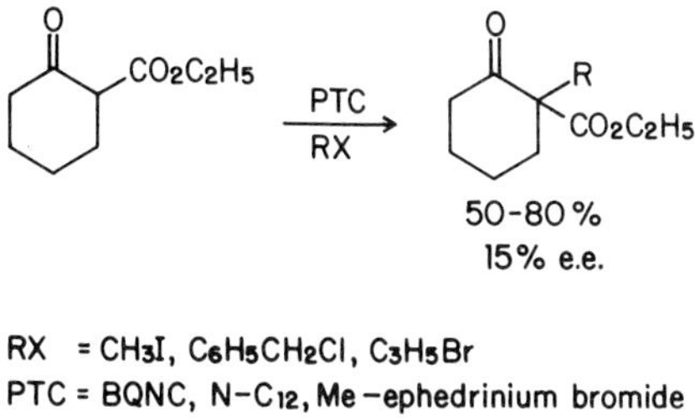

Fig. 4. Asymmetric alkylations with chiral phase-transfer catalysts.

II. CHIRAL PHASE-TRANSFER METHYLATION

The synthetic challenge we faced is shown in Figure 6. We were aiming for a phase-transfer catalyzed methylation upwards of 90% ee in which the base, the solvent, the alkylating agent and, most importantly, the chiral catalyst had to be defined. In this paper I would like to show you how we evaluated these parameters and how the knowledge of the function of each contributed to the understanding of this reaction as a whole thereby enabling us to succeed in our endeavor.

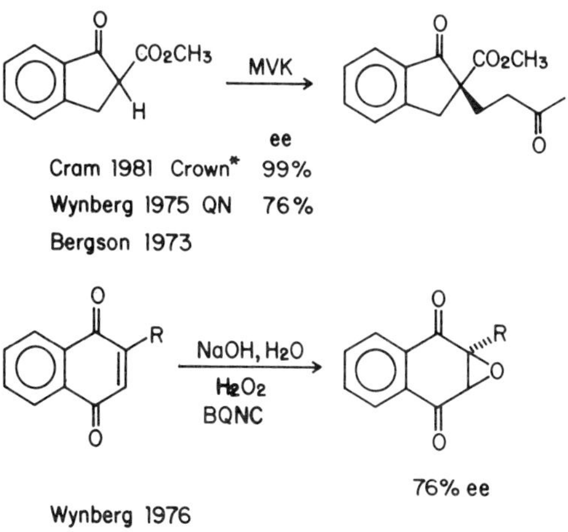

Fig. 5. Catalytic asymmetric Michael addition and epoxidation.

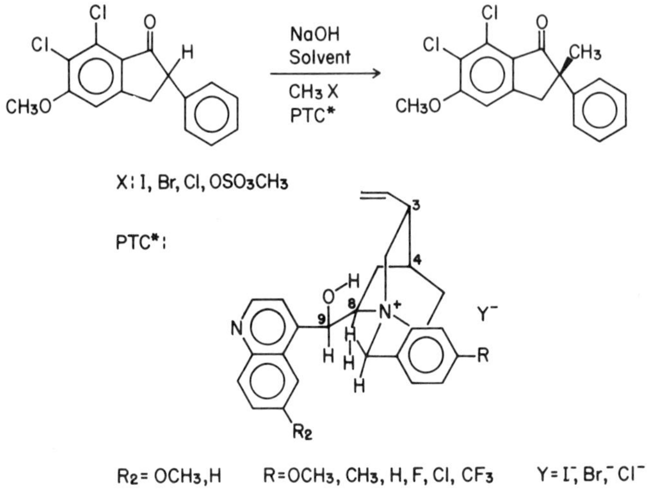

Fig. 6. Chiral phase-transfer methylation of indanone 5.

A. Important Reaction Parameters

Initially I would like to summarize some of the effects the reaction parameters had on the asymmetric phase-transfer methylation. Based on Fiaud's work[5] we screened the quaternized cinchona alkaloids as potential catalysts and found that the N-benzyl rather than N-methyl, butyl or octyl derivatized alkaloid was a better catalyst. The desired enantiomer was produced by the cinchoninium catalyst in higher ee than by the quinidinium catalyst, while quininium and cinchonidinium catalysts produced the opposite enantiomer. The ee at this point was in the 0-10% range.

The nonpolar solvents toluene or benzene produced higher ee's than polar solvents such as methylene chloride or methyl tert-butyl ether as previously observed by Wynberg.[9] The ee's increased with increasing NaOH concentrations, 50% NaOH giving the best results. Chloride and bromide as catalyst counterion produced similar results whereas iodide decreased the ee substantially. This brought the ee up to 20%. Finally, the less reactive alkylating agent methyl chloride gave by far the best selectivity (60% ee) relative to methyl bromide (30%), iodide (20%) or dimethyl sulfate.

B. Ion Pair Model

In a phase-transfer reaction obviously the ion pairing between the positively charged ammonium phase-transfer catalyst and the enolate substrate is very important. CPK molelcular models, the single crystal x-ray structure, and molecular modeling studies of the N-benzylcinchoninium ion suggest that the quinoline ring, the C-O bond, and the N-benzyl group all lie in one plane (Fig. 7). The enolate anion also has an almost planar structure with the negative charge delocalized into the 2-phenyl ring. Both molecules in their nearly planar conformations fit naturally on top of each other providing - interaction between the benzyl group of the catalyst and the 2-phenyl group of the enolate on the one side and between the quinoline and methoxydichlorobenzene moieties on the other. Hydrogen bonding between the C_9-hydroxyl and the enolate anion provides a directional handle and thereby, due to the asymmetry at C_8 and C_9 of the catalyst, completes an enantioselective three-point interaction between substrate and catalyst. Derivatization or removal of the C_9-hydroxyl group resulted in complete loss of enantioselectivity. Inversion of C_8 and C_9, as in the cinchonidinium derived catalysts, gives rise to the opposite enantiomer.

The ion pair can then only be alkylated by methyl chloride from the front side and forms (S)-(+)-6,7-dichloro-5-methoxy-2-methyl-2-phenyl-1-indanone. The solvent effect and the effect of increasing NaOH concentrations, which make the organic solutions increasingly drier, suggest that the tightness of the generated ion pair between substrate and catalyst is proportional to the observed ee.

This should make the asymmetric induction sensitive to the electronic effects of substituents on the N-benzyl group. A Hammett plot of log ee/ee_0 vs. the substituent constant of the para N-benzyl substituted catalysts (R = CH_3O, CH_3, H, F, Cl, CF_3) gave a reaction constant of ρ = 0.21 $\pm$ 0.02 with ee's in the range of 60% to 92% (Fig. 8). This demonstrates that substituents with increasing electron-withdrawing power enhance the interaction or tightness in the ion pair and thereby improve the catalyst selectivity. Clearly the N-(p-trifluoromethylbenzyl)cinchoninium bromide (p-CF_3BCNB) is the catalyst of choice in combination with CH_3Cl as methylating agent. The trifluoromethyl group is the most expensive part of

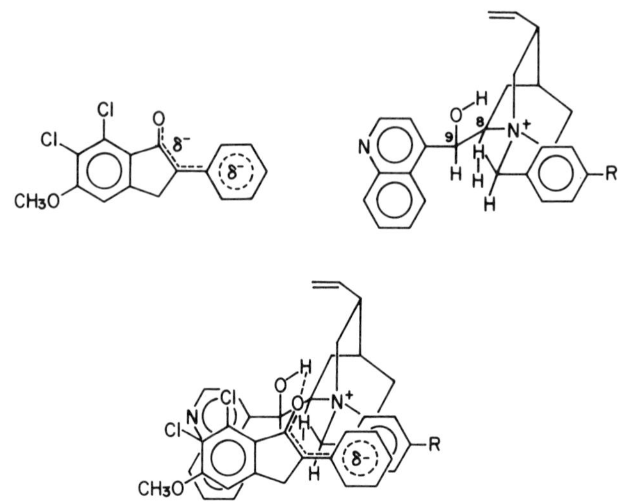

(a)

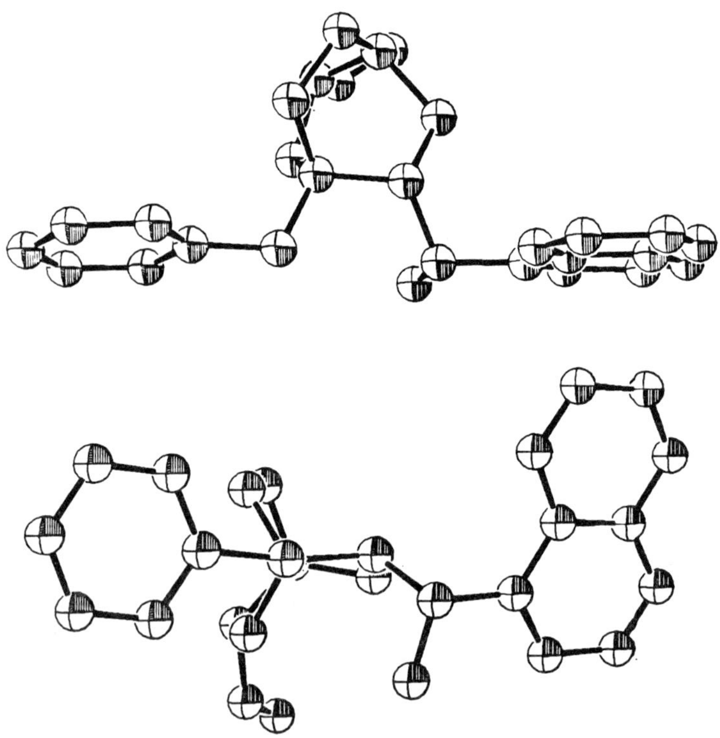

(b)

Fig. 7. (a) Ion pairing between indanone anion and benzylcinchoninium cation. (b) ORTEP structure of the catalyst cation rotated 90° to show the planar catalytic surface.

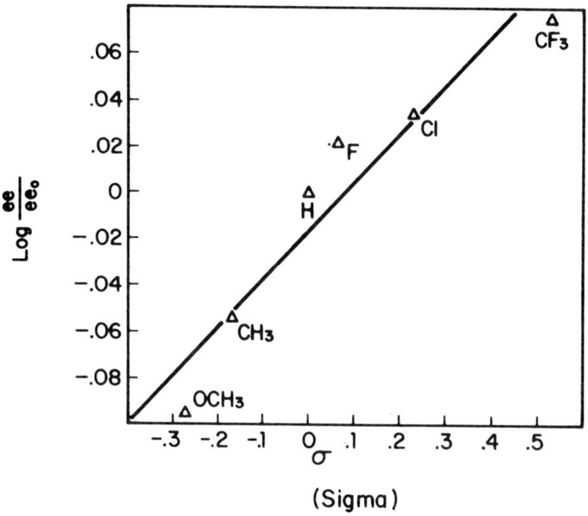

Fig. 8. Hammett plot.

the catalyst and in a commercial situation cost reduction becomes an important part in a synthesis. Based on the observed Hammett relationship we found that the 3-chloro and 3,4-dichlorobenzyl substituted catalysts are almost as effective as the p-trifluoromethylbenzyl catalyst.

C. Catalyst Decomposition

Of great concern to us was Dehmlow's cautionary note[10] in which he showed that under basic conditions chiral β-hydroxy ammonium catalysts undergo Hofmann elimination to give optically active oxiranes which in turn may contaminate the products and generate apparent optical activity. We determined the optical activity not only by rotation but also by specific NMR assays (Fig. 9) with a chiral shift reagent, (tris[3-((heptafluoropropyl)-hydroxymethylene)-d-camphorato]europium(III)), and HPLC using a Pirkle column (ionic phenylglycine). Having assured the validity of our results we investigated the catalyst decomposition and this led to a further understanding of the chiral alkylation. Indeed, we find small amounts of the corresponding oxirane **12** in our reaction mixtures (Fig. 10). However, under

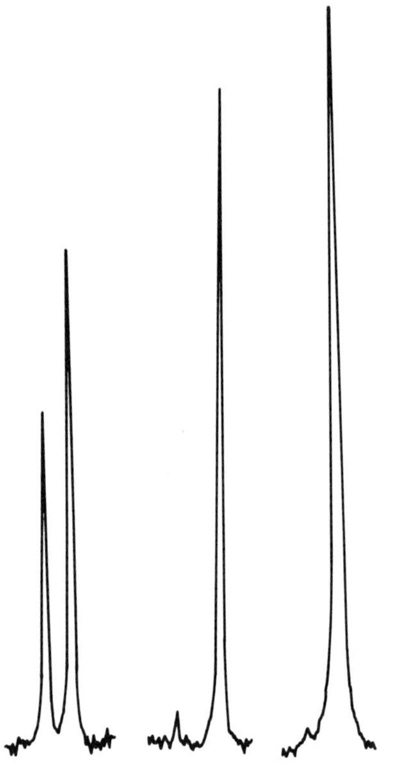

Fig. 9. NMR assays of a synthetic mixture (18% ee) of (+) and (-) enantiomers and reaction product mixtures (92% ee and 96% ee).

basic alkylating conditions the major decomposition product arises via alkylation of the alkoxide **9.** The resulting methyl ether **10** readily undergoes E2 elimination forming the enol ether **11.** Oxirane **12,** methyl ether **10** and enol ether **11** were independently prepared and characterized. Under reaction conditions the methyl ether **10** is rapidly converted to the enol ether. Both oxirane and enol ether are methylated with CH_3I under reaction conditions to form the new quaternary ammonium salts **13** and **14** which act as phase-transfer catalysts and produce racemic 2-methylindanone. This can account in part for the low asymmetric induction

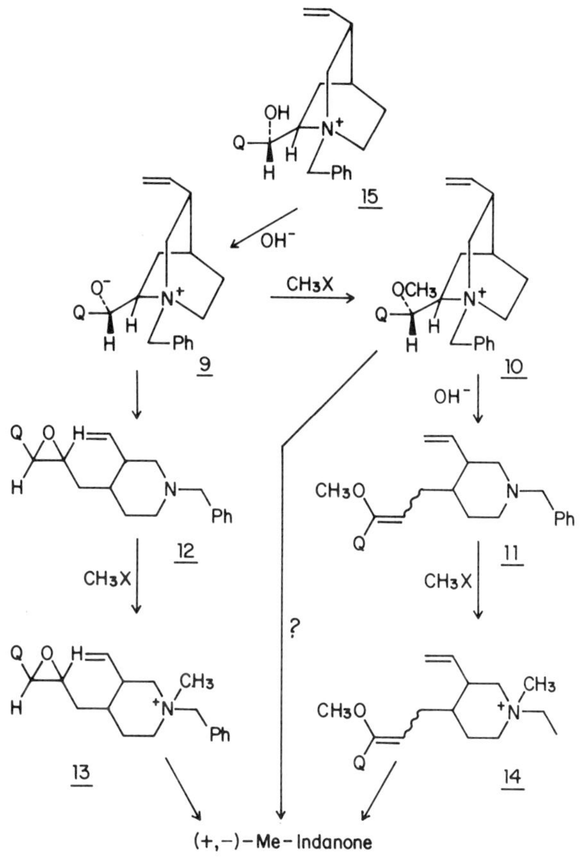

Fig. 10. Catalyst decomposition pathways.

using CH_3I as alkylating agent. With CH_3Br alkylation of tertiary amines **11** and **12** is slow relative to the alkylation of indanone **5**. CH_3Cl does not alkylate the amines under the reaction conditions and thereby preserves the integrity of the asymmetric alkylation.

D. Reaction Pathway

In a typical reaction the indanone, catalyst and CH_3Cl are charged to a 5 to 1 (volume) mixture of toluene and 50% aqueous NaOH. Stirring at 20°C for 18 hours completes the methylation. Experiments with p-CF_3BCNB and

TABLE I

Effect of the Molar Ratio of Indanone to Catalyst on the Asymmetric Induction.[a]

5 mmol	p-CF_3BCNB mmol	ee (+) %	5 / p-CF_3BCNB
2.5	0.25	90	10
5	0.25	69	20
10	0.25	0	40
10	0.50	70	20
10	1.0	82	10
5	0.50	87	10
5	0.125	0	40

CH_3Cl demonstrated some unusual behavior (Table I): at constant catalyst concentration the ee will dramatically decrease from 90% to 0% with increasing indanone concentration and at constant indanone concentration the ee will increase with increasing catalyst concentration. The table shows that the ee is directly related to the molar ratio of indanone to catalyst. To understand these findings we measured during the course of a reaction the concentration of indanone **5** and the catalyst in the toluene layer as well as the % of indanone remaining in the reaction mixture, the balance being the product 2-methylindanone (Fig. 11). The methylation proceeds slowly during the first 6 to 7 hours. During this "induction" period the concentration of the indanone in the toluene layer drops to less than 0.2 mg/ml. At this point the catalyst concentration, which remained so far very low in the toluene layer starts to increase which is accompanied by an increase in the rate of

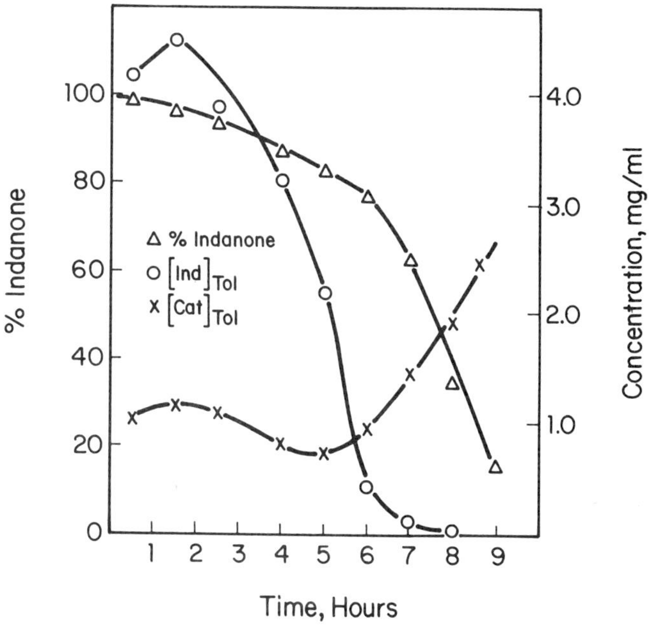

Fig. 11. Chiral phase-transfer methylation. Reaction profile.

methylation. An explanation of this unusual behavior was found in the following experiments. When a 16 mM solution of indanone **5** in toluene is stirred in the absence of any catalyst with 50% aqueous NaOH, the indanone is deprotonated at the interface to form the sodium enolate which precipitates thus reducing the concentration of **5** in toluene (Fig. 12). The precipitation of the sodium enolate is slower with 40% NaOH, and ceases with 30% NaOH. Just the opposite behavior was found for the catalyst which is almost insoluble in toluene or in 50% NaOH alone (<1 $^{-5}$M). But p-CF$_3$BCNB does dissolve when it is stirred in a mixture of toluene and 50% NaOH. The catalyst can be found in the toluene as a dimer of the catalyst **15** and its zwitterionic oxide **9** (Fig. 10). The dimer structure is supported by the following experimental evidence. The catalyst solution contains one mol of base and one mol of bromide per two mols of catalyst. Titration with acid does not generate an equivalent amount of water thereby excluding OH$^-$ as counterion. When the catalyst solution is quenched with

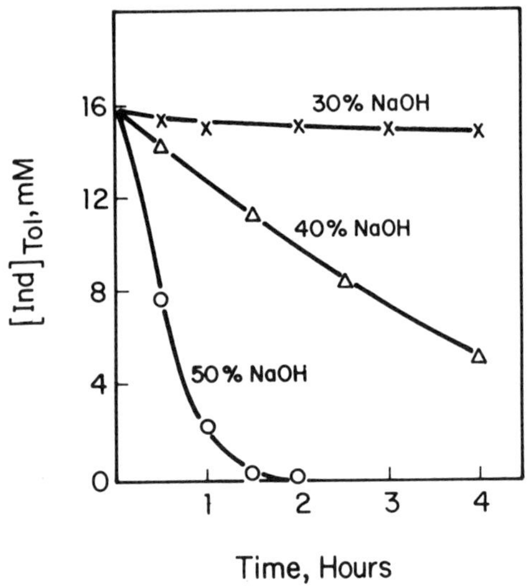

Fig. 12. Interfacial deprotonation of indanone 5 in toluene/aqueous NaOH.

excess CH_3I a 50/50 mixture of catalyst **15** and methyl ether **10** is generated. An NMR spectrum of a dimer solution of the dihydro catalyst in C_6D_6 shows the two cinchoninium ions to be equivalent. The protons on the quinoline rings are shifted downfield and those of the phenyl ring upfield relative to the chemical shifts of the soluble octylsulfonate salt (Fig. 13).

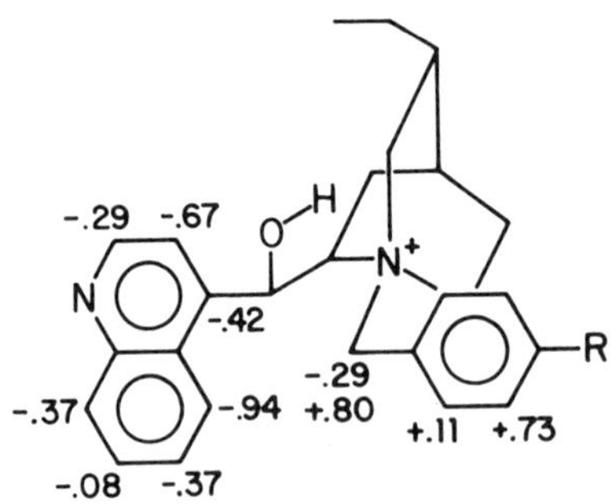

Fig. 13. Relative chemical shift differences between the catalyst dimer and the catalyst octylsulfonate salt.

This suggests π-π electron donation by the quinoline to the phenyl ring. Hydrogens on the face of the quinuclidine ring also experience large chemical shifts, whereas those in back remain essentially unchanged. Assuming that a symmetrical hydrogen bond forms between the hydroxyl of the catalyst and the alkoxyl of the zwitterion, a three dimensional structure for the dimer emerges that is remarkably similar to that proposed for the indanone anion-catalyst ion pair (Fig. 7).

Addition of indanone **5** to the catalyst dimer solution results in deprotonation of **5** by the zwitterion. Since the zwitterion is now protonated the catalyst is not any more completely soluble and its concentration in the toluene drops. In the reaction profiled in Fig. 11 only a low level of the catalyst enters the toluene layer as long as indanone is present but when all the indanone has been deprotonated the concentration of the catalyst increases.

Preformation of the indanone enolate in toluene/50% NaOH followed by addition of catalyst and CH_3Cl eliminates the induction period and most

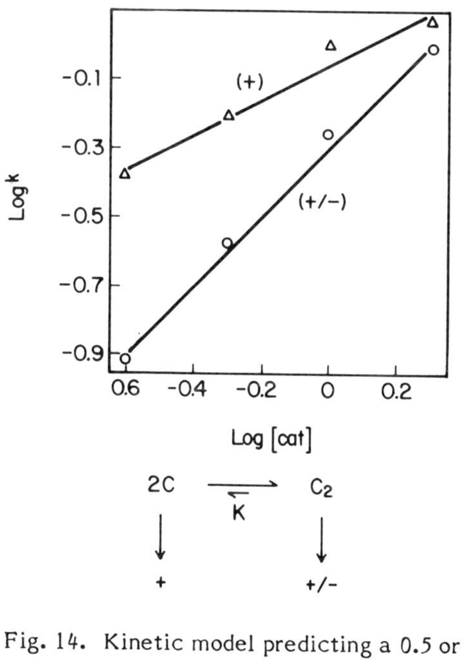

Fig. 14. Kinetic model predicting a 0.5 order in catalyst (30% NaOH).

importantly the high sensitivity of rate and ee to the indanone catalyst ratio. Under these conditions the reaction is 0.55 order in catalyst. A kinetic model in which the rate has a first order dependence on the monomeric catalyst, the amount of monomer is very small, and the equilibration between dimer and monomer is fast predicts the measured order in catalyst to be 0.5 (Fig. 14). This is consistent with our finding that the catalyst goes into solution as a dimer which must dissociate prior to complexation with the indanone anion. For comparison, with Aliquat 336 which cannot form such a dimer the order in the catalyst is 1.

III. CHIRAL ROBINSON ANNULATION

This chiral phase-transfer alkylation methodology was applied to the asymmetric synthesis of the tricyclic enone 20, an intermediate for a drug candidate under development.[11] The racemic compound was prepared via base catalyzed methyl vinyl ketone (MVK) addition to 2-propylindanone **17** followed by acid or base catalyzed aldol condensation (Fig. 15). Using 1,3-dichloro-2-butene (Wichterle reagent) as an MVK surrogate,[12] p-CF_3BCNB as catalyst and the conditions previously described produced (S)-(+)-6,7-dichloro-2-(3-chloro-2-butenyl)-2,3-dihydro-5-methoxy-2-propyl-1-inden-1-one **19** in 99% yield and 92% ee. Hydrolysis of the intermediate vinyl chloride (0°C, 1.5 h) and cyclization (70°C, 1h) in concentrated sulfuric acid produced **20** in 96% overall yield. However, as expected by the tight ion pair theory the enantiomer opposite to the desired one, shown in Fig. 15, was formed. The diastereomeric catalyst-N-(p-trifluoromethylbenzyl)cinchonidinium bromide (p-CF_3BCDB) which has only C_8 and C_9, the two chiral centers on the catalytic surface, inverted, produced a 97% yield at 76% ee. However, 30 mol % of the catalyst was necessary to completely alkylate the indanone. Study of space filling models suggests that the vinyl group in the cinchonine catalyst blocks the hydroxyl from the backside thus making it available for hydrogen bonding only from the front, the asymmetric catalyst surface. In the cinchonidine catalyst **16** this blocking is absent which may confer greater degrees of freedom to the system and hence lower ee's (Fig. 16).

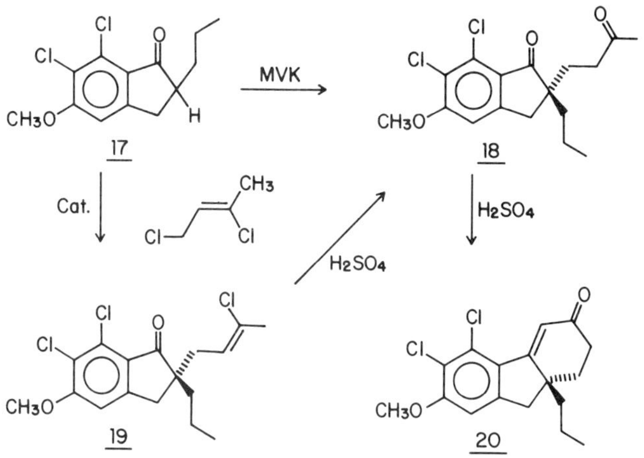

Fig. 15. Reaction scheme for the chiral Robinson annulation using the Wichterle reagent.

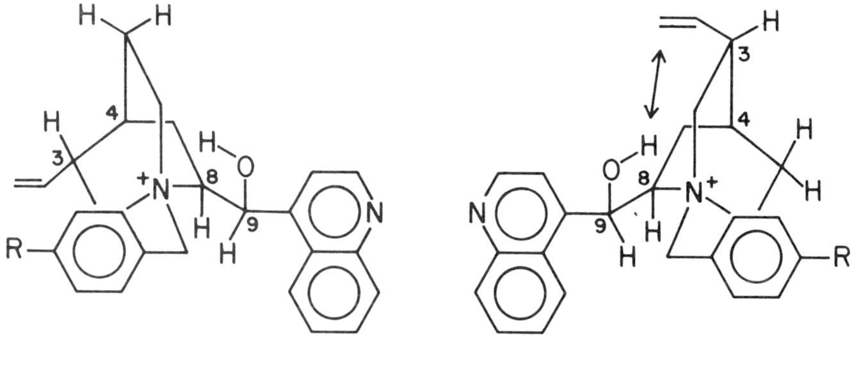

Fig. 16. Interaction between the C9-OH group and the C3-vinyl group in cinchonine 15 and cinchonidine 16 derived catalysts.

IV. DUAL CATALYSIS

In view of an economical process the need for 30 mol % of the chiral catalyst was not acceptable. The high catalyst levels were necessary due to the catalyst instability and the inefficient phase-transfer-catalyst properties relative to the common tetra-alkylammonium catalysts. What we needed was a way to facilitate the phase-transfer part of the reaction system without interfering with the chiral alkylation of the enolate-chiral catalyst ion pair. A surfactant, that is non-ionic to avoid interfering with the ion pairing, was hoped to facilitate the transport over the phase boundaries. By chance we had a bottle of such a surfactant, Triton X-405 (TX), a poly (ethylene glycol p-isooctylphenyl ether), on the shelf. Addition of a small amount of this surfactant enhanced the rate of reaction tremendously reducing the reaction time from 18 hours to less than 2 hours without decreasing the ee. Similar results were obtained with other poly (ethylene glycols) and their ethers (Fig. 17). Not surprisingly, closer examination showed that TX alone catalyses the racemic alkylation of indanone **17** (Table II) in a similar fashion as crown ethers.[13] In 18 hours with 0.5 mol % of TX 10% of racemic product is formed. An increase to 8 mol % produces 60% of product. No product was observed with 30% and 40% aqueous NaOH. However, solid KOH enhanced the racemic alkylation to 40% within 1 h. In contrast, with p-CF_3BCDB and toluene/solid KOH no alkylation is observed. But with 7.5 mol % p-CF_3BCDB and 0.1 mol % TX the reaction gives a quantitative yield, 80% ee. Surprisingly, the racemic reaction is completely suppressed.

TX extracts KOH or NaOH into the toluene layer, 5 mM solutions can readily be prepared. This is consistent with the recent observation of ion pair solvation by poly (ethylene oxide).[14] Addition of indanone to this solution immediately results in a yellow color characteristic of the indanone enolate anion. In contrast to the crown ethers and cryptands the poly (ethylene glycol) derived catalysts have a weaker binding thereby enabling the chiral catalyst to effectively compete for the enolate anion. Kinetic experiments show that in the presence of TX the order in the chiral catalyst is changed from 0.5 to 1. Thus a mechanistic picture emerges in which the TX transfers the base into the toluene layer where the indanone is deprotonated to form the TX-K^+-enolate ion pair (Fig. 18). Similarly TX

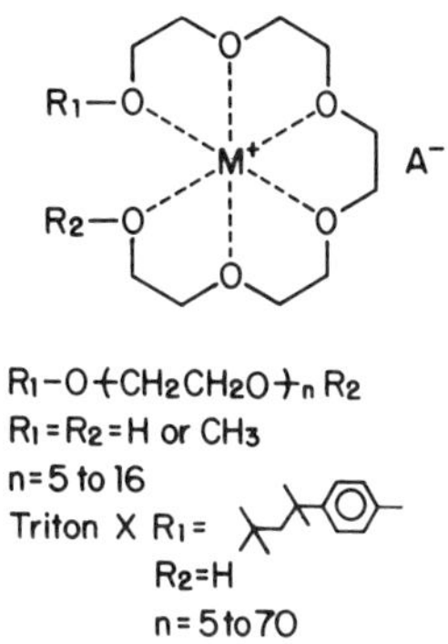

Fig. 17. Mechanistic scheme proposed for the dual catalysis.

solubilizes the chiral catalyst as monomeric ammonium salt eliminating the dimer formation in the toluene. The TX complexes are in equilibrium with each other and the enolate and chiral catalyst cation collapse to the tight ion pair due to their affinity for each other before racemic alkylation of the

TABLE II

Racemic Alkylation With Triton X-405[a]

NaOH %	Triton X-405 mol %	Racemic 19 % in 18 hr
30	0.5	0
40	0.5	0
50	0.5	10
50	8	60
KOH[b]	0.5	40[c]

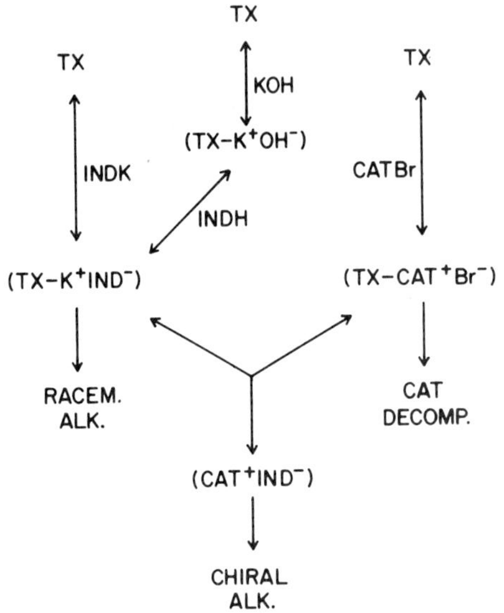

$$TX + KOH \longleftrightarrow [TX\text{-}K^{+}\text{-}OH^{-}]$$

$$[TX\text{-}K^{+}\text{-}OH^{-}] + INDH \longleftrightarrow [TX\text{-}K^{+}\text{-}IND^{-}] + H_2O$$

$$TX + BCDB \longleftrightarrow [TX\text{-}BCD^{+}\text{-}B^{-}]$$

$$[TX\text{-}K^{+}\text{-}IND^{-}] + [TX\text{-}BCD^{+}\text{-}B^{-}] \longleftrightarrow [TX\text{-}K^{+}\text{-}B^{-}] + [IND^{-}\text{-}BCD^{+}] + TX$$

$$[IND^{-}\text{-}BCD^{+}] + RY \longleftrightarrow IND^{*}R + BCDY$$

$$[TX\text{-}K^{+}\text{-}IND^{-}] + RY \longleftrightarrow INDR + [TX\text{-}K^{+}\text{-}Y^{-}]$$

Fig. 18. Mechanistic scheme proposed for the dual catalysis.

$TX\text{-}K^+$-enolate ion pair can occur. This process can obviously be influenced by mass action. Increasing amounts of KOH and co-catalyst, e.g. decreasing chiral catalyst/co-catalyst ratios, will decrease the observed ee. Molar ratios of chiral catalyst to co-catalyst larger than 25 are in general sufficient to preserve the high asymmetric induction. We usually use 5 mol powdered KOH (15% H_2O) per mole of indanone, 7 mol % chiral catalyst and

0.07 mol % co-catalyst. This represents a unique and novel dual catalysis in which the (inexpensive) racemic catalyst carries out the phase-transfer work while the efficiency of the (more expensive) chiral catalyst is enhanced by restricting its use to the selective complexation of the substrate in the organic solvent.

REFERENCES

1. U.-H. Doling, P. Davis and E.J.J. Grabowski, J. Am. Chem. Soc., **106**, 446 (1984). A. Bhattacharya, U.-H. Dolling, E.J.J. Grabowski, S. Karady, K.M. Ryan and L.M. Weinstock, Angew. Chem., Int., Ed., **25**, 476 (1986).
2. S.J. deSolms, O.W. Woltersdorf, Jr. and E.J. Cragoe, Jr., J. Med. Chem., **21**, 437 (1978). O.W. Woltersdorf, Jr., J. Labelled Compd. Radiopharm., **17**, 635 (1980). J.D. Irvin, P.H. Vlasses, P.B. Huber, J.A. Feinberg, R.K. Ferguson, J.J. Scrogie and O.R. Davies, Clin. Pharmacol. Ther., **27**, 260 (1980). A.G. Zacchei, M.R. Dobrinska, T.I. Wishowsky, K.C. Kwan and S.D. White, Drug Metab. Dispos., **10**, 20 (1982).
3. M. Pfau, G. Ravial, A. Guingant and J. d'Angelo, J. Am. Chem. Soc., **107**, 273 (1985). D. Enders, Chemtech. 504 (1981). A.I. Meyers, D.R. Williams, G.W. Erickson, S. White and M. Druelinger, J. Am. Chem. Soc., **103**, 3081 (1981). K. Tomioka, K. Ando, Y. Takemasa and K. Koga, J. Am. Chem. Soc., **106**, 2718 (1984). S. Hashimoto and K. Koga, Tetrahedron Lett., 573 (1978). D. Enders and H. Eichenauer, Chem. Ber., **112**, 2933 (1979).
4. P.G. Duggan and W.S. Murphy, J. Chem. Soc. Perkin I, 634 (1976). P. Duhamel, J.-Y Valnot and J. Jamal Eddine, Tetrahedron Lett., 2863 (1983).
5. J.-C. Fiaud, Tetrahedron Lett., 3495 (1975). K. Saigo, H. Koda and H. Nohita, Bull. Chem. Soc. Jpn., **52**, 3119 (1979). S. Julia, A. Ginebreda, J. Guixer and A. Thomas, Tetrahedron Lett., 3709 (1980).
6. H. Pluim and H. Wynberg, J. Org. Chem., **45** 2498 (1980). Y. Harigaya, H. Yamagucchi and M. Onda, Heterocycles, **15**, 183 (1981).

7. D.J. Cram and G.D.Y. Sogah, J. Chem. Soc., Chem. Commun., 625 (1981).
8. K. Hermann and H. Wynberg, J. Org. Chem., **44**, 2238 (1979). S. Colonna, A. Re and H. Wynberg, J. Chem. Soc. Perkin I, 547 (1981). H. Wynberg, Recl. Trav. Chem. Pays.-Bas, **100**, 393 (1981). H. Wynberg, Chemtech, 116 (1982).
9. H. Wynberg and B. Greijdanus, J. Chem. Soc., Chem. Commun., 427 (1978).
10. V.E. Dehmlow and J. Heider, J. Chem. Res, Synop., 292 (1981).
11. E.J. Cragoe, Jr., G.E. Stokker and N.P. Gould, U.S. Patent 4316043 (1982).
12. O. Wichterle, J. Prochaska and J. Hoffman, Coll. Czech. Chem. Commun., **13**, 300 (1948). M.E. Jung, Tetrahedron, **32**, 3 (1976). R.E. Gawley, Synthesis, 777 (1976).
13. E.V. Dehmlow and S.S. Dehmlow, Phase Transfer Catalysis, Monographs in Modern Chemistry, 11, (H.F. Ebel, ed.) Verlag Chemie, Weinheim (1983). C.M. Starks and C. Liotta, Phase Transfer Catalysis, Academic Press, New York (1978).
14. G.N. Arkhipovich, E.A. Ugolkova and K.S.Kozanskii, Polymer Bull., **12**, 181 (1984).

5

Catalytic Synthesis of Diesters, New Developments and Perspectives

Anne M. Gaffney, John J. Leonard, John A. Sofranko and Hsiang-ning Sun

ARCO Chemical Company
3801 West Chester Pike
Newtown Square, Pennsylvania

ABSTRACT

The syntheses of dicarboxylates, especially oxalates, with palladium on metal oxide catalysts are analyzed in detail. The origins of various products are examined. The possibility of involving palladium(I) intermediates in the reaction is discussed. Other newly developed chemistries and the utilization of CO_2 as the feedstock are also briefly presented.

I. INTRODUCTION

The syntheses of dicarboxylic acids and their derivatives have attracted significant attention for a long period of time. This is not unexpected considering the importance of many materials such as adipic acid, maleic anhydride, terephthalic acid, oxalic acid and others to the chemical industry. Due to the drastic changes in petrochemical feedstock prices during the past fifteen years, many processes have been re-examined, in terms of not only technology but also raw material availability and processing energy costs. It is therefore not surprising to witness that significant progress has been made in utilizing various feedstocks to produce olefins, oxygenates, gasoline components, and, among others, diacids.

TABLE I

Examples of Dicarboxylic Acid/Derivatives Syntheses

Reaction	Reaction Type
P-Xylene ⟶ Terephthalic Acid	Oxidation
Butane ⟶ Maleic Anhydride	Oxidation
Cyclohexane ⟶ Adipic Acid	Oxidation
Xylene ⟶ Trimellitic Anhydride	Carbonylation
Acrylonitrile ⟶ Adiponitrile	Dimerization
4-Pentenoate ⟶ 4-Octenedioate	Metathesis

A survey of one- to three-step syntheses of dicarboxylic acids and their derivatives reveals that there are four major type reactions, excluding many elegant reaction sequences for making complicated products. These include oxidations, carbonylations, dimerizations, and methathesis (Table I).

Oxidations have been used commercially to produce such important and large volume chemicals as adipic acid, maleic anhydride, p-terephthalic acid, etc.. Carbonylations, however, have not been used significantly in producing diacids in spite of the significant technical and commercial success in producing acetic acid from methanol. Recently, a three-step synthesis of trimellitic anhydride via a route involving an improved Gattermann-Koch reaction of xylene followed by oxidation/dehydration was revealed.[1] The original development was intended for p-terephthalic acid production.[2] Dimerization of unsaturated esters has been known for some time.[3] Hydro-dimerization of acrylonitrile to adiponitrile electrochemically is perhaps the most significant commercial success. Research has been quite active in exploring metathesis reactions with functionalized alkenes.[4] Both esters and nitriles have been investigated. The advances in the Re_2O_7/Al_2O_3-Lewis acid catalysts appear quite promising in producing a variety of long-chain diacid derivatives.[5] With the recent emphasis on producing specialty chemicals, it should not be surprising to see more applications and more new catalysts evolving around this unique reaction.

Oxidative carbonylation of alkenes to give carboxylic acid/esters has been known for some time[6-8] (eq. 1,2). Homogeneous palladium catalysts

$$CH_2{=}CH_2 + CO + 1/2\ O_2 \rightarrow CH_2{=}CH{-}COOH(R) \quad (1)$$

$$CH_2{=}CH_2 + CO + 2\ ROH + O_2 \rightarrow ROOCCH_2CH_2COOR + H_2O \quad (2)$$

along with copper or iron salts are used. These routes utilize CO and ethylene to produce more valuable acrylates or succinates. Oxalates become the major products in the absence of alkene (eq. 3). This is a very

$$2CO + 2ROH + 1/2\ O_2 \rightarrow RO\overset{O}{\overset{\|}{C}}{-}\overset{O}{\overset{\|}{C}}OR + H_2O \quad (3)$$

intriguing reaction because two CO molecules are coupled, an equivalent of "double carbonylation". A similar reaction in the presence of butadiene will produce a precursor for adipic acid[9] (eq. 4). This CO coupling reaction can

$$CH_2{=}CH{-}CH{=}CH_2 + CO + O_2 + ROH \rightarrow ROOCCH_2CH{=}CHCH_2COOR \quad (4)$$

be carried out in the presence of an alkyl nitrite as the oxidizing agent if a palladium on silica catalyst is used.[10]

A large number of one-electron and two-electron redox metal oxides have been studied as supports for the CO to oxalates reaction.[11] It was found that certain metal oxides, such as V_2O_5, V_2O_5-TiO_2, V_2O_5-P_2O_5 and Mn_2O_3, could significantly enhance oxalate formation, while others, such as WO_3, Nb_2O_5 and Sb_2O_5, were ineffective. This paper discusses the origins of various products and the reaction mechanism in view of recently published results from various laboratories.

II. EXPERIMENTAL

The details of experimental procedures, catalyst syntheses and product analyses are published elsewhere.[11]

III. DISCUSSION

The focus here will be placed on the sources of various products and on reaction mechanism. Quantitative data and kinetic analyses have been published elsewhere.[11] The V_2O_5 and V_2O_5-TiO_2 catalysts are used in most cases as examples, but the general conclusions should be applicable to other catalysts as well.

A. Origin of CO_2

Carbon dioxide is usually one of the major by-products formed during the reaction. It is known that CO oxidation on V_2O_5 alone is insignificant. Another possible source is oxidation of oxalates. The reaction between V_2O_5 and oxalic acid (eq. 5) was reported to occur in the presence of acetic acid at 110°C.[13] Since it was observed that longer reaction times produced more

$$V_2O_5 + 3H_2C_2O_4 \rightarrow 2VO(C_2O_4).2H_2O + 2\,CO_2 \qquad (5)$$

oxalic acid and CO_2, this secondary oxidation could be a major source of CO_2. However, a stability study of oxalic acid in methanol under reaction conditions without CO for one hour produced only trace amount of CO_2. This rules out the above oxalic acid - V_2O_5 reaction. The reaction probably requires the presence of acetic acid. In addition, methanol oxidation is also ruled out as a major source of CO_2. The bulk of CO_2 comes from direct CO oxidation on the palladium sites of the catalyst.

Comparing CO_2 formation on a 1% Pd/SiO_2 catalyst vs. a 1% Pd/V_2O_5 -TiO_2 catalyst shows a five-fold increase in CO_2 production for the latter (Table II). This is in qualitative agreement with the comparisons between Pd/SiO_2 and Pd/SnO_2 catalysts.[12] It shows that the Pd/V_2O_5-TiO_2 catalyst in solution provided not only a much higher oxalate formation rate, but also a higher CO oxidation rate. It has been reported that there is a significant enhancement of CO oxidation in the gas phase with a Pd/SnO_2 catalyst at 140°C, much higher than the 100°C used in our experiments.[12] There is an apparent synergistic effect between palladium and the SnO_2 support. It should be noted that the palladium on tin oxide catalyst was prepared in a reducing atmosphere (H_2) and used in a largely oxidizing environment (O_2:CO=10:3) while the palladium on vanadium oxide-titanium oxide was prepared in air and used in the presence of a somewhat reducing atmosphere of O_2:CO=1:5.

TABLE II

Comparison of CO Oxidation Rates

	$1\%\ Pd/V_2O_5\text{-}TiO_2$[1]	$1\%\ Pd/SnO_2$[2]
Reaction Temp. (°C)	100	140
O_2:CO	1:5	10:3
Reaction Medium	Methanol	Gas Phase
Relative Rates of CO Oxidation[3]	5	8

1 Reference 11.

2 Reference 12.

3 CO oxidation rates on metal oxide catalysts relative to those on 1% Pd/SiO_2 Catalyst under comparable reaction conditions.

CO oxidation kinetics show that the rate is nearly first order in CO, and a fractional order in O_2. It is difficult to compare our results with vapor phase data due to different diffusion characteristics. But qualitatively, our results are not inconsistent with published work on Pd/SnO_2 catalysts.

B. Dimethyl Carbonate

The selectivity to dimethyl carbonate is usually very low. A likely mechanism for DMC formation in a homogeneous liquid phase is shown in eq. 6.[13] It is probable that the same mechanism also operates for the

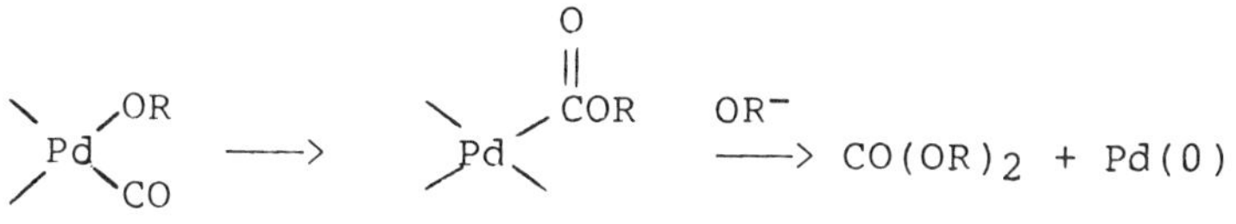

(6)

heterogeneous reaction. Higher CO pressure, thus higher CO concentration in solution and on the catalyst surface, would reduce the nucleophilic attack of OR^- on the carboalkoxy palladium intermediate (Table III). The

TABLE III

Dimethyl Carbonate Formation as a Function of CO Pressure

System	P CO (ATM)	DMC/Oxalate (Mole Ratio X 100)
A	6	16
A	40	3
B	144	1

A. $Pd(OAc)_2/PPh_3$ $CH_3OH/80^oC/2h$ Reference 17.

B. $Pd/V_2O_5\text{-}TiO_2/CH_3OH/100^oC/1h$ Reference 11.

possibility of the participation by the surface V-OR species due to CO pressure change/vanadium-palladium oxidation state change cannot be totally ruled out for the heterogeneous case. V-OR groups are very likely present on the catalyst surface.

C. Methyl Formate/Methylal Formation

The total selectivity to methyl formate and methylal varies in the 0-20% range.[11] Methyl formate can be produced via a simple reaction between CO and methanol with no redox reactions involved (eq. 7). Methylal could be

$$CH_3OH + CO \rightarrow HCOOCH_3 \qquad (7)$$

obtained through CH_3OH oxidation to formaldehyde followed by acetal formation. Although V_2O_5 is known to react with alcohols its involvement here in methylal formation may not be the most important factor. There are several pieces of evidence supporting this conclusion. First, 1% Pd/SiO_2 catalyst with no vanadium oxide component also gave substantial methylal in the products. Secondly, a number of recycle runs with the $Pd/V_2O_5\text{-}TiO_2$ catalyst did not produce methylal at all.

Work on well characterized complexes shows that thermal decomposition of $Pd_2(\mu\text{-}OCH_3)_2(F_6acac)_2$ gives methyl formate, methylal, methanol and a number of other species.[15] It is possible that bridged methoxy groups at certain specific pallaldium sites could be decomposed to give methylal directly. It is not clear why certain recycled catalysts gave no methylal at all. Changing the electronic environment near the palladium sites may be responsible for this observation.

D. Oxalate Formation

The kinetics for oxalate formation has been reported.[11] The rate of oxalate formation is nearly first order in CO, and a small fraction of order in O_2. It is also found that there is a reasonable correlation between the oxalate formation rates in different alcohols and corresponding tosylate solvolysis rates. This suggests that nucleopilic attack of the alcohol on the adsorbed CO is involved in the rate-limiting step or in a pre-equilibrium step. A mechanism based these data and results published on ethylene oxidation on similar catalysts[16] is as shown in Scheme 1.

Independent work on a number of well characterized phosphine containing complexes, $(Ph_3P)_2Pt(COOCH_3)_2$ and $(Ph_3P)_2Pd(COOCH_3)_2$,[17] and $(Ph_3P)_2ClPdCOCOPh$ and $(Ph_3P)_2ClPtCOCOPh$,[18] have essentially ruled out the consecutive double carbonylation route shown in eq. 8.[19]

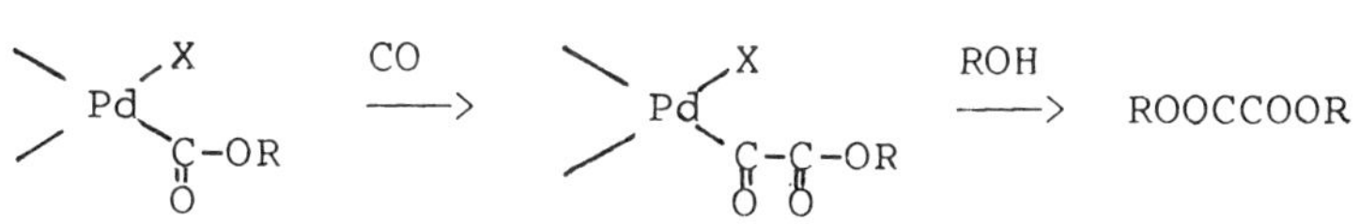

(8)

The possibility of a palladium (I) species during the reaction has not been seriously considered. This is perhaps due to the fact that most earlier reports on isolated Pd(I) complexes were not conclusive. There are quite a number of well characterized palladium complexes known now. Some of the examples are [PdX(CO)]X,[21] which is in a dimeric, and perhaps oligomeric form[20,21] containing bridged X and CO as in **1**.

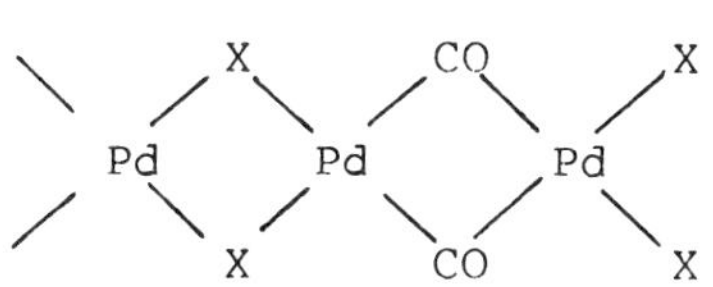

1

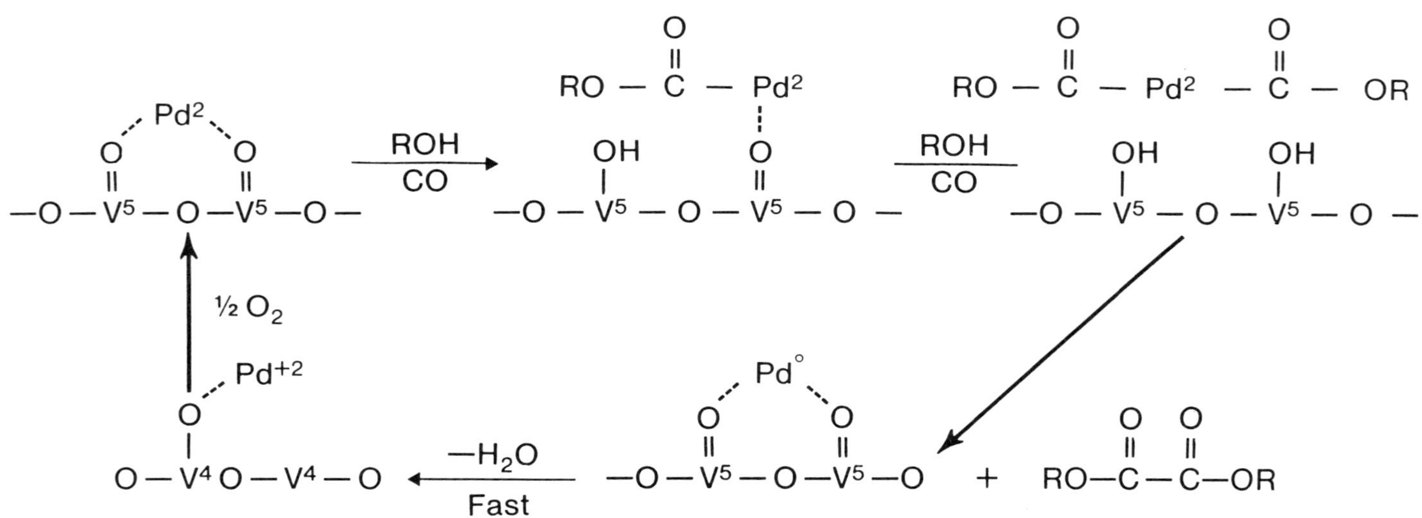

Scheme 1

A number of well characterized Pd(I) dimers have been reported.[22] CO is found to insert readily between the Pd-Pd bond. In addition, CO insertion into a Pd-R bond was also observed (eq. 9, 10). A few tetrameric

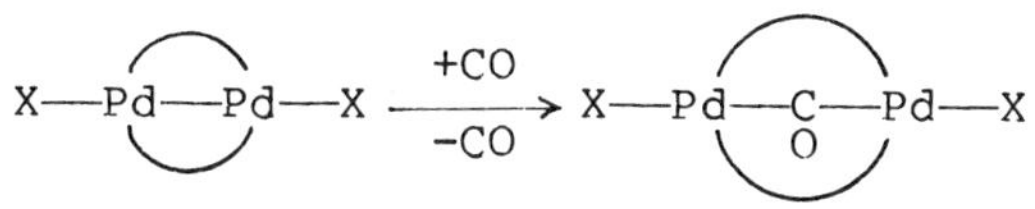
(9)

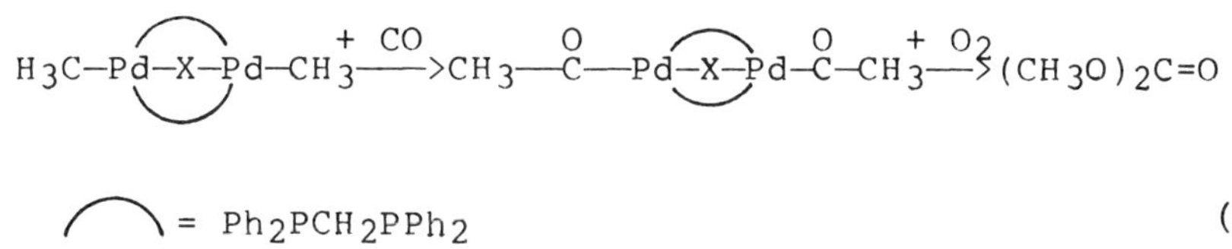
(10)

species, $Pd_4(OCOR)_4(CO)_4 \cdot 1/2RCOOH$, have been reported with their crystal structures.[23] They contain Pd-Pd bonds with bridging CO. One interesting common feature among these Pd(I) CO-containing complexes is that they all possess good stability toward oxygen. The presence of Pd(I)CO intermediates is therefore possible under the conditions for oxidative coupling of CO to give oxalates.

In the gas phase ethylene oxidation reaction, it was reported that addition of small amounts of TiO_2 has a significant effect on catalyst activity.[16] It is interpreted as a p-type dopant that lowers the Fermi level of the catalyst system, thus making the whole system more reducible. Similarly, the Pd/V_2O_5-TiO_2 catalyst in our case is found to be more active that the Pd/V_2O_5 catalyst. This is consistent with a feature observed with the palladium on metal oxide catalysts.[16] There is a reasonably good correlation between the catalyst activity and the one-electron redox potential of the metal oxide (Figure 1). There are, however, important differences between CO coupling to oxalates and ethylene oxidation to acetic acid and acetaldehyde. While palladium on copper oxide and palladium on cobalt oxide catalysts are not active for ethylene oxidation, they are good to mediocre catalysts for oxalate formation.

In a comparison between Pd/SiO_2 and $Pd/MgO/SiO_2$ for methanol synthesis from syn gas, the latter is much more active.[24] There is a change

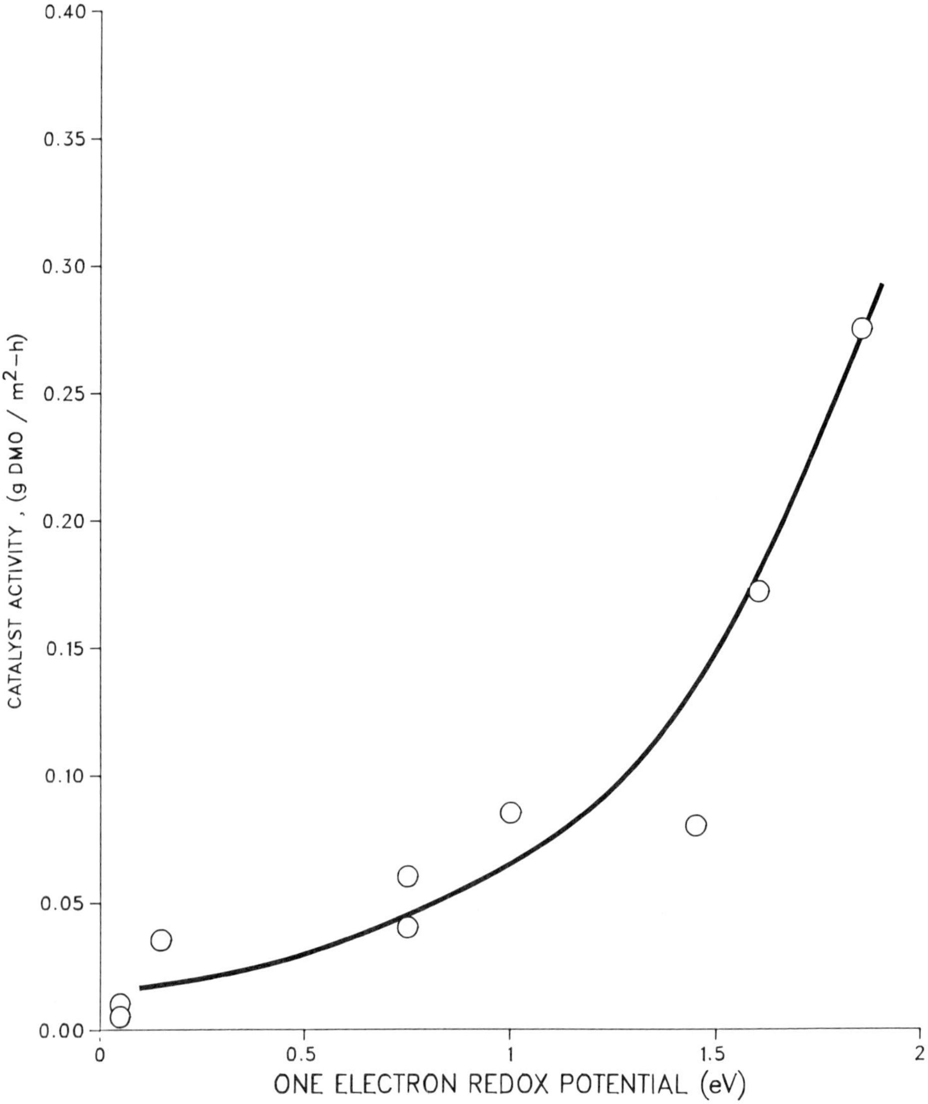

Fig. 1 Catalyst Activity as a Function of Metal Oxide Redox Potential

in CO IR bands. Also, ESR study detected the presence of Pd(I). Although the reaction conditions and environment are quite different, it is demonstrated that the existence of Pd(I) species can depend on the modification of the support.

Another mechanism involving Pd(I) intermediates can be envisioned. The key steps are similar to the Pd(O)-Pd(II) based route, except that two carboalkoxy groups on adjacent palladium sites couple to give the oxalate which is complexed to the Pd(I) sites. The Pd(I) species are then reoxidized by the support to Pd(II).

There are a few known oxalate complexes of palladium (II).[25] In the presence of different ligands, it is found that the oxalate could use two or four of its oxygens to bridge between two palladium centers. It can also use two oxygens to chelate on the same palladium center as shown in Scheme 2.

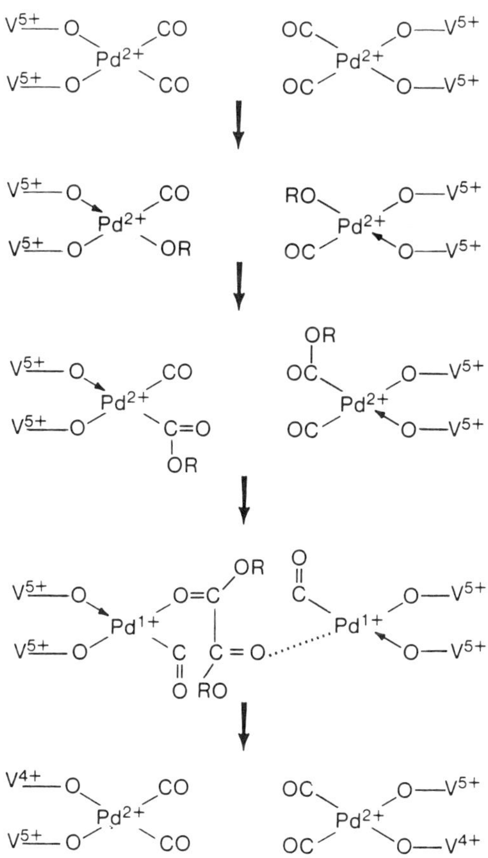

Scheme 2

IV. CONCLUSIONS

The reaction mechanism for oxalate formation from CO under oxidative dimerization conditions can be adequately explained by invoking the common Pd(O)-Pd(II) redox cycle. This cycle can be effectively coupled with a number of metal oxides to give high yields of oxalates. There appeared to be a correlation between the catalyst activity and the one electron redox potential of the metal oxide support. Metal oxides with primarily two-electron redox cycles are much less effective unless they can adopt the rutile type of crystal structure.

The involvement of palladium (I) species in the reaction provides another reasonable mechanism. There are many reported palladium (I) complexes containing bridging carbon monoxide. Many of them are rather stable in an oxidizing atmosphere, and all of them are present in the dimeric or oligomeric form. Well characterized dimeric or tetrameric complexes also contain chelating ligands. A "dimeric" form of palladium on the metal oxide surface is proposed as an alternative route to the popular Pd(O)-Pd(II) cycle. More experimental work is required to determine whether this is a viable mechanism.

V. PERSPECTIVES

There have been continuing efforts to bring about simple yet selective reactions for carboxylate syntheses. A recent report on electroreductive coupling of CO_2 to form oxalic acid appears to be especially interesting[26] (eq. 11). This reaction is catalyzed by Pd (II) or Ag (II) macrocyclic

$$CO_2 \xrightarrow{e} HOOCCOOH + (H_2) \qquad (11)$$

complexes. Although there are no details on yields, it is rather exciting that CO_2 coupling, rather than simple reduction takes place. A similar reaction catalyzed by light has been claimed in a Japanese patent.[27]

A very interesting synthesis of sebacic acid will be reported shortly.[28] The synthesis is also based on a homogeneous palladium catalyst.

Of particular interest is the recently reported reaction of direct carboxylation of aromatic compounds with CO_2 to give the corresponding carboxylic acids in the presence of a palladium catalyst and t-butyl hydroperoxide (eq. 12). Heteroaromatic compounds, such as furan and

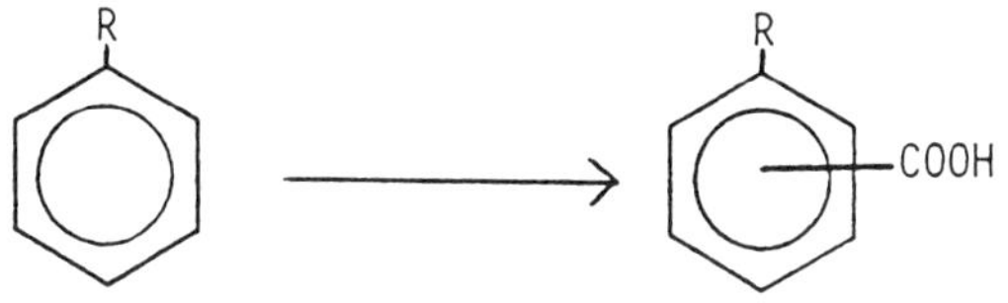

(12)

thiophene can be carboxylated similarly. A complex, ArCOOPdX is postulated as the reaction intermediate.

Undoubtedly there will be more new processes and new chemistry leading to the synthesis of a variety of dicarboxylates.

REFERENCES

1. SRI Process Economics Program Review, No. 85-1-1 December, 1985.
2. S. Fujiyama and T. Kasahara, Hydrocarbon Processing, **57**(11), 147 (1978).
3. For example, J. Shabatai and H. Pines, J. Org. Chem., **30**, 3854 (1965).
4. J.C. Mol, Chemtech, 250 (1983).
5. X. Xu and J.C. Mol, Chem. Comm., 631 (1985); E. Verkuijken, F. Kapteijn, J.C. Mol, and C. Boelhouwer, Chem. Comm., 198 (1977).
6. D.M. Fenton and P.J. Steinwand, J. Org. Chem., **37**, 2037 (1972).
7. D.M. Fenton and K.L. Oliver, Chemtech, 220 (1972).
8. D.M. Fenton and P.J. Steinwand, J. Org. Chem., **39**, 701 (1974).
9. H.S. Kesling, ACS Nat. Meeting, Chicago, Ill., Sept. 1985.
10. K. Nishimura, K. Fuju, K. Nishihura, M. Matsuda, and S. Uchiumi, U.S. Patents 4,229,589 and 4,229,591 (Ube Industries).
11. A.M. Gaffney, J.J. Leonard, J.A. Sofranko and H.N. Sun, J. Catal., **90**, 261 (1984).
12. G.C. Bond, M.J. Fuller, and L.R. Molloy in **Proceedings, 6th International Congress on Catalysis**, (G.C. Bond, P.B. Wells and F.C. Tompkins, Eds.) p. 356, The Chemical Society, London, 1976.

13. W.N. Smith, U.S. Patent 3,689,515 (Foote Mineral Company).
14. P.M. Henry **Palladium Catalyzed Oxidation of Hydrocarbons**, Reidel, Dordrecht, 1980.
15. A.R. Siedle and L.H. Pignolet, Inorg. Chem., **21**, 3030 (1982).
16. A.B. Evnin, J.A. Rabo and P.H. Kasac, J. Catal., **30**, 109 (1973).
17. F. Rivetti and V. Romano, J. Organomet. Chem., **154**, 323 (1978) and **174**, 221 (1979).
18. J-T. Chen and A. Sen, J. Amer. Chem. Soc., **106**, 1506 (1984).
19. R. Ugo in **Catalysis in C_1 Chemistry** (W. Keim, Ed.) Reidel, Dordrecht 1983).
20. R. Colton, R.H. Farthing, and M.J. McCormick, Aust. J. Chem., **26**, 2607 (1973).
21. P.L. Goggin and J. Mink, J. Chem. Soc. Dalton, 534 (1974).
22. L.S. Benner and A.L. Balach, J. Amer. Chem. Soc., **100**, 6099 (1978). C.-L. Lee, C.T. Hunt and A.L. Balch, Organometallics, **1**, 824 (1982).
23. T.A. Stromnova, L.G. Kuzmina, M.N. Vargaftik, G. Ya. Mazo, Yu. T. Struchkov, and I.I. Moiseev, Izv. Akad. Nauk SSSR, **3**, 720 (1978).
24. E.K. Poels, Faraday Discus. Chem. Soc., **72**, 194 (1981).
25. J. Chatt, F.G. Mann and A.F. Wells, J. Chem. Soc., 2086 (1935).
26. J.Y. Becker, B. Vainas, R. Eger, and L. Kaufman, Chem. Comm., 147 (1985).
27. S. Tafu and N. Kitamura, Japan Kokai 79 112,813 (1979).
28. H.S. Kesling, ACS Nat. Meeting, New York, April 1986.
29. H. Sugimoto, I. Kawata, H. Taniguchi and Y. Fujiwara, J. Organomet. Chem., **266**, C44 (1984).

Preparation of Amines and Anilines

6

The Production of o-Hydrazotoluene by the Selective Catalytic Hydrogenation of o-Nitrotoluene

Charles G. Coe and James W. Brockington

Air Products & Chemicals, Inc.
Corporate Science Center
Allentown, PA

ABSTRACT

The catalytic hydrogenation of o-nitrotoluene to produce high yields of o-hydrazotoluene is reported. The reduction is carried out at 60°C and 60 psig over a palladium catalyst in the presence of sodium hydroxide and a quinone in either a homogeneous liquid phase or a two-phase solution. The effect of solvent, catalyst charge, quinone and hydroxide concentration, and reaction conditions on rate and product selectivity will be discussed.

This study has provided a clearer understanding of the multistep reaction sequence required to produce high selectivities to o-hydrazotoluene. The presence of hydroxide affects both the rate and selectivity of the hydrogenation while the quinone inhibits hydrogenolysis of the desired product. Under optimum conditions, it was found that the small amount of o-toluidine co-produced is due primarily to hydrogenation of coupled intermediates and not the direct reduction of single ring compounds. Evidence presented indicates that the dielectric constant of the organic liquid phase can affect the selectivity of the hydrogenation.

I. INTRODUCTION

o-Tolidine is currently a specialty diaryl diamine with a world-wide market of approximately ten million pounds. Most of the o-tolidine is marketed to the organic dye industry for the production of azo dyes. Small amounts of o-tolidine are phosgenated to tolidine diisocyanate which is used to make high performance, specialty polyurethanes. Less costly o-tolidine might significantly expand its use in these applications. A potential new application for o-tolidine is its use as a polymer crosslinker, particularly in urethane and epoxy systems.

To achieve a more cost effective route to o-tolidine, the efficient conversion of o-nitrotoluene (ONT) to o-hydrazotoluene (OHT) by a catalytic process is needed. Once produced, OHT can be quantitatively converted via the benzidine rearrangement to o-tolidine under mild conditions using acidic catalysts.[1] Various catalytic processes have been reported for the preparation of aromatic hydrazocompounds, but in most cases yields vary greatly depending on the nature of substituents on the aromatic ring. Hydrogenation of ONT using reported catalytic methods failed to produce high yields of OHT (vide infra).

This paper describes a study in applied catalysis and reports an improved catalytic route for the preparation of OHT. In addition, this work has provided a clearer understanding of the multistep reaction sequence required to produce high yields of OHT and a simplified way of recovering the product.

II. EXPERIMENTAL

A. Chemicals

All the chemicals used in the hydrogenation studies were reagent grade or better and used without further purification. The supported metal catalysts were obtained from Engelhard Corporation.

Compounds used as calibration standards for the analyses were prepared in pure form using known procedures. o-Hydrazotoluene was prepared by alkaline reduction using zinc following the procedures of

Vogel.[2] o-Nitrosotoluene (Aldrich Chemical Co.) was recrystallized from ethanol and azoxytoluene was isolated from the base catalyzed condensation of nitrosotoluene. Azotoluene was prepared using a modification of the synthetic procedures described for azobenzene.[3] o-Toluidine and o-nitrotoluene (Aldrich Chemical Co.) were purified by vacuum distillation. A procedure reported by Bamberger[4] was used to prepare o-N-hydroxytolylamine which was found to be very unstable and rapidly disproportionated. All the standards were stored at 3°C in amber bottles under a nitrogen atmosphere.

B. Procedures

The hydrogenations were carried out in a 600 cc Parr stainless steel, stirred autoclave. In a typical experiment the autoclave was charged with 27.4 g o-nitrotoluene (0.20 moles), 17.0 g of 50% aqueous sodium hydroxide, 0.8 g 2,3-dichloro-1,4-napthoquinone, 0.40 g sodium dodecylbenzene sulfonate, 60 g tetrahydrofuran, 41 g 95% ethanol and 0.534 g 5% Pd/carbon. Typical hydrogenation conditions were 60°C and 60 psig. The reaction mixture was stirred at 1000 rpm using two four-blade turbine-type impellers. The hydrogen uptake was continually monitored by observing the pressure drop in the discharge vessel feeding the reactor. Generally, the run was terminated when gas uptake ceased. The course of the hydrogenation could be followed by periodically removing samples and measuring the concentration of reactants and products using liquid chromatography.

After the run the reactor contents were discharged directly into a tube funnel maintained under a nitrogen atmosphere and separated from the catalyst. The filtrate (containing the OHT) was added to a two-fold excess of water to precipitate the OHT product. Isolation of the OHT product followed by drying under vacuum typically resulted in 80% OHT yield (MP = 158-159°C). Liquid chromatographic analysis and H' NMR both indicated the OHT was better than 99% pure.

If THF is not used in the reaction mixture then the OHT is intimately mixed with the heterogeneous catalyst. In this case the OHT can be recovered by dissolving the crude OHT in THF, filtering off the catalyst, and adding water to the precipitate the OHT.

III. RESULTS AND DISCUSSION

A. Simplified ONT Reduction Network

The reduction of nitroaromatics to hydrazo compounds is not a simple one-step process, but consists of a number of parallel and consecutive reactions involving the generation of several intermediates. To understand the critical pathways necessary for the production of the desired hydrazo compound, it's useful to review the ONT reduction network shown in Fig. 1. In neutral or acidic media reduction of ONT produces high yields of o-toluidine. This is generally believed to proceed by formation of two reactive intermediates, nitrosotoluene and N-hydroxytolylamine, which are reduced further to o-toluidine. Preparing hydrazo compounds requires the formation of an N-N bond. This occurs through coupling of reactive intermediates to form azoxytoluene. The key to producing azoxytoluene selectively, is the interception of the highly reactive intermediates before they are reduced further. It is well known that in the presence of base these intermediates readily combine to form the corresponding azoxy compounds. The exact mechanism by which this coupling occurs is still unresolved but believed to proceed through the formation of highly reactive radical anion intermediates.[5] High selectivities to OHT require that most of the

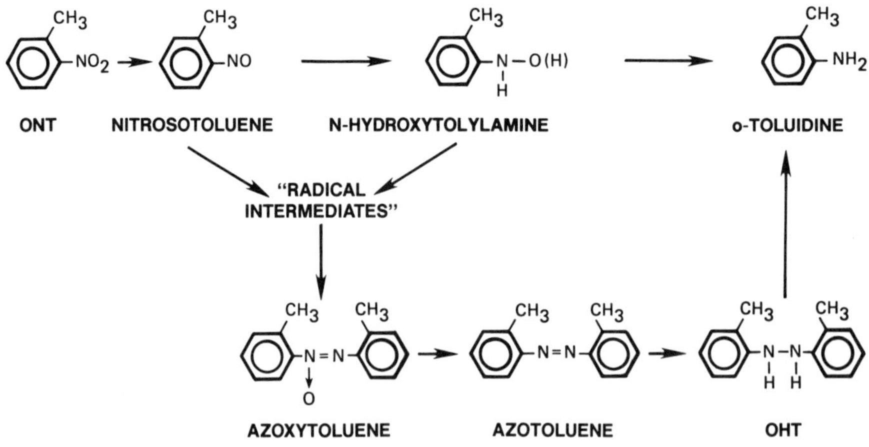

Fig. 1. ONT reduction network.

reduction is diverted to the formation of coupled precursors. Once the azoxytoluene has been produced, further reduction to the azo and hydrazo compounds is straightforward. The rates for the production of azotoluene are significantly slower than the previous portions of the reduction sequence but nevertheless proceed under mild conditions. In order to successfully produce OHT, the product must be stable under reaction conditions. In other words, the OHT cannot undergo further reduction via hydrogenolysis to produce the unwanted by-product, o-toluidine. This is not a trivial problem and occurs readily under most conditions.

B. Literature Review

A large amount of literature exists for the reduction of nitroaromatics; however, reports specifically teaching a selective catalytic process to prepare hydrazocompounds are limited. The Germans in the late 19th century first reported the use of a palladium on carbon catalyst in the presence of aqueous potassium hydroxide to produce hydrazobenzene from nitrobenzene. Much later it was shown that the hydrogenation of ONT to a mixture of 60% OHT and 40% o-toluidine can be carried out in an alkaline medium using alcohol solvent, over an active noble metal catalyst.[6] The use of a similar system in the presence of a quinone has been reported to produce good yields of o-chlorohydrazobenzene from o-nitrochlorobenzene.[7] A two-stage hydrogenation process using Pd/C and at least 20% NaOH in methanol is reported to give 80-90% yields of OHT without the addition of a quinone.[8] In this process the temperature is controlled from 0°C to 30°C in the first stage and raised to 90°C in a second stage. Carrying out this process on a commercial scale would be unattractive and require expensive refrigeration. Generally, the prior art describing the production of hydrazo compounds uses a combination of a precious metal catalyst on an activated carbon support in combination with an alkali base. For the most part, these reductions are carried out in a multiphase system containing two liquid phases, one consisting of an aqueous alkaline media and the other the organic nitrocompound being reduced. The addition of various quinones has been shown for certain nitrocompounds to increase the yields of hydrazo aromatics.

REACTION:

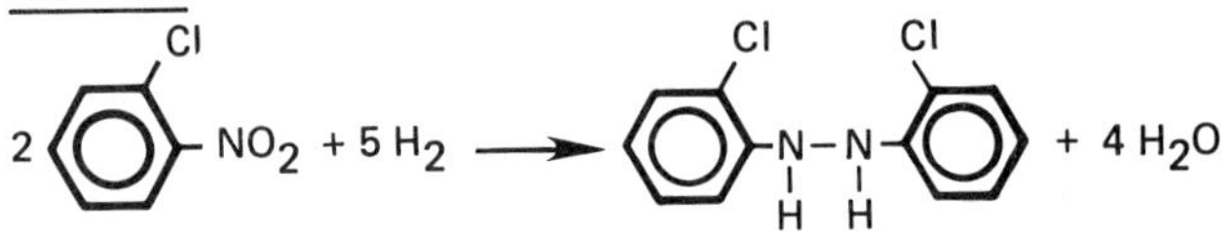

CONDITIONS:

SOLVENT:	12% AQUEOUS NaOH
CATALYST:	5% Pd/C POWDER
INHIBITOR:	2, 3–DICHLORO–1, 4–NAPHTHOQUINONE
TEMPERATURE:	60°C
PRESSURE:	60 PSIG

RESULTS:

90% YIELD OF 2, 2'DICHLOROHYDRAZOBENZENE

Fig. 2. The quinone assisted reduction of o-chloronitrobenzene (US 3,156,724).

The technology represented in Fig. 2 summarizes the best method available for preparing substituted hydrazoaromatics on a large scale. o-Chloronitrobenzene can be reduced to the corresponding chlorohydrazobenzene in approximately 90% yield, using the conditions shown in Fig. 2.[7] This reduction is carried out in a multiphase reaction medium containing two liquid phases, and a solid catalyst. However, one cannot always draw direct analogies between various substituted or unsubstituted nitrocompounds. Aside from solubility factors, what may be advantageous for achieving high selectivities in one case may not work for another. Despite repeated attempts using the procedures described for obtaining high yields of o-chlorohydrazobenzene, similar yields of OHT could not be efficiently produced from ONT.

C. Key Reaction Variables

Screening experiments indicated that the following variables were critical to obtaining high yields of OHT:

- reaction medium
- sodium hydroxide concentration
- noble metal catalyst
- quinone inhibitor

Varying any one of the above significantly alters the course of the hydrogenation. Each of these variables was studied to determine its effect on the hydrogenation and its role in controlling the selectivities.

1. Reaction Media

Table I gives the OHT yield and turnover number for hydrogenations using a variety of reaction media. (Throughout this work all turnover numbers are defined as moles of ONT converted per mole of metal per hour.) Repeated attempts to reduce ONT to OHT, using a two liquid phase reaction media with aqueous sodium hydroxide as one phase, failed to produce acceptable yields of OHT. Changing to an alkaline ethanolic medium which solubilizes both the ONT and sodium hydroxide in the reactive phase resulted in a dramatic increase in the yield of OHT. The identical reduction using methanol in place of ethanol is not nearly as selective. This is somewhat surprising since it has been reported that, for the production of hydrazobenzene from nitrobenzene, methanol is the preferred solvent.[9]

The hydrogenation of ONT in ethanol containing sodium hydroxide and 2,3-dichloro-1,4-napthoquinone produced high yields of OHT; however, the OHT is precipitated and intimately mixed with the catalyst. A commercial process based on this system would require costly separations for catalyst recovery and water removal. A reaction medium containing two liquid

TABLE I

Effect of Reaction Medium on the Selectivity and Rate of ONT Hydrogenation

REACTION MEDIUM	OHT YIELD (%)	TURNOVER ($HR.^{-1}$)
o-NITROTOLUENE/ AQUEOUS	< 5	750
ETHANOL	78	1,587
METHANOL	54	1,632
THF: ETHANOL/ AQUEOUS	87	650
THF: METHANOL/ AQUEOUS	83	620

phases, consisting of tetrahydrofuran (THF), ethanol, and ONT as one phase and 50% aqueous NaOH as the other phase, provides a system in which the water of reaction can be easily removed and the product OHT remains soluble in the orgnic phase for further processing. Using this reaction medium, 80-90% yields of OHT have been obtained in a system that appears amenable to continuous processing. No examples of this type of reduction medium for nitroaromatics have been previously reported.

The data in Table I shows that the addition of THF to methanol not only provides an improved solvent facilitating separation but unexpectedly affects the actual selectivity of the hydrogenation resulting in an increased OHT yield. (The addition of THF to the ethanol alters the OHT yield less significantly.)

A possible explanation for the effect of THF on selectivity could be related to the lowering of the dielectric constant of the solvent mixture. The predicted dielectric constant of the organic phase for the 60/40 mixture of THF and either methanol or ethanol is similar and approximately half that of pure methanol. A solvent's dielectric constant can drastically alter the rate of electron transfer in homogeneous systems and could have a similar effect here. Reduction of single-ring intermediates to o-toluidine is thought to occur by an electron transfer pathway and will be accelerated as the dielectric of the solvent increases. Some recent work exhibits just this effect for the nickel catalyzed reduction of nitrobenzene.[10] It is reasonable to postulate that the selectivity to the hydrazocompound arises from the rate of coupling to produce azoxytoluene being much faster than the rate of further hydrogenation (via electron transfer) to produce o-toluidine.

2. Addition of DMSO

Rylander has shown that in neutral or acidic media high yields of N-arylhydroxylamines are achieved during the hydrogenation of nitroaromatics in the presence of small amounts of dimethyl sulfoxide (DMSO).[11] The preferred catalyst for this process is platinum on carbon; palladium on carbon in combination with DMSO is claimed to be much less effective for producing hydroxylamines. The condensation of the hydroxylamine intermediate with its precursor nitrosocompound is a key step to generating the coupled OHT precursors, therefore it was reasonable to determine the

TABLE II

Effect of DMSO Additive on OHT Yield

CATALYST (g)	2,3-DICHLORO-1,4-NAPHTHOQUINONE ADDED (g)	RATIO OF QUINONE TO CATALYST	INITIAL RATE (10^{-2} MOL MIN^{-1})	OHT YIELD
0.4	-0-	-0-	6.2	48
0.4	0.2	0.5	5.6	65
0.4	0.4	1.0	4.2	86
0.4	0.8	2.0	3.1	82
0.4	1.2	3.0	3.3	83

*ALL RUNS CARRIED OUT OVER A 5% Pd/C CATALYST IN ETHANOL SOLVENT CONTAINING 6% NaOH AT 60°C AND 60 PSIG.

influence of DMSO in the presence of base for ONT hydrogenations. As seen in Table II in most cases the addition of 1% DMSO to runs using 5% Pd/C improved the OHT yield, however, the rate decreased significantly. A single run using 1% Pd/C with DMSO present showed that the rate can be improved while maintaining a high OHT yield. This rate enhancement is discussed in Section 4.

3. Sodium Hydroxide Concentration

The effect of sodium hydroxide concentration in these reductions has been carefully studied. The results of some of these studies are shown in Fig. 3.

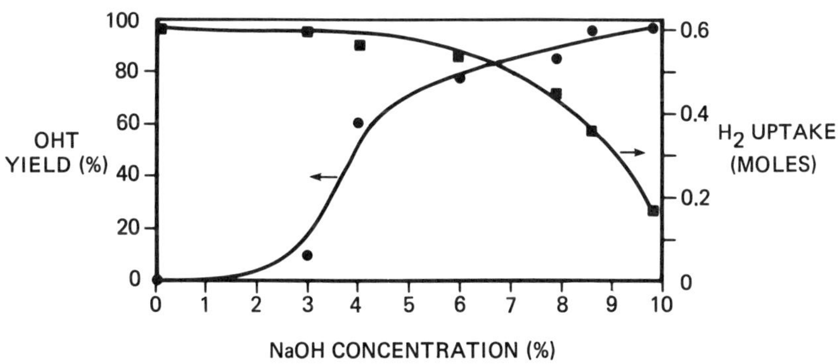

Fig. 3. Effect of sodium hydroxide concentration on ONT reduction.

The addition of small amounts (less than 1%) of sodium hydroxide are totally ineffective in diverting the reduction of ONT to coupled products and instead gives high yields of o-toluidine. Reductions performed under identical conditions in ethanol employing 1%, 0.37% (pH = 13) and no base all resulted in the formation of large amounts of o-toluidine and less than 1% OHT. Curiously, at pH = 13 both o-nitrosotoluene and o-hydroxytolylamine rapidly react to produce azoxytoluene. Therefore, the addition of larger amounts of sodium hydroxide is not required to generate the azoxytoluene, presumably the OHT precursor; however, the reduction of ONT under these conditions does not give any of the desired product. The data in Fig. 3 is for the experiments in ethanolic sodium hydroxide, but the same general trends were found in the THF/ethanol medium as well. Significant OHT yields are achieved only after a 4% sodium hydroxide concentration is maintained in the organic phase of the reaction medium. Eighty percent yields of OHT require at least 6% NaOH. Essentially quantitative yields of OHT are obtained by using 8% or greater concentrations of sodium hydroxide and considerably longer reaction times. However, a large portion of the OHT comes from a stoichiometric reduction rather than a catalytic one as evidenced by the dramatic decrease in the amount of hydrogen consumed. Under these conditions, besides consuming sodium hydroxide, some of the alcohol solvent is converted to intractable by-products which complicate downstream processing. It is well known that high concentrations of alkali in alcohols are powerful reducing agents for aromatic nitrocompounds.[12] This study shows that a catalytic process producing high yields of OHT requires a sodium hydroxide concentration of about 6% in the reactive phase.

Using the preferred THF/alcohol solvent, the actual amount of aqueous NaOH initially added is not critical. The THF/alcohol ratio controls the NaOH concentration present in the active phase. In these media the aqueous phase simply provides a reservoir for excess NaOH.

A control experiment was carried out to be certain that the OHT is not being produced by a stoichiometric reduction from the alkaline alcohol. Combining all the reagents generally employed in the catalytic reduction including the quinone and 6 wt.% NaOH under a nitrogen atmosphere at reaction temperature for two hours gave no OHT. Only 12% of the ONT was converted during the two hour period. Of this, approximately 50% selectivity to o-toluidine and 30% selectivity to azoxytoluene was observed. These results suggest that under noncatalytic conditions, the quinone is having little effect in directing the production distribution. At 6% NaOH, the competing stoichiometric reduction is not a significant factor in these hydrogenations. Higher amounts of base appear to accelerate the stoichiometric reduction since the rate and amount of hydrogen consumed is greatly reduced yet high OHT yields are obtained.

4. Metal Loading

Under the conditions studied both the concentration and metal loading of the catalyst affect these hydrogenations. Varying the catalyst (5% Pd/C) to ONT ratio gives a linear correlation with the observed rate, suggesting that gas to liquid mass transfer resistances are not influencing the reduction. As shown in Fig. 4 an unusual effect was noted, however, when the metal loading on the catalyst was varied. Running the hydrogenation in

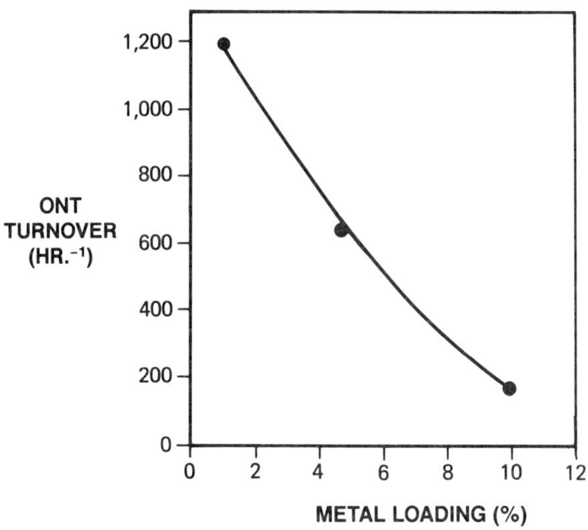

Fig. 4. Influence of Pd loading on the ONT reduction rate (in these runs the amount of Pd and ONT was held constant).

THF/ethanol with a constant total metal concentration but at different loadings on the carbon support, resulted in a wide variation in the turnover frequency. As the Pd loading decreased the reaction rate was accelerated. It is likely that this difference in turnover frequency is due to an increase in the metal dispersion as the metal loading on the catalyst decreases. [Due to the high surface area of the carbon supports no attempt was made to measure the metal dispersions via standard hydrogen chemisorption.] The increased activity of the 1% Pd/C catalyst compared to the 5% loading is important since faster rates can be achieved in the preferred THF/alcohol media. Turnover frequencies approaching those obtained with 5% Pd/C using the less desirable ethanol solvent are observed without sacrificing any OHT selectivity.

5. Quinone Inhibitor

The addition of a quinone is essential to achieving high OHT yields. In its absence approximately half of the hydrogenation products are single-ring compounds, mainly o-toluidine. In the presence of a quinone and using the preferred reaction media, > 80% OHT yields are routinely achieved. The data in Table III shows that a quinone to catalyst weight ratio of one is

TABLE III

Effect of Catalyst/Quinone Ratio on the Selective Reduction of ONT

REACTION MEDIUM	1% DMSO	OHT YIELD (%)	TURNOVER NO. (Hr^{-1})
ETHANOL	NO	82	1690
	YES	93	226
METHANOL	NO	54	1632
	YES	44	320
ETHANOL/THF	NO	83	660
	YES	96	177
		92*	690*
METHANOL/THF	NO	83	620
	YES	93	<160

*1% PD/C INSTEAD OF 5% PD/C

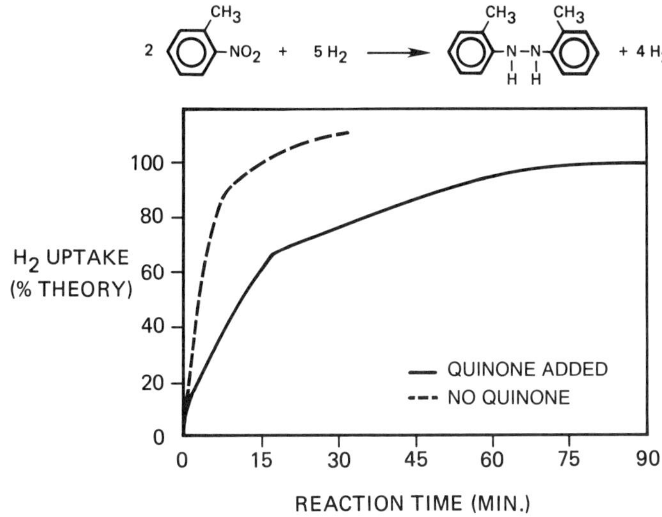

Fig. 5. Effect of quinone on the hydrogenation of ONT.

optimum for high OHT selectivity. Additional quinone does not improve the yield and adversely influences the rate. Thus it acts as an inhibitor, yet results in higher selectivity to OHT. In this chemistry the quinone lowers the overall hydrogenation rate. Fig. 5 shows the influence quinone has on the hydrogen consumption curves for the ONT hydrogenation. The 100% hydrogen uptake shown on this graph is the amount required to convert all the ONT to OHT. The inhibition of the hydrogenation can be easily seen by comparing the quinone assisted reduction to the run without quinone. Without quinone, despite the presence of sodium hydroxide, the amount of hydrogen consumed increases above the theoretical amount required for OHT production and approaches the amount necessary for complete reduction to o-toluidine. However, with the addition of quinone the theoretical amount of hydrogen is generally consumed. The rate of hydrogenation of azoxytoluene to OHT was not affected by the presence of quinone.

In all cases where sodium hydroxide and a quinone are present there is an abrupt change in the rate of hydrogenation after 60% of the theoretical amount of hydrogen is consumed. The reaction profiles for hydrogenations

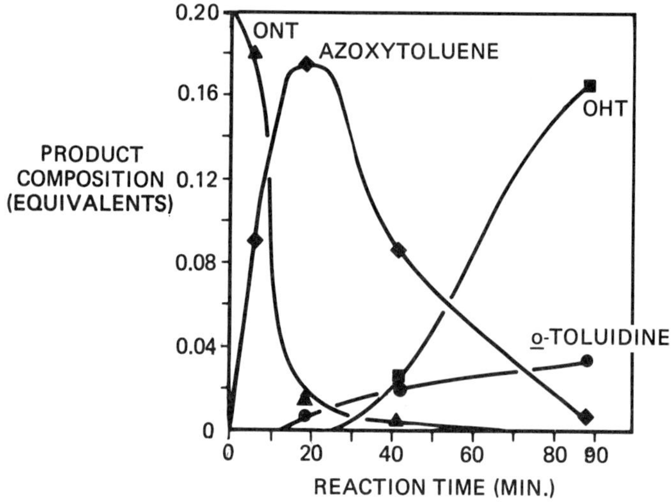

Fig. 6. ONT hydrogenation profile in the presence of quinone.

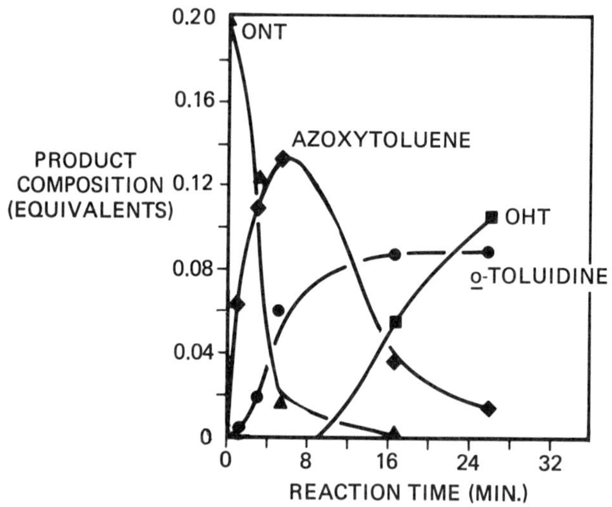

Fig. 7. ONT hydrogenation profile without quinone.

under identical conditions with and without quinone are shown in Figs. 6 and 7. With quinone the abrupt change in rate coincides with the majority of the ONT being converted to a large amount of azoxytoluene. The hydrogen consumption after the break in the rate curve is associated with the hydrogenation of azoxytoluene to OHT. This evidence supported the hypothesis that high selectivities to azoxytoluene are required to produce good OHT yields. The rate for the reduction of azoxytoluene is identical to that observed for the rate of ONT reduction after the sharp break. Therefore, the reduction of ONT occurs in a somewhat consecutive rather than concerted fashion. Note that only a small amount of o-toluidine is produced during the hydrogenation.

In the absence of quinone the reaction profile is quite different (Fig. 7). The ONT is more rapidly depleted compared to the run having quinone. The reduction without quinone affords less azoxytoluene and immediately starts producing significant amounts of o-toluidine. It is also important to realize that the amount of OHT produced is lower than the amount of azoxytoluene formed. This suggests that some of the coupled product is being converted to o-toluidine due to further reduction.

The intermediates and products shown in the network (Fig. 1) are known to react under various conditions to either produce or consume the coupled precursors necessary for OHT production. The reactions represented by dashed lines in Fig. 8 were screened and found not to occur at significant rates under the conditions used for this work. Also, addition of 10 wt% o-toluidine to the reaction mixture didn't alter the observed rate, conversions or selectivities. This indicates that o-toluidine is not a reactive intermediate under the conditions used.

In addition to generating the required coupled precursors it is necessary to prevent the OHT from undergoing further reduction. The effect of NaOH concentration and quinone on the further reduction of OHT to the unwanted by-product was investigated and the results are shown in Fig. 9. In these studies the reactor was charged with OHT in the THF/ethanol reaction medium and maintained at 60°C and 60 psig. The formation of o-toluidine was followed for each case. The graph in Fig. 9 shows that the presence of quinone, whether at low or high base concentration, inhibits the hydrogenolysis of OHT and thus greatly retards

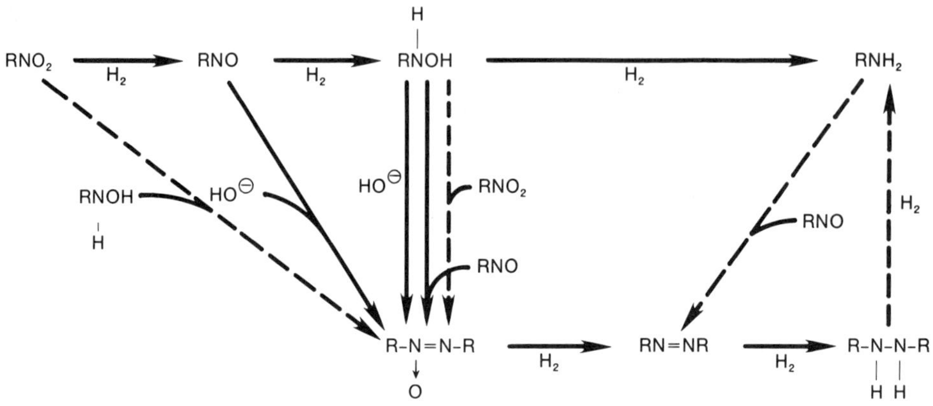

$$2RNHOH \xrightarrow[\Delta]{} RNO + RNH_2$$

Fig. 8. Competing reactions which could contribute to OHT formation.

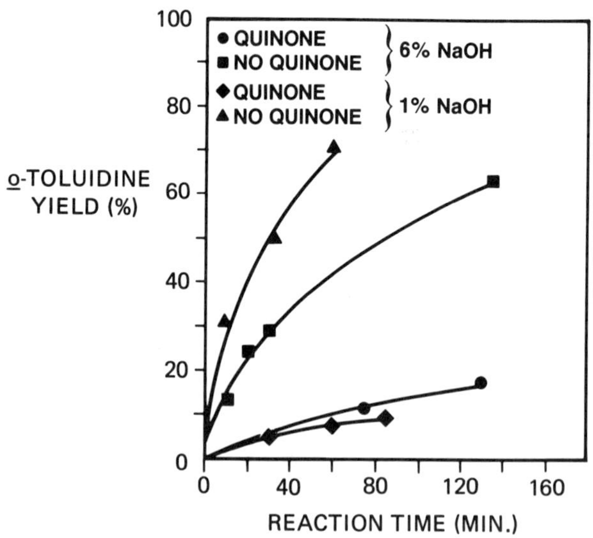

Fig. 9. Effect of quinone on OHT stability under hydrogenation conditions.

o-toluidine formation. Even in the presence of quinone the OHT is reduced cleanly to o-toluidine if the NaOH is omitted; however, 1% NaOH is sufficient to inhibit this further reduction. Charging the reactor with azoxytoluene in place of OHT produced similar results. Intermediate samples taken from a run in this study not containing quinone showed the presence of large amounts of OHT which were further reduced to o-toluidine. These studies indicate that once a coupled OHT precursor is present the only route to o-toluidine is through the formation of OHT. The quinone plays a key role in these hydrogenations by protecting the OHT product from further reduction.

D. Pathway for OHT Formation

From the results presented, it is reasonable to suggest that the quinone selectively poisons the Pd catalyst and blocks the most active sites. Such a poisoned catalyst may still be effective for hydrogenating easily reduced substrates but be ineffective for compounds that are more difficult to reduce. All the data presented is consistent with the hypothesis that the catalyst contains two different types of active sites. Those active sites responsible for the hydrogenolysis of OHT and hydrogenation of the hydroxylamine may be selectively poisoned by the quinone. The preferred adsorption of the quinone over that of the N-hydroxytolylamine and nitrosotoluene allows these reactive intermediates to move away from the catalyst surface and undergo the base catalyzed coupling to form azoxytoluene. The coupled products containing an unsaturated functionality (N=N bond) can be easily adsorbed on the less active catalytic site until the unsaturated site has been reduced. The OHT produced desorbs from the catalyst since it no longer has any unsaturation and consequently has a lower heat of adsorption; thus, preventing hydrogenolysis of the N-N bond. In the absence of quinone o-toluidine is formed over the more active catalyst sites by both the further reduction of the aromatic hydroxylamine and the hydrogenolysis of the OHT product.

IV. SUMMARY

The catalytic hydrogenation of o-nitrotoluene (ONT) to give high yields of o-hydrazotoluene (OHT) is carried out at 60°C and 60 psig using a supported palladium catalyst and a quinone co-catalyst. The reduction can be conducted in either a homogeneous liquid phase with alcohol as the solvent or preferably in a two phase solution containing alcohol/THF and aqueous base. A reaction medium containing two liquid phases, consisting of THF, ethanol, and ONT in the organic phase and concentrated sodium hydroxide as the aqueous phase provides a system from which the water of reaction can be easily removed. Also, the THF solubilizes the OHT product allowing it to be easily separated from the catalyst.

From a mechanistic viewpoint this study has confirmed several hypotheses concerning the probable pathway for the formation of OHT. High selectivities for OHT require (1) diverting N-hydroxytoluylamine and nitrosotoluene to azoxytoluene, (2) reduction of azoxytoluene to OHT, and (3) blocking the hydrogenolysis of the OHT product.

ACKNOWLEDGEMENTS

The authors gratefully acknowledge Louise Schoonmaker and Anthony Palmisano for carrying out a large portion of the experimental work described in this paper, and Air Products and Chemicals, Inc. for permission to publish this work.

REFERENCES

1. N.V. Sidgwick, The Organic Chemistry of Nitrogen, Oxford University Press, London, p. 512 (1966).
2. A.I. Vogel, Laboratory Text in Organic Chemistry, Prentice-Hall Inc., New York, 1950, p. 277.
3. "Organic Synthesis", Coll. Vol. III, John Wiley & Sons, New York, p. 103 (1955).
4. E. Bamberger, Annalen, **441**, 207 (1925).

5. G.A. Russell, E.J. Geels, F.J. Smentowski, K.Y. Chang, J. Reynolds, and G. Kaupp, J. Am. Chem. Soc., **89**, 3821 (1967).
6. C. Henke, et. al, US 2,233,129 (1941).
7. R. Werner, et. al, US 3,156,724 (1964).
8. R. Sayigh, et. al, US 3,260,750 (1966).
9. Dynamit Nobel Aktiengesellschaft, GB 1,560,004 (1977).
10. R.S. Berman and J. Kochi, Inorg. Chem., **19**, 248-254 (1980).
11. P. Rylander, US 3,694,509 (1972).
12. L. Anderson, US 2,804,453 (1957).

7

The Selective Hydrogenation of Bromonitrobenzenes to Bromoanilines

William Pascoe

Research Laboratories - Manufacturing Technology
Eastman Kodak Company
Rochester, New York

ABSTRACT

The selective hydrogenations of o-, m-, and p-bromonitrobenzenes to the corresponding bromoanilines were accomplished with less than 2% debromination using commercially available catalysts. Ethanol and tetrahydrofuran were used for solvent comparisons and the reductions were conducted at two pressures, 60 and 1500 psig. There was no direct hydrogen pressure correlation with catalyst selectivity. However, tetrahydrofuran was preferred to ethanol as the solvent for the reductions. No one catalyst-solvent-pressure system emerged as the best system for the selective reduction of all three isomers. The reduction mechanism apparently followed two parallel pathways. The first was hydrogenation of the nitro group through nitroso and hydroxylamine intermediates to the bromoaniline. The second pathway was the condensation of the nitroso and hydroxylamine intermediates to form an azoxy intermediate, followed by hydrogenation through the azo and hydrazo intermediates to bromoaniline.

I. INTRODUCTION

The hydrogenation of halogen-substituted aromatic nitro compounds to the corresponding anilines often results in partial dehalogenation of the product. Frequently, catalytic systems recommended in the literature for limiting dehalogenation during these types of reductions fail to give high yields of the halogenated aniline. The objective of this work was to develop a better understanding of the best catalyst-solvent-pressure systems for providing high yields of halogenated aromatic amines via hydrogenation of halogenated aromatic nitro compounds.

Kosak[1] reported that the order of susceptibility of aromatic halogens to hydrogenolysis was I > Br > Cl > F, and by ring position was ortho > para > meta. The hydrogenations of ortho; meta; and para-substituted bromonitrobenzenes were studied as model reactions for determining the best catalyst for this class of hydrogenations. Most interest in the hydrogenation of halonitroaromatics involves either bromo or chloro substituents. Bromine is more susceptible to hydrogenolysis than chlorine, and catalyst systems that selectively hydrogenate bromonitrobenzenes to bromoanilines with minimum hydrogenolysis of the bromo group generally can be applied to the selective reduction of chloronitrobenzenes.

In recent years, two types of catalyst modifications have been described which are reported to prevent dehalogenation during hydrogenation of halonitroaromatic compounds. One involves adding an inhibitor to a conventional catalyst, and the other is a prepared poisoned catalyst, usually containing sulfur. Table I lists a brief survey of the literature on the selective hydrogenation of bromonitrobenzenes.

We studied the selective reduction of all three bromonitrobenzene isomers over four commercially available catalysts. These were 5% Pt/C, 5% Pt/C-sulfided, and 5% Rh/C purchased from Engelhard Industries, and No. 28 Raney nickel purchased from W.R. Grace. We also examined the effects of two solvents, ethanol and tetrahydrofuran (THF), and two pressures, 60 and 1500 psig, on the selectivity of the reactions. In order to limit the number of variables studied, the reactions were run at room temperature.

TABLE I

Hydrogenation of Bromonitrobenzenes - Literature Review

Isomer	Cat.	Additive	Temp (°C)	Press. (psig)	Solv.	BNB[a] (mole %)	Dehal.[b] (mole %)	Ref.
para	Pt/C	morpholine	70-80	500	MeOH	—	10	1
	Pt/C	phorphorous acid	70-80	500	MeOH	—	1.6	1
meta	Rh/Al	Ca(OH)	22-40	20-40	MeOH	86	—	2
para	Pt/C	MgO	30-60	500	MeOH	87	2.2	3
para	Pt/C-S	—	100-130	800	MeOH	99.5	—	4
para	Rh/C-S	—	105	800	MeOH	100	trace	4
para	Co/K-S	—	115	800	MeOH	99	trace	4
para	ReS	—	60-100	3000	—	100	—	5
	Re O	—	124	3000	—	100	—	5
meta	Pt/C	morpholine	85	500	MeOH	91	1.5	6

[a]BNB - Bromonitrobenzene
[b]Dehal. - Dehalogenated product

II. EQUIPMENT

All high-pressure reductions (1500 psig) were run in an Autoclave Engineers 1-liter stainless steel autoclave equipped with a turbine type impeller. The low-pressure reactions were run on Parr 3911 shakers. The reaction mixtures for the high-pressure reactions contained 500 ml of a 2% by wt. solution of bromonitrobenzene in solvent and 1.0 g of catalyst (10% by wt. of the reactant). The low-pressure reaction mixtures contained 250 ml of the 2% solution and 0.5 g of catalyst. During the reductions, samples were taken of the reaction solution at desired intervals and were analyzed by gas chromatography and/or gas chromatography/mass spectrometry (GC/MS). Gas chromatographic analyses were run on a Hewlett-Packard 5880A gas chromatograph equipped with a 12-meter crosslinked methyl silicone capillary column and a flame-ionization detector. The peaks were identified by comparison with known samples and GC/MS. The gas chromatograph was calibrated to calculate concentrations of intermediates and product on a mole percent basis.

III. RESULTS AND DISCUSSION

As expected, the rate of the hydrogenations of bromonitrobenzenes showed a clear dependence on the reaction pressure, increasing with increasing pressure (Table II). The solvent effect was not as well defined, although for most reactions the rate in ethanol was greater than the rate in THF. An interesting exception was the rate of para isomer hydrogenation, which was faster in THF than in ethanol. The rate of hydrogenation of all three isomers in THF at 1500 psig was greater than the rate of hydrogenation in ethanol at 60 psig, indicating that the effect of pressure on the rate was greater than the solvent effect.

The effect of the isomer on the reaction rate appears to be more dependent on the solvent than the pressure. In ethanol, the rate was the same for all three isomers at 60 psig. At 1500 psig, the rate decreased in the order of ortho > meta > para. In THF at 60 psig and at 1500 psig, the rate decreased in the order of para > meta > ortho, the opposite of the ethanol effect.

TABLE II

Rate of Hydrogen Absorption [(mole H_2/mole-min BNB[a]) X 10^{-3}]

Isomer	Rate at 60 psig EtOH	Rate at 60 psig THF	Rate at 1500 psig EtOH	Rate at 1500 psig THF
ortho	.45	.31	1.35	.75
meta	.48	.39	1.18	.83
para	.47	.41	1.00	1.25

Effect of pressure, solvent, and isomer on rate of hydrogenation.
[a]BNB - bromonitrobenzene

The effect of the catalyst on the reaction rate is influenced by the pressure more than by the solvent (Table III). At 60 psig, in both THF and ethanol, the rate decreases in the following order: Pt/C > Rh/C > Pt/C-S > Raney nickel. At 1500 psig in ethanol, the rate of hydrogen consumption was the same for each catalyst. The rates were also the same for each catalyst in THF at 1500 psig. When the rate of formation of amine product (both bromoaniline and aniline) was measured, the rate was higher in ethanol

TABLE III

Rate of Hydrogen [(mole H_2/mole-min BNB[a]) X 10^{-3}]

Catalyst	Rate at 60 psig EtOH	Rate at 60 psig THF	Rate at 1500 psig EtOH	Rate at 1500 psig THF
5% Pt/C-S	0.5	0.4	1.1	1.0
5% Pt/C	0.8	0.7	1.2	1.0
5% Rh/C	0.7	0.4	1.3	0.9
Raney Ni	0.1	0.05	1.2	1.0

Effect of catalyst on the rate of hydrogenation.
[a]BNB - bromonitrobenzene

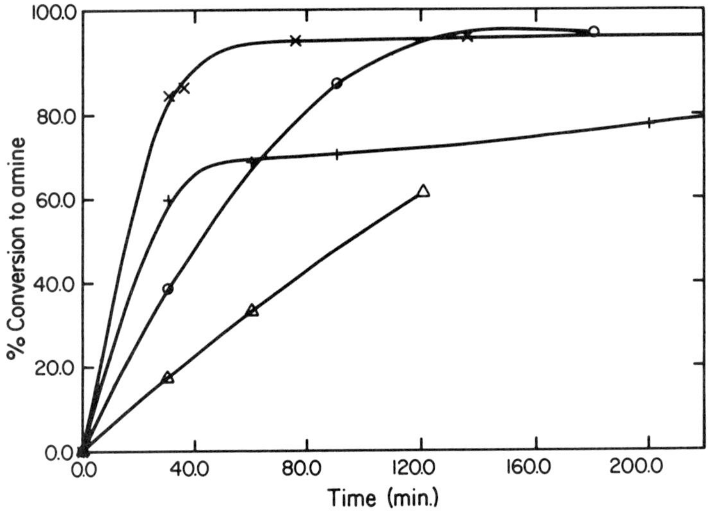

Fig. 1. Hydrogenation of 1-bromo-4-nitrobenzene at 1500 psig. o - Raney nickel/THF, Δ - Raney nickel/EtOH, + - 5% Pt/C-THF, x - 5% Pt/C-S/EtOH.

than in THF for the noble metal catalysts, and higher in THF for Raney nickel. This was true for all isomers at both 60 and 1500 psig. This is a reversal for Raney nickel from the rate of hydrogen consumption, which was greater in ethanol. Fig. 1 shows typical curves for the formation of amine products over both Pt/C-S and Raney nickel. Similar results were observed for all isomers with the rate of amine formation faster in ethanol for the noble metal catalysts, and faster in THF for Raney nickel.

The effect of solvent, pressure, isomer, and catalyst on the level of dehalogenation during the reductions was less clear. For the ortho and para isomers, the ratio of bromoaniline to aniline in the final product was higher when THF was the solvent. However, for the meta isomer, less dehalogenation was observed when ethanol was used with either 5% Pt/C or 5% Rh/C at 1500 psig. The rhodium catalyst also demonstrated greater selectivity toward m-bromoaniline at 60 psig in ethanol. When either 5% Pt/C-S or Raney nickel was used, the selectivity toward m-bromoaniline was greater in THF at both pressures (Table IV). The effect of pressure, isomer,

TABLE IV

Selectivity of Catalyst for Bromoaniline

Catalyst	Pressure (psig)	Ratio of Bromoaniline to Aniline (mole% bromoaniline/mole% aniline)					
		Ortho		Meta		Para	
		EtOH	THF	EtOH	THF	EtOH	THF
5% Pt/C-S	60	7	38	12	83	14	53
	1500	3	38	7	53	28	77
5% Pt/C	60	3	23	5	29	1	29
	1500	8	143	150	37	5	16
5% Rh/C	60	11	53	63	38	53	63
	1500	2	45	100	—	1	23
Raney Ni	60	14	12	—	—	—	—
	1500	7	13	3	40	56	30

and catalyst on selectivity were interdependent. The degree of dehalogenation over 5% Pt/C-S was less for the ortho and meta isomers at 60 psig, and less for the para isomer at 1500 psig. When 5% Pt/C was used as catalyst, the opposite selectivity was observed. Higher pressure resulted in less dehalogenation for the ortho and meta isomers, and greater dehalogenation for the para isomer. Using 5% Rh/C resulted in less dehalogenation for the ortho and para isomers at 60 psig, and less dehalogenation of the meta isomer at 1500 psig. The rate of reaction over Raney nickel was very low at 60 psig, so accurate data on high conversions to amine products were not obtained at this pressure.

Table V lists the best systems found for the hydrogenation of the various bromonitrobenzene isomers to the corresponding bromoanilines with minimum dehalogenation to aniline. The mole percentages do not add up to 100% as some intermediates are still found in the product solutions. Further hydrogenation of these solutions led to a product higher in aniline and lower in bromoaniline.

TABLE V

Hydrogenation of Bromonitrobenzenes

Isomer	Catalyst	Solvent	Pressure (psig)	Bromoaniline (mole %)	Aniline (mole %)
Ortho	5% Pt/C	THF	1500	83	0.6
	5% Rh/C	THF	1500	77	1.7
	5% Pt/C	THF	60	84	3.8
Meta	5% Pt/C	EtOH	1500	91	0.0
	5% Rh/C	EtOH	60	96	1.5
	Raney Ni	THF	1500	96	2.4
Para	5% Pt/C-S	THF	1500	93	1.4
	5% Pt/C-S	THF	60	91	1.7
	5% Rh/C	THF	60	89	1.4

TABLE VI

Rearrangement of Para Isomer[a]

Catalyst	Wt% Rearranged EtOH	Wt% Rearranged THF
5% Pt/C-S	2.0	2.1
5% Pt/C	2.2	0.0
5% Rh/C	2.2	0.0
Raney Ni	3.0	0.3

[a]Rearrangement to o-, and m-isomers at 1500 psig

IV. REARRANGEMENT

Some rearrangement to the ortho and meta positions occurred during the reduction of para-bromonitrobenzene. All four catalysts caused some rearrangement in THF. Pt/C-S and Raney nickel also caused rearrangement in ethanol (Table VI). Trace amounts of rearranged hydroxylamine intermediate were detected during the para isomer reductions, indicating that rearrangement either occurred during the reduction of the nitroso intermediate or to the hydroxylamine. There was no detection of either o- or m-bromonitrobenzene contamination in the starting p-bromonitrobenzene solutions.

V. MECHANISM

Fig. 2 shows the mechanistic pathways proposed by Haber[7] to which a pathway has been added for the disproportionation of the arylhydroxylamine intermediate to arylnitroso and arylamine products. This pathway was discussed by Kosak.[1]

Of the four possible pathways described, we believe only two are of significance. The primary pathway (I) has been widely reported in the literature. This involves direct addition of hydrogen to the nitro group, reducing it through the nitroso and hydroxylamine intermediates to the

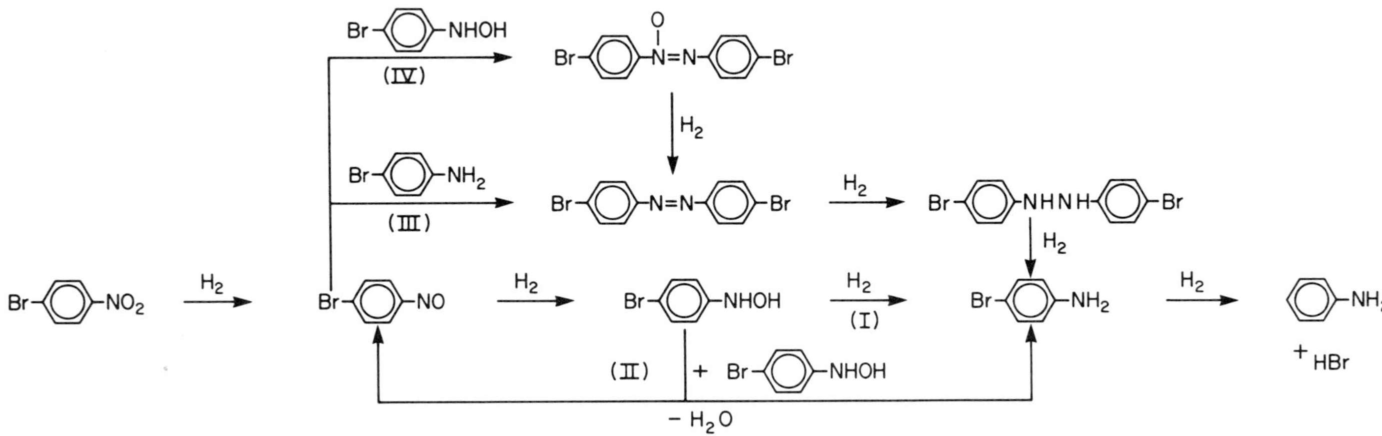

Fig. 2. Proposed pathways for the hydrogenation of bromonitrobenzenes.

bromoaniline product. The second pathway of importance (IV) is the condensation of the hydroxylamine and the nitroso intermediates to form the azoxy intermediate, which is then reduced through the azo and hydrazo intermediates to the bromoaniline product. At no time during our experiments were we able to detect any intermediates in the reaction solution that were either partly or completely dehalogenated. This indicates that dehalogenation occurs after the bromoaniline product is formed. That none of these dehalogenated intermediates were found also indicates that pathway III, involving the condensation of the bromoaniline product with the nitroso intermediate to the azo intermediate, is insignificant. Otherwise, one would expect some dehalogenated azo intermediate to be formed from the condensation of aniline with the nitroso intermediate.

The disproportionation of the hydroxylamine intermediate to nitroso and aniline products (II) would be a very exothermic reaction and would also occur when there was a significant buildup of hydroxylamine. Only a mild exotherm occurred during any of the reactions. Also, at the low temperatures at which the reactions were run, it is unlikely that there was enough energy present for disproportionation to occur.

The rate and amount of azoxy intermediate formation was dependent on the isomer being reduced. As can be seen in Fig. 3, the fastest rate and largest amount of azoxy was formed during the hydrogenation of the para isomer, while the ortho isomer gave the slowest rate and lowest amount of azoxy. This is easily explained by the stereochemistry of the compounds. The large bromo group in the ortho position would sterically hinder the reaction between the nitroso and hydroxylamine intermediates. The steric hindrance would be lowest in the para isomer. One would, therefore, expect to see a higher level of hydroxylamine intermediate during the ortho isomer reduction and lower levels during the para isomer reductions. This was observed, and Fig. 4 shows typical hydroxylamine formation over 5% Pt/C-S.

VI. SUMMARY

The three bromonitrobenzene isomers can be selectively hydrogenated to the corresponding bromoanilines with $\leq$ 2% debromination using commercially available catalysts. The best catalyst for reducing all three

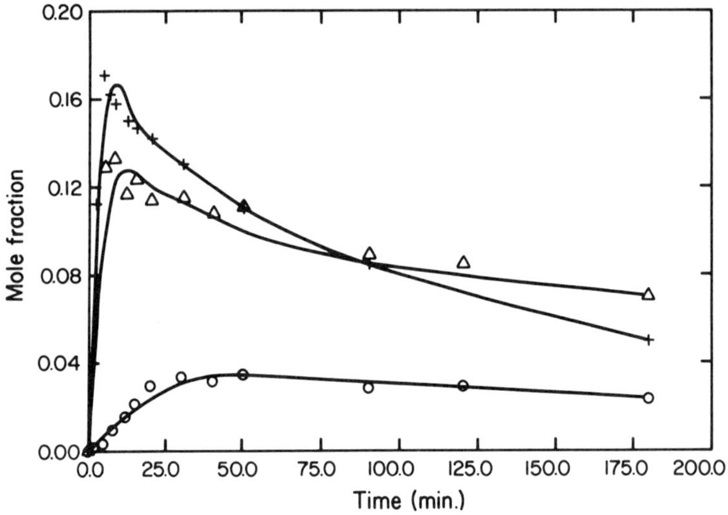

Fig. 3. Formation of the azoxy intermediate during the hydrogenation of bromonitrobenzenes over 5% Pt/C-S in THF at 1500 psig. o - ortho, Δ - meta, + - para isomer.

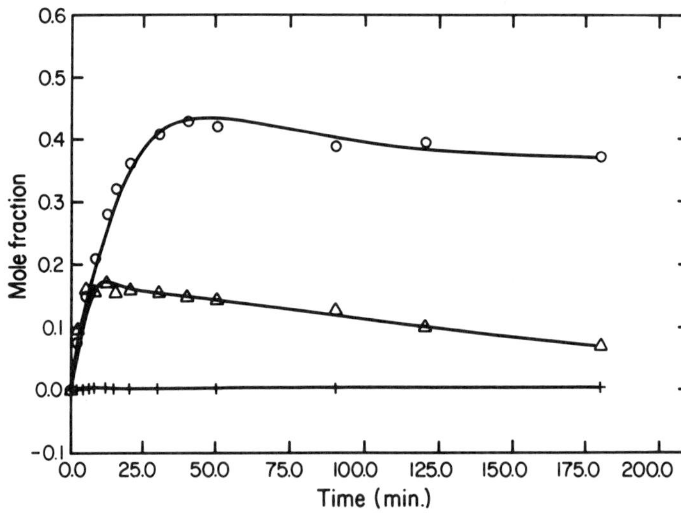

Fig. 4. Formation of hydroxylamine intermediate during the hydrogenation of bromonitrobenzenes over 5% Pt/C-S in THF at 1500 psig. o - ortho, Δ - meta, + - para isomer.

isomers in good yields with minimum debromination proved to be 5% Rh/C. Five percent Pt/C was effective for reduction of the ortho and the meta isomers, and 5% Pt/C-S was effective for the para isomer. THF was the preferred solvent for the ortho and para isomers, while ethanol was the better solvent for reduction of the meta isomer. The effects of pressure, isomer, and catalyst on debromination were interdependent. However, the amount of debromination generally was greater in ethanol than in THF, with the exception of the reduction of the meta isomer over either 5% Pt/C or 5% Rh/C.

The rate of reaction was greater in ethanol and at the higher pressure. Also, the rate of reaction was greater over the noble metal catalysts than the Raney nickel 28 catalyst. There was no detectable effect on the rate by the different isomers.

The reduction mechanism apparently follows two parallel pathways. These are, direct hydrogenation of the hydroxylamine intermediate to bromoaniline and condensation of the nitroso and hydroxylamine intermediates to an azoxy intermediate that is subsequently reduced to bromoaniline. The degree to which the azoxy intermediate pathway was followed was dependent on the particular isomer reduced. The para isomer follows that pathway to the greatest degree and the ortho isomer follows it to the least degree. Debromination occurred only to the bromoaniline product, and not to any of the intermediates.

This work has demonstrated that the reduction of halonitroaromatics with minimum dehalogenation is a difficult reaction dependent on several reaction variables. In this paper we have only investigated a few of those variables, but the indications are that each individual halonitroaromatic may require very specific reaction conditions, solvents, and catalysts in order to maximize yield of the haloaniline and minimize dehalogenation.

REFERENCES

1. J.R. Kosak, in Catalysis in Organic Syntheses (W.H. Jones ed.), Academic Press, New York, p. 107, 1980.
2. A.J. Dietzler and T.R. Kiel, U.S. Patent 3,051,753 (Aug. 28, 1962).
3. L. Spiegler and N.J. Woodbury, U.S. Patent 3,073,865 (Jan. 15, 1963).

4. F.S. Dovell and H. Greenfield, U.S. Patent 3,350,450 (Oct. 31, 1967).

5. H.S. Broadbent, Ann. N.Y. Acad. Sci., **145**, 58 (1967).

6. J.R. Kosak, Ann. N.Y. Acad. Sci., **172**, 175 (1970).

7. F.Z. Haber, Elektrochem., **22**, 506 (1898).

8

Hydrogenation of Nitroarenes—The Hydroxylamine Intermediate

John R. Kosak

E.I. Du Pont de Nemours and Co.
Deepwater, NJ

ABSTRACT

Catalytic hydrogenation of nitroarenes to the corresponding arylamines can proceed through intermediate species (nitroso, hydroxylamino) which may or may not be detected by interrupting the hydrogenation mid-way through the reduction. The arylhydroxylamine, because of its ability to disproportionate thermally (35 kcal/mole), can be potentially hazardous if present in significant quantity during the course of a hydrogenation. Hydrogenation processes for several substituted nitroarenes were investigated to determine their propensity for arylhydroxlamine formation. The influence of catalyst modifiers on formation/accumulation of hydroxylamine intermediates was also investigated. The sensitivity of p-tolylhydroxylamine to temperature was shown and its disproportionation products were identified.

I. INTRODUCTION

Catalytic hydrogenation of nitroarenes* to the corresponding aromatic amines is conducted routinely in the laboratory on a few grams at a time

*Nitroarenes are defined here and throughout this paper as nitrobenzenes and with one or more hydrogen atoms of the ring replaced with other atoms or groups.

and on the industrial scale where many tons per day are produced. The reduction is said to proceed by way of a number of intermediate species.

$$R\text{–}NO_2 \longrightarrow R\text{–}NO \longrightarrow RNHOH \longrightarrow R\text{–}NH_2$$

Nitro is reduced to nitroso which in turn is reduced to hydroxylamino, and this is reduced to amine. If one interrupts the hydrogenation before completion one may or may not detect these transient species depending on handling of the reaction mass and the mode of detection. However, bimolecular reduction species are usually found. These are the azoxy, azo and hydrazo compounds. Haber in 1898[1] showed that all the intermediate species could lead to the amine product.

Since that time a number of investigators have studied this reduction process and based on their findings have proposed various schemes to explain their results. Some researchers claim the reduction proceeds stepwise through nitroso and hydroxylamino intermediates to the final amine. Others claim intermediacy of the bimolecular species (azoxy, azo, hydrazo) as the pathway to producing the amine.

Our study of the reduction of nitroarenes was directed toward learning more about the hydroxylamine intermediate since it is capable of reacting in two ways (Fig. 2) (disproportionation, rearrangement) that do not require the use of hydrogen. Specifically, we were interested in the disproportionation

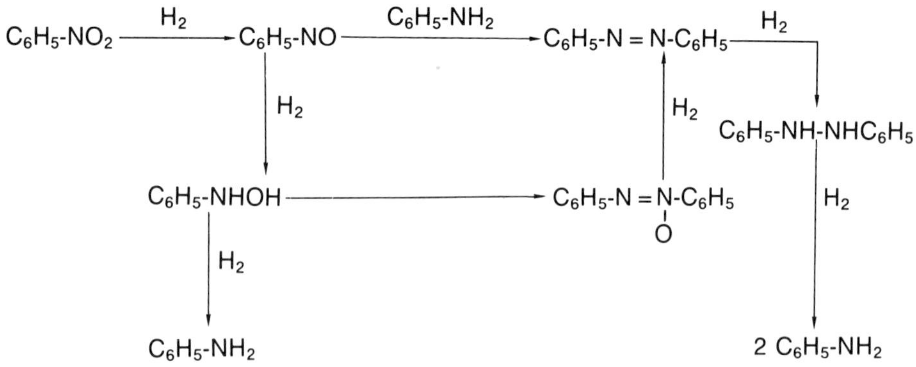

Fig. 1. Haber Reduction Scheme for Nitrobenzene.

$$\text{RNHOH} \nearrow \text{R-NO} + \text{RNH}_2 \text{ (disproportionation)}$$
$$\text{RNHOH} \searrow \text{HO-R-NH}_2 \text{ (rearrangement)}$$

Fig. 2. Hydroxylamine Chemistry.

reaction which is exothermic (~ 35 kcal/mole). If the hydroxylamine is present in substantial quantity it can lead to a runaway reaction during the hydrogenation. One such case involved the destruction of a plant scale autoclave used for the manufacture of 3,4-dichloroaniline.[2] Thus, we were interested in learning what caused it to form and accumulate.

We have studied hydrogenation processes for a number of nitroarenes over different palladium and platinum catalysts, and under a variety of conditions. Procedures were developed to identify the various intermediate species that could form. In particular, a procedure was developed to identify and measure the concentration of the relatively unstable aryl hydroxylamine species.

II. EXPERIMENTAL

A. Hydrogenation

All hydrogenations were done in a Parr Instrument Co. 1 liter titanium autoclave equipped with internal cooling coils and an automatic sampling valve so that samples could be taken during the course of the hydrogenation without interrupting the hydrogenation. Hydrogen was fed to the autoclave from a 2-liter pressurized cylinder and the volume was measured to determine H_2 consumption vs time. (Fig. 3)

B. Sample Handling

Samples taken while the hydrogenation was in progress were transferred to a -30°C refrigerator and held until they could be analyzed. Samples taken at the completion of hydrogenation were treated similarly if they were to be analyzed for the hydroxylamine. If not they were stored at ambient temperature.

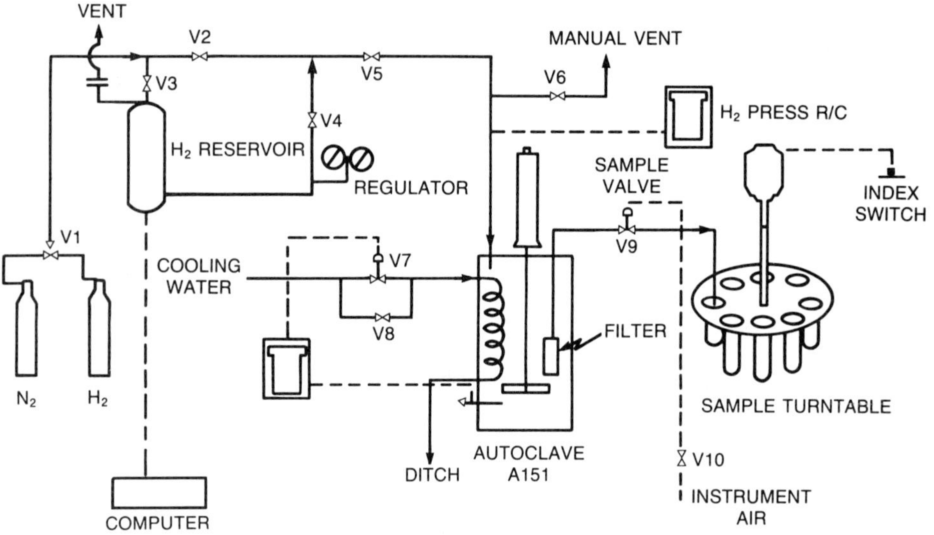

Fig. 3. 1 liter batch autoclave for hydrogenation, remote control sampler, automatic cooling water control, computer monitored hydrogen feed.

C. Analysis for Arylhydroxylamine

Absorption at 1.544 microns in the near IR is specific identification for substituted arylhydroxylamines. The limit of detection is 2%. Where there was specific interest in a given arylhydroxylamine, that compound was synthesized and standards were established.

D. Gas Chromatography

Analyses of 3,4-dichloroaniline were done on a Perkin-Elmer 900 Gas Chromatograph equipped with a flame ionization detector, a 6-ft. by 1/4 in. O.D. glass column packed with 10% SE 30 on Gas Chrom Z (60-80) mesh. Helium flow is 60 ml./min. The injection port temperature is 300°C and the initial column temperature is 150°C. The initial column temperture is held for 3 minutes and then is programmed to rise at a rate of 10°C/min. to a final temperature of 280°C. The sample is prepared by dissolving ~ 0.25 g.

of molten 3,4-dichloroaniline in 10 ml of methylene chloride. A 1 microliter sample size is injected. Retention times, in seconds, for the various components are: aniline - 73, o-chloroaniline - 135, p-chloroaniline - 192, 3,4-dichloronitrobenzene - 350, 3,4-dichloroaniline - 387, and 3,3',4,4'-tetrachloroazo - (and hydrazo)benzene - 1100.

Other amines were done with the same system and the components were identified by their corresponding retention times.

III. RESULTS AND DISCUSSION

A. 3,4-Dichloroaniline

Hydrogenation of 3,4-dichloronitrobenzene to 3,4-dichloroaniline requires the use of special conditions to minimize concurrent dehydrochlorination during the reduction. Two different processes were investigated: morpholine + 5% platinum/carbon catalyst (Ni-Cr modified)[3] and, phosphorous acid + 5% platinum/carbon catalyst (purchased from Engelhard Industries). With the morpholine process a number of intermediates were detected from samples taken during the course of the hydrogenation. Two of these are shown in Fig. 4.

The highest concentration of 3,4-dichlorophenylhydroxylamine (3,4-DCPHA) occurs at about 2 moles hydrogen consumed, which is what one might expect. Formation of 3,3',4,4'-tetrachlorohydrazobenzene (TCAB) appears to reach a maximum of about 1.25% at about 60% conversion and then decreases to a final value of 0.50%. However, this is an artifact caused by some disproportionation of the arylhydroxylamine (on standing and prior to refrigeration) to primarily the azo compound, thereby giving higher TCAB concentrations. Under these reduction conditions the hydrazo compound cannot be reduced to the amine. This is shown in Table I. A 3,4-dichloroaniline reduction mass was analyzed at the end of a hydrogenation and then the mass was subjected to higher temperatures for 30 minute intervals, sampled and analyzed for the hydrazo component.

Thus, it is apparent that the hydrazo compound is not reduced to the amine at or below 125°C.

Other chloronitroarenes tested under morpholine process conditions also showed accumulation of the corresponding hydroxylamine when

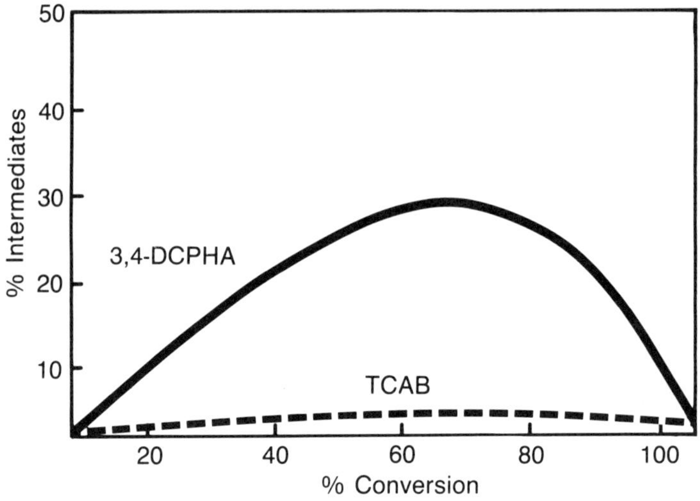

Fig. 4 Morpholine Process. Accumulation of intermediates vs. % conversion of 3,4-dichloronitrobenzene. Hydrogenation conditions were: 2M 3,4-dichloronitrobenzene, 1 wt % morpholine, 10 ppm Pt as 5% Pt on carbon (Ni-Cr modified), 95-100°C, 200 psig H_2.

TABLE I

Hydrogenation of 3,3',4,4'-Tetrachlorohydrazobenzene[a,b]

Hydrogenation Temp., °C	Time, Min.	Hydrazo %
95	0[b]	0.49
95	30	0.53
125	30	0.50
150	30	0.34
175	30	0.08

[a]A standard laboratory hydrogenation of 3,4-dichloronitrobenzene was carried out as in Fig. 4.

[b]Reduction mass was sampled 15 min. after active hydrogenation ceased and analyzed for TCAB.

3,4-Dichloronitrobenzene	(35%)
o-Nitrochlorobenzene	($\sim$ 15%)
2-Chloro-4-nitrotoluene	($\sim$5.0%)
p-Chloronitrobenzene	1.9%)

Fig. 5. Order of hydroxylamine accumulation.

hydrogenation was interrupted before completion. However, the amount accumulated varied with the structure of the arene. At 67% conversion the greatest amount was obtained from 3,4-dichloronitrobenzene (35%) and the least from p-chloronitrobenzene (1.9%). The order is shown in Fig. 6.

When the phosphorous acid process was used to hydrogenate 3,4-dichloronitrobenzene, formation of the hydroxylamine intermediate was much lower and reached a maximum of only about 8% after the reduction was $\sim$ 10% complete. However, it was found that a marked increase in hydroxylamine accumulation occurs when 1% by weight of sodium nitrate is added to the substrate before hydrogenation (Fig. 5). Also, the maximum of about 40% is reached at 80% conversion.

B. o-Anisidine

Two o-nitroanisole samples of different quality (Table II) were hydrogenated using a palladium and a platinum catalyst. The results are shown in Table III.

It had been shown that the presence of o-chloronitrobenzene and o,o'-dichloroazoxybenzene in sample B enhances the ability of the palladium catalyst, but not the platinum catalyst, to produce hydroxylamine.[4] When the hydroxylamine intermediate is formed in significant amounts then a large amount of o,o'-hydrazoanisole is found in the final product, and again the hydrazo compound is not reduced to the arylamine.

C. Effect of Hydrogenation Variables

Variables encountered in batch hydrogenation technology were compared in one model system to determine whether some generalizations could be made

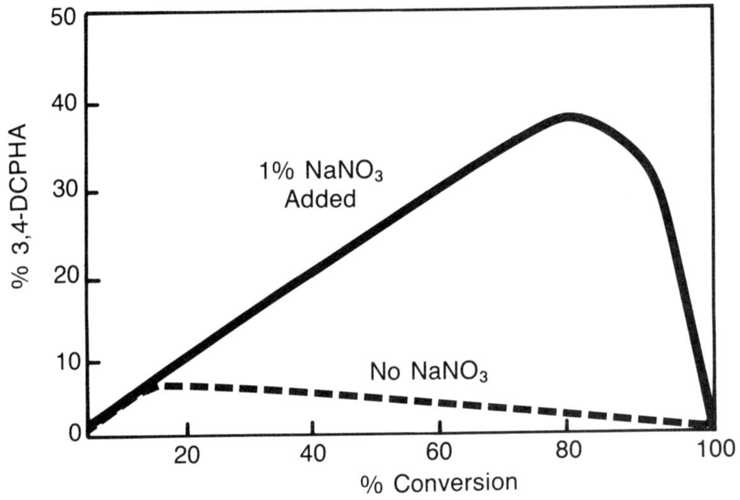

Fig. 6. Phosphorous Acid Process. 3,4-DCPHA accumulation. Hydrogenation conditions were: 2M 3,4-dichloronitrobenzene, 0.1% H_3PO_3, 20 ppm Pt as 5% Pt on carbon, 110-115°C, 200 psig H_2.

TABLE II

o-Nitroanisole Compositions

	Composition, %[a] A	B
o-Chloroanisole	.05	.09
o-Chloronitrobenzene	.11	.60
o-Nitroanisole	99.82	98.87
o,o'-Dichloroazoxybenzene	.01	.33
o-Nitrophenol, ppm[b]	<10	<10

[a]Area % determined by GC.
[b]Determined spectrophotometrically.

TABLE III

Platinum Versus Palladium Catalysts[a]

Example No.[b]	Catalyst Type	o-Nitroanisole Quality[d]	Product Composition (%) o-Anisidine	o,o'-Hydrazoanisole	o-Anisyl-Hydroxylamine[f]
1	Pd-Pt-Fe[c]	A	94 - 97	3-1	2
2	Pt-Ni-Cr[e]	A	99 - 99.2	0.1	2
3	Pd-Pt-Fe	B	82 - 85	14-12	18
4	Pt-Ni-Cr	B	97 - 98	2-1	2

[a]Reduction condtiions: 1M o-Nitroanisole. Precious metal usage was 10 ppm, 95 $\pm$ 5 °C, 300 psig H_2.
[b]Minimum number of experiments per example is 2.
[c]4.5% Pd-0.5% Pt-5.0% Fe on carbon prepared according to U.S. Patent No. 2,823,235.
[d]See Table II.
[e]5% Pt-2.5% Ni-2.5% Cr on carbon prepared according to U.S. Patent No. 3,546,297.
[f]Samples analyzed by near IR at 65% conversion.

about the factors governing hydroxylamine formation, accumulation and its final fate in the reduction process. Hydrogenation of p-nitrotoluene to p-toluidine was selected as the model system. Variables compared were platinum vs. palladium, morpholine vs. sodium hydroxide, and phosphorous acid vs. phosphoric acid (Table IV).

Both palladium and platinum (Examples 1, 3 & 9) appeared to give a detectable (by near IR) but very small amount of p-tolylhydroxylamine. Morpholine (Examples 5 & 6) and phosphorous acid (Examples 7 & 10) both promoted a significant increase in hydroxylamine accumulation, whereas sodium hydroxide (Example 4) and phosphoric acid (Example 11) had no effect on the process. In addition to being a base and an acid respectively, morpholine and phosphorous acid are catalyst modifiers which inhibit hydrogenolysis of aromatic carbon-halogen bonds when hydrogenating nitrohaloarenes. Thus pH alone is not a factor in causing hydroxylamine formation. It is evident from these data that arylhydroxylamine accumulation results from the presence of catalyst modifiers or poisons which alter the catalyst surface and inhibit reduction of arylhydroxylamines to the corresponding amines. Accumulation of the arylhydroxylamines can then lead to subsequent reactions - decomposition, rearrangement, etc. - which are dependent on conditions such as temperature, pressure, pH, and other reactants.

A recent patent[5] claims the use of mixtures of an organic amine in combination with an organic phosphorous compound to enhance conversion of nitroarenes to the corresponding hydroxylamines in high yield.

D. p-Tolylhydroxylamine Decomposition

The stability of p-tolylhydroxylamine (PTHA) and its degradation products was examined at various temperatures (Table V). It was found that PTHA in p-toluidine is stable at -30°C for 13 days (Example 2). Decomposition at room temperature (Example 4) occurs at a significant rate and at 150-160°C (Example 3 & 6) PTHA is dissipated in about half an hour or less. At ambient temperature decomposition leads only to azo formation. However, at the elevated temperature range (Examples 3 & 6) both the azo and azoxy derivatives are formed.

TABLE IV

p-Tolylhydroxylamine (PTHA) Accumulation

Example	Catalyst[a] Type	Additive Type	Additive %	Reduction (%) H_2 Consumed	Reduction (%) G.C.	PTHA (%) IR[b]
1	Pd	None	—	65.6	69.6	2
2	Pt-Ni-Cr	None		100	99.8	—
3	"	None		67	67.2	3
4	"	NaOH	0.1	64	69.3	3
5	"	Morpholine	1.0	40.7	36.7	9.3
6	"	Morpholine	1.0	63.2	61.6	13.5
7	Pt	Phosphorous Acid	0.05	63.2	65.2	11.6
8	Pt	None		100	99.7	0
9	Pt	None		64	67.3	3
10	Pt	Phosphorous Acid	0.1	64	64	6
11	Pt	Phosphoric Acid	0.1	64	66.7	3

[a]Precious metal catalyst concentration in all experiments was 20 ppm; hydrogenations were conducted at 60°-65°C and 400-500 psig H_2.

[b]Limit of detectability is about 2%.

TABLE V

p-Tolylhydroxylamine (PTHA) Decomposition

Example	pH	PTHA (by IR)	PT	PNT	Azoxy	Azo	Temp., °C	Hold Time, hrs.
1	Alkaline[a]	13.5	61.6	31.9	0.1	2.8	-30°	24
2	"	13.5	61.3	31.9	0.1	3.1	-30°	13 (days)
3	"	—	57.3	32.6	3.9	6.0	150-160°	0.5
4	"	—	58.0	32.4	0.1	8.3	22	16
5	Acid[b]	11.6	65.0	28.2	0.2	3.2	-30°	24
6	"	3.1	62.8	28.6	1.9	4.8	150-160°	0.25

[a]Sample from Table IV, Example 6.
[b]Sample from Table IV, Example 7, initial PTHA - 11.6.

IV. SUMMARY

Hydrogenation of nitroarenes (under the stated conditions) may or may not lead to accumulation of the corresponding hydroxylamine.

Certain catalyst modifiers will enhance accumulation of arylhydroxylamines. In the cases cited, morpholine, phosphorous acid, sodium nitrate, o-chloronitrobenzene, and o,o'-dichloroazoxybenzene all caused increased arylhydroxylamine accumulation under the stated conditions.

p-Tolylhydroxylamine is stable for at least 13 days at -30°C, disproportionates at a moderate rate at ambient temperature and rapidly at 150-160°C.

Hydroxylamines present in partially hydrogenated reduction masses will disproportionate to the corresponding azoxy and azo compounds depending on temperature.

Hydrogenation of nitroarenes discussed herein appear to follow a direct reduction pathway to the amine.

REFERENCES

1. F. Haber, Z. Electrochem., **4**, 506 (1898).
2. W.R. Tong, R.L. Seagraves and R. Wiederhorn, 3,4-Dichloroaniline Autoclave Incident, Chemical Engineering Progress, Loss Prevention, Vol. II, pp. 71-75 (1977).
3. J.R. Kosak, US 3,546,297 (1970), issued to DuPont.
4. J.R. Kosak, in **Catalysis of Organic Reactions,** (W.R. Mosher, ed.), Marcel Dekker, New York, NY, p. 461-471 (1981).
5. P. Hope and A.H. Sharma, European 0147879, issued to Akzo N.V.

9

Synthesis of Aliphatic Amines Over Methanol Synthesis Catalysts

Gaminī A. Vedage, Richard G. Herman, and Kamil Klier

Air Products and Chemicals, Inc.
Allentown, PA
and
Center for Surface and Coatings Research
Lehigh University
Bethlehem, PA

ABSTRACT

It has now been demonstrated that copper-based and palladium-based methanol synthesis catalysts can be effectively used as catalysts for aliphatic amine synthesis. Methylation of aliphatic amines with H_2/CO synthesis gas was carried out over Cu/ZnO catalysts at moderate temperatures and pressures. Under the experimental conditions with H_2/CO/amine feeds, both methanol and methylamine synthesis simultaneously occurred. Over these binary catalysts, higher temperatures favored higher yields, while lower temperatures favored higher selectivities of methylamines. The reduction of aliphatic amides was another route for amine synthesis, and it was shown that diethylmethylamine was the major product in the hydrogenation of N,N-diethylformamide over Cu/ZnO catalysts. Based on a contact time dependence study, a mechanism for the hydrogenation of amides was proposed. Finally, synthesis of aliphatic amines from alcohol/amine reactant feed streams was studied over Cu/ZnO, Fe/Cu/ZnO, and Pd/SiO_2 catalysts, and the results have been compared with those reported in the literature. Only Cu-based catalysts gave good results with respect to both activity and selectivity. Catalysts containing Group VIII metals generally showed high activity, but the selectivities were much lower than the Cu-based catalysts because of their strong tendency for carbon-carbon bond cleavage of the alcohols. This C-C bond cleavage was not observed over the Cu/ZnO catalyst. A mechanism for alcohol/amine coupling over methanol synthesis catalysts was proposed.

I. INTRODUCTION

The synthesis of aliphatic amines is of particular interest since they are widely used as corrosion inhibitors, epoxy hardeners, textile additives, explosives, and for the manufacture of rubber chemicals. Some known processes utilizing aliphatic alcohols are given in Table I, from which it can be seen that catalysts of very different chemical compositions are reported to be effective for the amination reaction. However, with few exceptions, these catalysts give poor selectivities or they tend to deactivate rapidly in the absence of hydrogen.

Despite the fact that many catalysts have been proposed in the patent literature (Table I), only Cu-based catalysts gave good results with respect to both activity and selectivity. Group VIII metal catalysts, such as Ni or Co, also showed good activity but the selectivity was much lower than the Cu based catalysts[6,7] due to their strong tendency for carbon-carbon bond cleavage of the alcohols. This carbon-carbon bond cleavage was not observed with Cu-catalysts and, thus, most of the intermediates undergo condensation instead of decarbonylation.

Catalytic methods for the preparation of aliphatic amines have been extensively reviewed by Baiker et al.[7] Nekrasova et al.[8] and Herman.[6] In the present work, it will be demonstrated that catalysts that can be utilized for low temperature/pressure methanol synthesis can also be used for alkylation of amines with alcohols, methylation of amines with CO/H_2 synthesis gas and for the hydrogenation of amides to amines.

TABLE I

Group VIII and Cu Catalysts for the Alkylation of Amines with Alcohols or Aldehydes

Catalysts	Reactants[a]	Product Selectivity[b]	Total Conversion	Temp., Pressure	Reference
Co-Kieselguhr	NH_3, 1.1 EtOH, 1.0 H_2, 1.4	~45%	64.9%	170°C 16 atm	1
Co-Al_2O_3	NH_3, 1.1 EtOH, 1.0 H_2, 1.4	~20%	63.7%	170°C 16 atm	1
Ni-Kieselguhr	NH_3, 1.1 EtOH, 1.0 H_2, 1.4	~45%	44.0%	170°C 16 atm	1
Ni-Al_2O_3	NH_3, 1.1 EtOH, 1.0 H_2, 1.4	~45%	43.4%	170°C 16 atm	1
Pd-Al_2O_3	NH_3, 9.2 PhOH, 1.0 H_2, 9.2 H_2O, 1.1	71.8%	43.0%	250°C 1 atm	2
Pd-Al_2O_3	NH_3, 9.2 PhOH, 1.0 H_2, 9.2 H_2O, 1.1	18%	63%	200°C 1 atm	2
Rh-Al_2O_3	NH_3, 9.2 PhOH, 1.0 H_2, 9.2 H_2O, 1.1	37.1%	28.3%	250°C 1 atm	2
Cu-Mo-Al_2O_3	Me_2NH, 3 BuOH, 1 H_2	80.9%	89.3%	214°C 26 atm	3
Cu-Zn-Al_2O_3	$MeNH_2$, 2 $Me(CH_2)_{11}OH$, 1 H_2	65.9%	83.3%	192°C 26 atm	3
Cu-Na-Sn-Al_2O_3	Me_2NH $Me(CH_2)_{11}OH$ H_2	78%	95.4%	250°C 26 atm	4
Cu-Mo Cu-W	Me_2NH, 3 $Me(CH_2)_{11}OH$, 1 H_2	80.8%	96.3%	193- 203°C 27 atm	5

[a] Mole ratio given.
[b] Mole %.

II. EXPERIMENTAL

A. Catalyst Preparation

The Cu/ZnO (30/70 mol %) catalysts were coprecipitated from a nitrate solution by Na_2CO_3, calcined, pelletized, and sieved to 10-20 Mesh, according to a procedure previously described in detail.[9,10] A 4.8 wt % Pd on SiO_2 catalyst was prepared by adding 200 g of 10-20 Mesh silica gel to 200 ml of concentrated HCl containing 16.7 g of $PdCl_2$, as described elsewhere.[11] The Fe/Cu/ZnO catalyst was coprecipitated from Fe(III), Cu(II), and Zn(II) nitrate solution (total metal ion concentration was approximately 0.25 M) heated to 80-90°C by the dropwise addition of 1.0 M sodium carbonate until a pH of 6.8 was achieved. After digesting for 1 hr, the resultant solid was filtered, washed, and air dried.

B. Testing Procedure

Approximately 2.45 g (3 ml) of calcined catalyst mixed with 7 ml of 3 mm diameter Pyrex beads were centered in the fixed bed single pass tubular reactor (1.8 cm i.d.) that was used to test the catalysts. Reduction was carried out at 250°C and 1 atm with a H_2/N_2 (2/98 vol %) gas mixture flowing at a rate of 60 ml/min (STP) until water was no longer detected in the exit stream by gas chromatography. The catalyst was then cooled to ambient temperature prior to pressurizing the reactor to 75 atm with H_2. The gases used for testing were ultra-high purity grade hydrogen, chemically pure carbon monoxide that was passed through activated charcoal, and certified standard CO/CO_2 mixtures. The isopropylamine that was utilized was 98% pure, with n-propylamine and diisopropylamine as the main impurities.

A schematic of the catalytic reactor systems has been presented,[6,9] and this testing unit has a high pressure pump, located at the top of the reactor preheater section, for injecting liquids into the synthesis gas stream at reaction pressure, wherein the liquids are vaporized in the preheater section of the reactor. The exit gas was reduced to ambient pressure by a back-pressure regulator and was sampled in intervals using a heated automatic on-line sampling valve on a Hewlett-Packard 5730A gas chromatograph that was coupled with a HP Model 3388jA

Integrator/Controller. Analyses was typically carried out with a Poropak Q or Chromosorb 101 column. Response factors for the thermal conductivity detector were taken from Ref. (12) except for the following: methanol, 45; isopropylamine, 90; methylisopropylamine, 115; dimethylisopropylamine, 130; ethylisopropylamine, 130; methylethylisopropylamine, 145; diethylisopropylamine, 160; and n-propylisopropylamine, 175. Condensables (at 0°C) were collected and further analyzed by direct liquid injection into the HP GC and a Finnigan 4021 GC/MS/ Nova system. The analyses resulted in less than 2% error in the carbon atom balance, and over the binary Cu/ZnO catalyst, this procedure gave carbon and oxygen balances within 0.5 and 1.6%, respectively. Testing was usually allowed to run for at least 24 hr for each set of experimental conditions.

C. Catalyst Characterization

The catalysts were characterized by BET surface area determinations[13] using argon (16.8×10^{-20} m^2/Ar atom) and by standard X-ray powder diffraction. Auger analyses were carried out with a Physical Electronics Auger spectrometer (Phi Model 50-220), and X-ray photoelectron spectroscopy (XPS) was performed using a Physical Electronics ESCA-Auger spectrometer (Model 548). Samples were prepared by briefly grinding a few catalyst pellets to a fine powder, suspending the powder in absolute ethanol, and evaporating the ethanol with the aid of an infrared lamp so that the dispersed solid settled onto a 1 cm^2 gold-plated sample holder. Catalyst samples, subjected to the standard reduction procedure, produced no detectable signal corresponding to sodium.

III. RESULTS AND DISCUSSION

A. Alkylation of Amines by Alcohols Over Pd/SiO_2 and Cu/ZnO Methanol Synthesis Catalysts

In this section it will be demonstrated that the catalysts utilized for methanol synthesis can also be used as amine alkylation catalysts. Both Cu/ZnO and Pd/SiO_2 methanol synthesis catalysts were chosen for this investigation.[14,15] The investigated catalytic amination was carried out in a simple one-step procedure, and it can be represented by the reaction scheme shown in Eq. 1. where R and R' can be either alkyl groups or a

$$R_2CHOH + R'NH_2 \rightarrow R_2CHNHR' + H_2O \quad (1)$$

hydrogen atom.

A 2.45 g portion of the Cu/ZnO catalyst (4.0 g of 4.8 wt % Pd/SiO_2 catalyst), after reduction, was pressurized with nitrogen to 75 atm, and the GHSV was set at the desired value. Under these conditions, the reactor was heated to the reaction temperature, and a mixture of isopropylamine and isopropanol was injected to the gas stream at the top of the reactor. To study the disproportionation reactions only isopropylamine was injected at the top of the reactor. The product composition exclusive of nitrogen, is compiled in Table II. Over the Cu/ZnO catalyst at 190°C, over 70% of the isorpopylamine was alkylated to diisopropylamine. No ammonia was detected in the exit gas, indicating that disproportionation of issopropylamine was not occurring under the present experimental conditions. When the isopropylamine/isopropanol mixture was replaced by

TABLE II

Results[a] of the Alkylation of Isopropylamine with Isopropanol and of the Disproportionation of Isopropylamine in Nitrogen at 190°C, and a Pressure of 75 atm

Catalyst	GHSV		Iso-propanol	Isopropyl-amine	Ammonia	Water	Diisopropyl-amine
4.0 g of	3750	mmol/hr In	9.4	5.7	0.0	0.0	0.0
4.8 wt%	$\ell(N_2)$/	mmol/hr Out	8.5	2.5	1.3	0.7	2.0
Pd/SiO_2	kg cat/	mmol/hr In	0.0	7.1	0.0	0.0	0.0
Catalyst	hr	mmol/hr Out	0.0	3.4	1.9	0.0	1.8
2.45 g of	6120	mmol/hr In	8.6	6.4	0.0	0.0	0.0
30/70	$\ell(N_2)$/	mmol/hr Out	4.6	1.9	0.0	4.2	4.3
Cu/ZnO	kg cat/	mmol/hr In	0.0	7.1	0.0	0.0	0.0
Catalyst	hr	mmol/hr Out	0.0	5.5	0.7	0.0	0.9

[a] ± 0.1 mmol/hr

pure isopropylamine, about 22% of the isopropylamine disproportionated into ammonia and diisopropylamine. Vedage et al.[14] further showed that decreasing the temperature from 190°C to 159°C decreased the degree of disproportionation to approximately 3%.

When a mixture of isopropanol/isopropylamine was injected into the N_2 carrier gas stream over the Pd/SiO_2 catalyst, the observed total amine conversion was about 56% (see Table II). The ammonia and water yields, 1.3 and 0.7 mmol/4g cat/hr, respectively, indicated that approximately two-thirds of the substituted amine product was formed by disproportionation of isopropylamine, while about one-third arose by alkylation of the reactant amine with isopropanol. Replacing the injected liquid mixture by pure isopropylamine led to nearly the same degree of amine conversion as observed in the previous test.

The results compiled in Table II show that the alkylation of a primary amine by a primary alcohol occurred at a faster rate with a higher selectivity over the Cu/ZnO catalyst than those obtainable with the Pd/SiO_2 catalyst. The Cu/ZnO catalyst was also an excellent catalyst for the shift reaction, and the water produced by process 1 can be combined with reaction 2 so that the oxygenated product is carbon dioxide instead of water.

$$H_2O + CO \rightarrow CO_2 + H_2 \tag{2}$$

The Pd/SiO_2 catalyst, on the other hand, is a poorer shift catalyst and will not produce CO_2 from CO and H_2O under conditions at which reaction 1 takes place.

The principal undesirable reaction occurring over the Pd/SiO_2 catalyst is disproportionation of the reactant substituted amine, e.g. one given by Eq. 3, and this reaction can be monitored by the yield of ammonia.

$$2(CH_3)_2CHNH_2 \rightarrow [(CH_3)_2CH]_2NH + NH_3 \tag{3}$$

Disproportionation requires an activation and breaking of an N-alkyl bond and the N-H bond. In the present work, the disproportionation of isopropylamine occurred over both catalysts in the absence of isopropanol, and it was shown that the disproportionation of isopropylamine occurred at a faster rate over the Pd/SiO_2 catalyst than over the Cu/ZnO catalyst. In fact, this reaction is completely inhibited by isopropanol over Cu/ZnO catalyst, while for the Pd/SiO_2 catalyst, addition of isopropanol did not completely suppress the disproportionation of isopropylamine. The possible reasons for the difference in the behavior of these two catalysts are discussed below.

It is known that SiO_2 is an acidic material[16] and that ZnO is slightly basic. Further, it has been documented that acidic materials such as silica-alumina promote disproportionation of amines.[17] Therefore, the disproportionation reaction over the Pd/SiO_2 catalyst can be partly attributed to the acidic nature of the catalyst support. However, Cu/ZnO, which is a basic catalyst, also promotes the disproportionation of amines. Therefore, there has to be another kind of site responsible for this disproportionation reaction. The ZnO portion of Cu/ZnO catalyst will not disproportionate amines, since it is not generally accepted that two bases such as ZnO and amine would interact with one another. Therefore, the sites for disproportionation of amines might also include the hydrogenation-dehydrogenation sites on the catalysts, as previously proposed.[18] It has been demonstrated that the Cu/ZnO is a very active hydrogenation catalyst,[19] and when isopropanol is added to the amine gas stream over this catalyst, the disproportionation of amine is inhibited due to the ability of alcohols to preferentially compete for the hydrogenation-dehydrogenation sites. Pd^o is also a well-known hydrogenation-dehydrogenation site;[20] hence, the Pd/SiO_2 catalyst should contain both the acidic and hydrogenation-dehydrogenation sites for the disproportionation of amines. Therefore, when isopropylamine by itself is added, the disproportionation to diisopropylamine and ammonia would occur on both sites. In the presence of isopropanol, the disproportionation occurring on hydrogenation-dehydrogenation sites would be inhibited, but the disproportionation of isopropylamine over acid sites would continue without interruption because acid/base (amine) interactions are much greater than acid/alcohol interactions. This accounts for the lowering, but not inhibition, of isopropylamine disproportionation rate in the presence of isopropanol over Pd/SiO_2 catalyst.

Alkylation reactions (Eq. 1) and disproportionation reactions (Eq. 3) would be expected to proceed without hydrogen as a reactant. However, H_2

is often utilized as a carrier gas,[21-23] and it has been reported that the presence of H_2 promotes the disproportionation of dimethylamine over a Cu/γ-alumina catalyst at 237°C and atmospheric pressure.[24] The experiments of Baiker[24] showed that H_2 also acts as an inhibitor for catalyst deactivation during the reaction involving the alkylation of dimethylamine with 1-dodecanol. in fact, as soon as hydrogen was replaced by N_2 in their experiments, both the conversion and the selectivity began to decrease. In the present work, Cu/ZnO (30/70) catalyst has been used in the absence of hydrogen for the alkylation of primary amines with primary or secondary alcohols at pressures of 26-75 atm and temperatures of 159-215°C for up to 150 hr without catalyst deactivation.[14] When hydrogen was added to the reactant gas stream, a moderate increase in catalytic activity and conversion was observed.

It should be noted that after 150 hr of continuous testing in the absence of hydrogen, the catalyst was observed to be pitch black when removed from the reactor under a nitrogen atmosphere. This indicated that the catalyst had not suffered sintering such that metallic copper would agglomerate. Indeed, the black color has been assigned to the presence of highly interdispersed, electronically interacting catalyst components that produce the active state of the catalyst.[10]

The previously observed[24] deactivation of copper-based catalysts in the absence of hydrogen has been reported to be caused by surface nitride formation, e.g. as represented by Eq. 4. The ammonia, in this case, would be formed by disproportionation of primary alkylamines.

$$NH_3 + 3Cu \rightarrow Cu_3N + 3/2\ H_2 \tag{4}$$

The evidence for such a nitride formation was obtained by temperature programmed desorption (TPD) and differential scanning calorimetric (DSC) measurements.[24] Such a nitride formation is indicated to not occur over the presently used Cu/ZnO catalyst because catalyst deactivation was not observed even after 150 hr of catalyst testing in the absence of hydrogen and XPS analyses of the tested catalysts did not detect surface nitrogen. In any case, it would be desirable to suppress the disproportionation reaction of the amines in the system. Since acidic catalysts promote the disproportionation reaction, less acidic catalyst supports should be utilized or the strongly acidic groups on the catalyst should be neutralized prior to use.

The mechanism of the amination of alcohols has been considered by a number of investigators,[24-29] and it was agreed that the reaction proceeds via an aldehydic or ketonic type of intermediate produced by the dehydrogenation of alcohols. For example, the amination of[27] αα-dideuterated 1-octanol with ammonia over a molten iron catalyst and with dimethylamine over Cu-γ alumina catalyst[24-29] showed no dideuterated amines in the product mixture. This observation excluded the possibility of a SN2 type of mechanism and supported an aldehydic-type mechanism for alkylation of amines with alcohols. On the other hand, Omote et al.[26] studied different stages of the amination of cyclohexanol over Cu/ZnO catalyst by trapping appropriate intermediates; he suggested a reaction path involving cyclohexanone and cyclohexanone imine.

The results of the investigation of Vedage et al.[15] on the Cu/ZnO catalyst also suggest an aldehydic/ketonic-type intermediate, originating from the dehydrogenation of alcohols, for the reaction of primary or secondary alcohols with isopropylamine. Over Cu/ZnO catalyst, isopropylamine was alkylated at a faster rate by secondary alcohols than the corresponding primary alcohol. This fact excludes the possibility of a SN2 type mechanism since such a mechanism would give a reverse order than the observed one. It was concluded that the first step involved in amine/alcohol coupling is the dehydrogenation of alcohol to the aldehyde or the corresponding ketone. Accordingly, the mechanistic scheme (Eqs. 5-7) was proposed,[30] based in part on reference 28 for the amine/alcohol reaction over Cu/ZnO catalyst:

$$R_2CHOH \rightarrow R_2CO + H_2 \tag{5}$$

$$R_2CO + R'NH_2 \rightarrow R_2C{=}NR' \; H_2O \tag{6}$$

$$R_2C{=}NR' + H_2 \rightarrow R_2CH{-}NHR \tag{7}$$

where R and R' = alkyl group or a hydrogen atom. According to this scheme, the first step is the dehydrogenation of alcohol to the corresponding aldehyde, which, in turn, would react with the amine to produce an imine. Subsequently, the imine would undergo hydrogenation to produce an alkylamine.

B. Synthesis of Methylalkylamines from CO/H_2 Synthesis Gas and Alkylamines Over a Cu/ZnO Catalyst

Addition of primary or secondary amine to CO/H_2 synthesis gas under methanol synthesis conditions produces methylalkylamines according to Eq. 8 and methanol according to Eq. 9.[30]

$$R_1R_2NH + CO + H_2 \rightarrow R_1R_2NCH_3 + H_2O \quad (8)$$
$$CO + 2H_2 \rightarrow CH_3OH \quad (9)$$

Water produced by reaction 8 was converted to CO_2 via the water gas shift reaction (Eq. 2). Methylation of amines with synthesis gas (CO/H_2) was carried out over Cu/ZnO (30/70) catalyst under the methanol synthesis conditions at 75 atm and GHSV = 6120 ℓ/kg cat/hr (Table III). A 2.45 g

TABLE III

Results of the Reaction of Synthesis Gas (CO/H_2 = 30/70) with Isopropylamine at 75 atm and a GHSV = 6120 /kg cat/hr Over 2.45 g of Cu/ZnO (30/70) Catalyst

	Temperature	CO	CO_2	H_2O	CH_3OH	Isopropyl-amine	Methyliso-propylamine	Dimethyliso-propylamine
mmol/hr In	150°C	184.0				7.1		
mmol/hr Out		182.0	0.2	0.1	1.5	6.8	0.3	0.0
mmol/hr In	159°C	184.0				7.1		
mmol/hr Out		180.7	0.7	0.0	1.9	6.6	0.5	0.0
mmol/hr In	180°C	184.0				7.1		
mmol/hr Out		175.6	1.6	0.0	5.3	5.8	1.3	0.0
mmol/hr In	215°C	184.0				7.1		
mmol/hr Out		140.9	9.4	0.6	23.9	1.3	2.9	2.8

portion of the catalyst was centered in the reactor and was reduced in a manner described in reference.[9] After cooling to room temperature, the reactor was pressurized to 75 atm with a H_2/CO (70/30 vol %) synthesis gas. Maintaining a gas flow of 6120 ℓ/kg cat/hr, the reactor was heated to the reaction temperature and an amine was injected continuously into the gas stream at the reactor inlet.

The results summarized in Table III show that at 215°C about 82% of the injected amine was methylated and half of the amine product was dimethylisopropylamine. At the same time when methylation was occurring, methanol synthesis was also observed. The yield of methylisopropylamine was higher at 215°C than at 180°C, indicating that higher temperatures favor higher yields of methylisopropylamine. However, it is evident that higher tempertures lead to the formation of dimethylisopropylamine, thus lowering the selectivity to methylisopropylamine.

As shown by Eq. 8, the methyl group originating from synthesis gas replaced a hydrogen atom from the amine nitrogen atom. This methylation reaction can be viewed as a reaction between an intermediate [I] formed during methanol synthesis (Eq. 10) and isopropyl amine, or a reaction between the product methanol and isopropylamine.

$$CO + 2H_2 \rightarrow [I] \rightarrow CH_3OH \qquad (10)$$

$$RNH_2 + [I] \rightarrow RN(H)CH_3 + H_2O \qquad (11)$$

$$RNH_2 + CH_3OH \rightarrow RN(H)CH_3 + H_2O \qquad (12)$$

To differentiate between the possible pathways represented by Eqs. 11 and 12, first isopropylamine was injected into CO/H_2 synthesis gas under the conditions specified in Table IV. After Experiment A was completed, the reactant CO was replaced by nitrogen and methanol was injected into the inlet gas stream. The quantity of methanol injected was approximately equivalent to the number of methyl groups as methanol plus methylisopropylamine product in Experiment A. The results for this Experiment B in Table IV show steady state concentrations of methanol reacted with isopropylamine over the catalyst to produce methylisopropylamine and dimethylisopropylamine. Comparison of these two experiments in Table IV demonstrates that the degree of methylation was lower when methanol was the source of methyl groups.

TABLE IV

Results of the Catalytic Synthesis of Methylamines at 190°C, GHSV = 6120 /kg cat/hr and at 75 atm Over 2.45 g of Cu/ZnO (30/70) Catalyst

	Isopropyl-amine	CO	H_2	CO_2	H_2O	CH_3OH	Isopropyl methylamine	Isopropyldi-methylamine
Experiment (A)								
mmol/hr In	42.3	184.0	429.0					
mmol/hr Out	38.3	169.9	n.d.	4.0	0.3	5.6	4.0	0.0
Experiment (B)								
mmol/hr In	41.3[a]		429.0			9.9		
mmol/hr Out	39.5	0.0	n.d.	0.0	1.7	8.2	1.7	0.1

[a]Plus 184.0 mmol/hr nitrogen.
n.d. = Not determined.

The results presented here show that isopropylamine reacts with the synthesis gas CO/H_2 at rates comparable to the methanol synthesis rate. Clearly the carbon source for the methyl group of the amine product is CO in the synthesis gas, but the key mechanistic problem resolved[30] was which carbon species, gaseous or adsorbed, reacted with the injected amine. The following paths were considered for the methylated amine.

1. Direct reaction of the amine with the methanol product

$$RNH_2 + CH_3OH \rightarrow RNHCH_3 + H_2O \quad (13)$$

2. Routes via formamide intermediate

$$RNH_2 + CO \rightarrow H(CO)NHR \quad (14)$$

$$RNH_2 + HCOO^-_{(s)} + H^+_{(s)} \rightarrow HCOO^- \, H_3^+NR \quad (15)$$

$$HCOO^- \, H_3^+NR \rightarrow H(CO)NHR + H_2O \quad (16)$$

$$H(CO)NHR + 2H_2 \rightarrow RNHCH_3 + H_2O \quad (17)$$

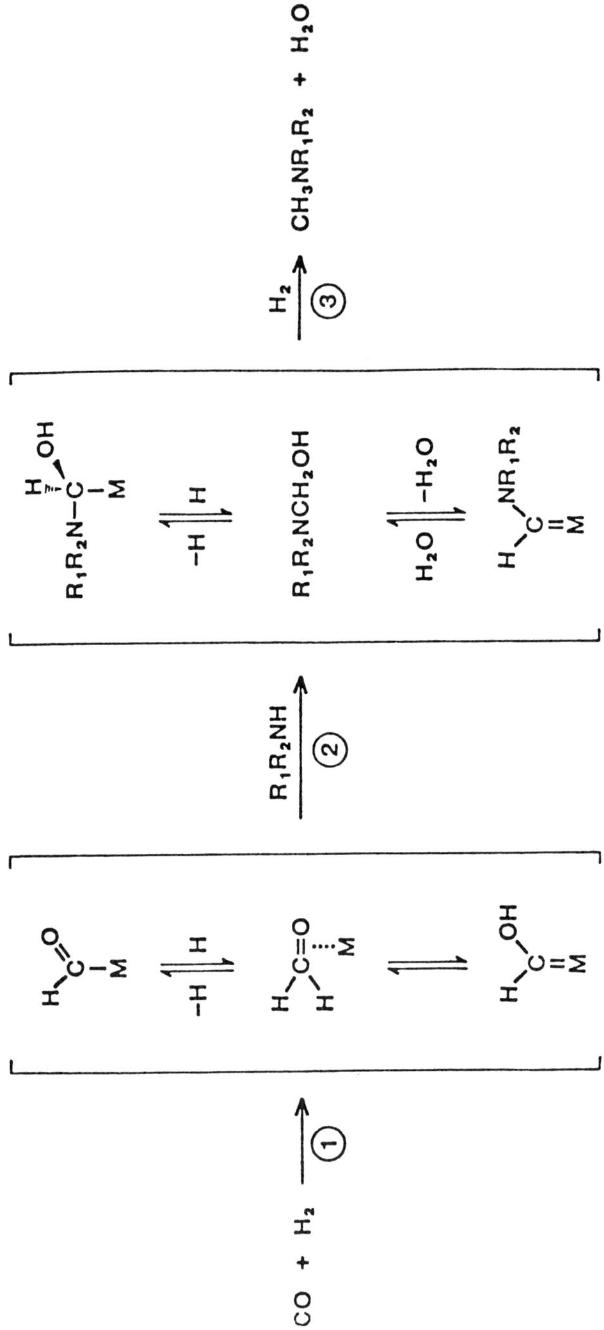

(23)

3. Route via carbamic acid and carbamates

$$RNH_2 + CO_2 \rightarrow RNHCOOH \quad (18)$$
$$RNHCOOH + RNH_2 \rightarrow RNHCOO^- \, H_2^+NR \quad (19)$$
$$RNHCOOH + 3H_2 \rightarrow RNHCH_3 + 2H_2O \quad (20)$$

4. Routes via aldehydic intermediates

$$H_2CO \text{ (or HCOH, HCO)} + RNH_2 \rightarrow H_2C(OH)NHR \quad (21)$$
$$H_2C(OH)NHR + H_2 \rightarrow RNHCH_3 + H_2O \quad (22)$$

Based on the comparison of the rates of the methyl group synthesis on the amine, of methanol synthesis, and of formamide hydrogenation, routes 1-3 were ruled out as major pathways under the present conditions using the Cu/ZnO catalyst. Route 4, involving the amine trapping of aldehydic intermediates, was consistent with the observed product composition and formation rate. In this route, formaldehyde H_2CO, adsorbed formyl HCO, and surface-bonded hydroxycarbene HCOH formed from synthesis gas are expected to undergo similar reactions (reaction 23). A detailed analysis underlying the above conclusions is given in Ref. 30.

C. Synthesis of Methylamines Over a Cu/ZnO Catalyst

A pelletized low density CuO/ZnO (30/70 mol %) catalyst (2.45 g = 4.8 ml) was reduced using the standard procedure and was then exposed to H_2/CO = 2.0 synthesis gas at 57 atm and 120°C. Heated ammonia gas was added to the synthesis gas stream, and the temperature was gradually increased to 300°C. A description of the resultant reactant gas mixture and the slate of products formed is given in Table V. With the high rate of ammonia addition ($NH_3/CO/H_2$ = 40.9/19.6/39.5 mol %), the small amount of CO converted to products (4.0%) was divided in equal proportions between methylamine and carbon dioxide. No methanol was detected in the exit gas, nor was dimethylamine formed under these experimental conditions.

Maintaining the temperature and pressure, the flow rate of ammonia was decreased in a step-wise manner. This led to a decrease in the gas hourly space velocity and an increase in the CO conversion to products. As

TABLE V

Synthesis of Methylamines from Ammonia and H_2/CO = 2.0 Synthesis Gas at 300°C and 57 atm Over a Cu/ZnO = 30/70 mol% Catalyst

$NH_3/CO/H_2$ (mol%)	NH_3/CO Molar Ratio	GHSV (hr^{-1})	% CO Conversion	Yield of Products (mol/l/hr)					Methylated Products (mol/kg/hr)
				CH_3OH	H_2NCH_3	$HN(CH_3)_2$	$N(CH_3)_3$	CO_2	
40.9/19.6/39.5	2.09	1040	4.0	0	0.19	0	0	0.19	0.37
28.2/23.8/47.9	1.18	870	7.2	0	0.31	Trace	0	0.29	0.61
12.7/29.0/58.3	0.44	700	33	0.04	0.04	0.66	0	1.27	1.37
7.8/30.6/61.6	0.25	670	68	0.27	0.01	1.75	Trace	1.92	3.45
5.2/31.5/63.3	0.16	650	63	0.27	0.01	0.02	0.88	2.23	1.78

indicated in Table V, the observed selectivity changed in a systematic way so that the predominant product shifted from monomethylamine to dimethylamine and then to trimethylamine as the ammonia content of the reactant gas mixture decreased. For example, with the $NH_3/CO/H_2$ = 7.8/30.6/61.6 mol % gas stream, 68% of the CO reactant was converted to products and about 80% of the ammonia was converted to methylated products with over a 99% selectivity to dimethylamine. With this lower level of inlet ammonia, the formation of methanol was observed and an appreciable quantity of CO was converted to CO_2 by the water gas shift reaction.

D. Effect of Doping the Cu/ZnO Catalyst with Fe on the Synthesis of Amines

It has previously been shown[31] that under methanol synthesis testing conditions of 75 atm and 225°C with 2.45 g (3.0 ml) of CuO/ZnO = 30/70 mol % catalyst, utilization of H_2/CO = 70/30 vol % synthesis gas at GHSV = 5000 hr^{-1} resulted in the production of 9.0 mmol methanol/hr (4.9% carbon conversion), while under the same conditions with a $H_2/CO/CO_2$ = 70/24/6 vol % synthesis gas, the production of methanol increased to 33.5 mmol/hr (18.2% carbon conversion to methanol). Addition of isopropylamine to the

TABLE VI

Alkylation of Isopropylamine with H_2/CO = 70/30 mol% Synthesis Gas Over Cu/ZnO = 30/70 mol% and Fe/Cu/ZnO = 1.2/31.8/67.0 mol% Catalysts at 75 atm and GHSV = 5000 hr^{-1}. Catalyst Charge = 2.45 g.

	Cu/ZnO (225°C)	Fe/Cu/ZnO (235°C)
Input Excluding H_2 (mmol/hr)		
CO	184.0	184.0
Isopropylamine	28.2	14.1
Output Excluding H_2 (mmol/hr)		
CO	137.2	175.0
CO_2	14.6	0.92
H_2O	0.9	5.8
NH_3	0.0	1.3
Acetone	0.0	1.3
CH_3OH	16.5	0.0
C_1-C_7 Hydrocarbons	0.0	4.3
Isopropylamine	15.4	11.6
Methylisopropylamine	10.1	0.10
Dimethylisopropylamine	2.7	0.13
Other Alkylisopropylamines	0.0	0.91
	H2/CO/Amine 2.33/1.00/0.15	H2/CO/Amine 2.33/1.00/0.08

H_2/CO = 70/30 vol % synthesis gas resulted in an enhanced yield of methanol and in the formation of the mono- and di-methylated isopropylamine products, as shown in Table VI. Indeed, the number of methyl groups synthesized in the presence of isopropylamine corresponded to 32.0 mmol/hr, which is approximately the same yield of methyl groups formed from the $H_2/CO/CO_2$ = 70/24/6 vol % synthesis gas. As deduced from previous studies on the effects of CO_2 [31] and H_2O [32] in the synthesis gas on the yield of methanol over these binary catalysts, the water (and resultant carbon dioxide formed via the water gas shift reaction) produced during the methylation of the amine, e.g. Eq. 8, increased the rate of formation of the C_1 intermediate, i.e. [I] in Eq. 10.

In the presence of an iron pentacarbonyl catalyst in a methanol solution, it has been demonstrated that tertiary amines act as methyl transfer agents and induced the formation of ethanol.[33]

In the mechanistic sequence of reaction steps, it was proposed that a methyl group was transferred from a $N(CH_3)_4^+$ species to an iron carbonyl moiety, where CO insertion subsequently occurred to yield an aldehyde that was easily hydrogenated. It has previously been shown that incorporation of small amounts of iron into the Cu/ZnO catalyst resulted in decreased yields of methanol and a shift in selectivity toward hydrocarbons.[34] In the present work, addition of isopropylamine to H_2CO = 70/30 mol % synthesis gas over a Fe/Cu/ZnO = 1.2/31.8/67.0 mol % catalyst resulted in complete inhibition of alcohol synthesis, as indicated in Table 6. However, the carbon conversion ($\sim$4.4 mol %) to products was not altered by the injection of the amine to the reactant gas stream. Instead, a wide variety of higher alkyl-substituted amines were produced, in contrast with the undoped Cu/ZnO catalyst for which only methylisopropylamine and a smaller quantity of dimethylisopropylamine were formed from the reactant amine. Table VI also shows that some decomposition of the isopropylamine occurred over the Fe/Cu/ZnO catalyst and that the water gas shift reaction did not proceed as readily over this catalyst.

Of the hydrocarbons formed over the Fe-doped catalyst, the selectivity was shifted from paraffins to olefins by the addition of the amine to the inlet stream. This shift in hydrocarbon selectivity could be induced by (i) competitive adsorption of the amine on olefin adsorption sites or hydrogen activation sites or (ii) the formation of secondary α-olefins via the Hofmann elimination, which can be represented by the following equation:

$$RH_2N^+\!-\!\overset{|}{\underset{|}{C}}\!-\!\overset{|}{\underset{|}{C}}H \xrightarrow{OH^-} RH_2N + {>}C{=}C{<} + H_2O \qquad (24)$$

The latter possibility was explored by pumping n-butanol and isopropylamine in a CO-free feed mixture over the Fe/Cu/ZnO catalyst under the conditions listed in Table VII. It is shown in this table that a large number of products

TABLE VII

Reaction of n-Butanol and Isopropylamine Over the 1.2% Fe Catalyst. Catalyst at 235°C, 75 atm and 5000 hr^{-1}. Liquid Pumping Rate = 40 l/min.

N_2/H_2 (vol%)	100/0	70/30	100/0
Liquid Feed (mole ratios)			
n-butanol	1	1	1
isopropylamine	1	1	1
water	0	0	0.5
Reactor Effluent[1] (%)			
n-butanol	28.6	11.7	37.8
isopropylamine	15.3	6.0	16.3
water	3.7	6.7	9.3
acetone	3.5	5.0	5.5
butane nitrile	5.3	3.0	1.7
N-isopropylpyrrolidine	9.7	0	8.8
n-butylisopropylamine	21.9	61.1	12.9←
dibutylamine	3.5	6.5	2.2
C_9+ compounds	8.7	0	5.5
butene + hydrocarbons	0	0	0

[1] Percent of all products excluding N_2 as recorded by the inline GC.

was observed, but no olefins were produced. In addition, the expected coupled product, n-butylisopropylamine, was formed in the pure N_2 carrier gas, but the activity and selectivity toward this product were greatly enhanced by the addition of hydrogen to the carrier gas. On the other hand, addition of water to the N_2 carrier gas decreased both the activity and selectivity toward the dominant coupled product. These experiments indicate that the Hofmann elimination is not operative.

E. Synthesis of Amines by the Reduction of Amides

The reduction of amides over the Cu/ZnO (30/70) catalyst was of interest because formamide was a potential intermediate considered during the synthesis of methylamines from CO/H_2 synthesis gas and amine. As seen in Table 8, the formamide hydrogenation products are quite complex. The formamide hydrogenation study was conducted with N,N-diethylformamide at 215°C, 75 atm, and hydrogen GHSV = 4290 ℓ (STP)/kg cat/hr over Cu/ZnO catalyst.[30] The hydrogenation of formamide was also studied as a function of contact time to obtain mechanistic information. As indicated in Table VIII, when the flow rate of N,N-diethylformamide was 10.8 mmol/2.45 g cat/hr, 62% of it was converted to diethylmethylamine and the remaining 38% was distributed among methanol, ethanol, methylethylamine, diethylamine, and triethylamine. When the flow rate of N,N-

TABLE VIII

Hydrogenation of N,N-diethylformamide (NNDEF) at 215°C, 75 atm Over 2.45 g of Cu/ZnO (30/70) Catalyst

Flow Rate of Substrate	Flow Rate of Hydrogen		H_2O	CH_3OH	C_2H_5OH	MEA	DEA	DEMA	TEA	NNDEF	CO	CO_2
Experiment (A)												
20 μl/min of NNDEF (10.8 mmol/hr)	10.5 ℓ (STP)/hr	Exit gas composition[a] (mol %)	34.70	13.24	2.28	6.39	8.68	30.59	4.11	<0.01	<0.01	<0.01
Experiment (B)												
40 μl/min of NNDEF (21.6 mmol/hr)	21.0 ℓ (SIP)/hr	Exit gas composition[a] (mol %)	30.37	17.76	1.64	3.97	14.49	29.44	2.34	<0.01	<0.01	<0.01
Experiment (C)												
60 μl/min of NNDEF (32.4 mmol/hr)	31.5 ℓ (STP)/hr	Exit gas composition[a] (mol %)	29.25	19.91	1.23	3.06	16.85	28.02	1.68	<0.01	<0.01	<0.01

[a] Exit gas composition excluding hydrogen.

MEA = methylethylamine DEMA = diethylmethylamine
DEA = diethylamine TEA = triethylamine

diethylformamide and hydrogen was doubled and tripled at 215°C and 75 atm, the yields of diethylamine and methanol increased while the yields of all other products decreased (Table V (B) and (C)). Under all conditions studied, the conversion of N,N-diethylformamide to products was 100% with no observable amounts of CO, CO_2 and N,N-diethylformamide within the analytical detection limit of 0.01 mol %. The exit gas composition is plotted in Fig. 1 as a function of the reciprocal flow rate of hydrogen.

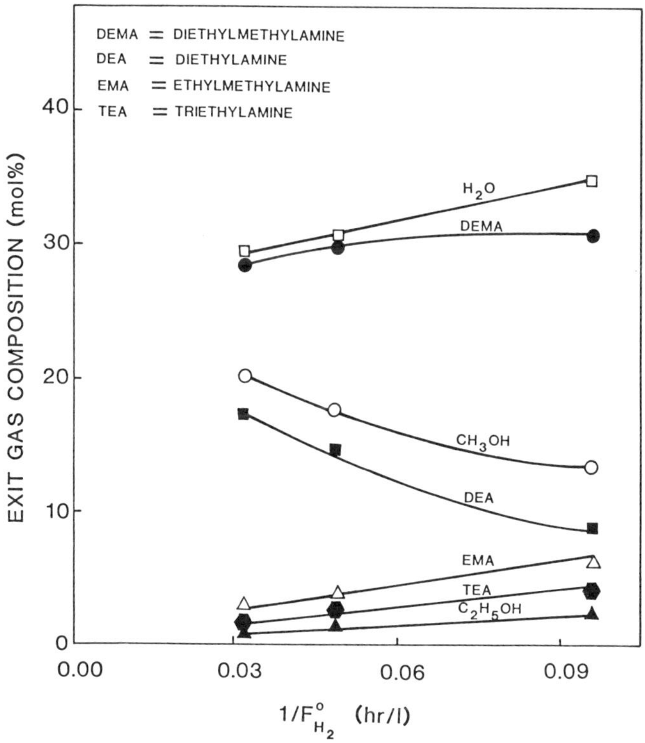

Fig. 1. The products of the hydrogenation of N,N-diethylformamide as a function of the reciprocal flow rate of hydrogen at 215°C, 75 atm and over 2.45 g Cu/ZnO (30/70) catalyst. The molar feed rate of N,N-diethylformamide was kept at 0.025 x flow rate of hydrogen in mol/hr.

As seen in Fig. 1, the formamide hydrogenation products are quite complex, and for this reason it was deemed necessary to determine which are the primary products and what are the rates of their formation. It is observed in Fig. 1 that as the flow rate increased, the yields of EMA, TEA and ethanol decreased; extrapolated to infinite flow rate or zero contact time, their yield approached zero. This shows that EMA, TEA, and ethanol were secondary products of the hydrogenation of NNDEF. The reactions given by Eq. 23-26 were formulated from balancing the CH_3 and C_2H_5 groups and water, and the method is briefly described below. Since Eqs. (25) and (26) are the primary reactions, the amount of methanol produced has to be equal to the amount of DEA produced from NNDEF by reaction 26. Because the amount of methanol was found equal to the sum of DEA and TEA in the exit gas stream, however, TEA has to be produced by a secondary reaction of DEA.

$$(C_2H_5)_2N\text{-}\overset{\overset{\displaystyle O}{\|}}{C}-H + 2H_2 \rightarrow (C_2H_5)_2NCH_3 + H_2O \tag{25}$$

$$(C_2H_5)_2N\text{-}\overset{\overset{\displaystyle O}{\|}}{C}-H + 2H_2 \rightarrow (C_2H_5)_2NCH_3NH + CH_3OH \tag{26}$$

$$(C_2H_5)_2NCH_3 + (C_2H_5)_2NH \rightarrow C_2H_5NHCH_3 + (C_2H_5)_3N \tag{27}$$

$$(C_2H_5)_3N + H_2O) \rightarrow (C_2H_5)_2NH + C_2H_5OH \tag{28}$$

In addition, ethylamine was not observed in the exit gas stream within the analytical detection limit of 0.01%. Therefore, TEA and MEA would originate by the reaction given by Eq. (27). The amount of MEA was equal to the sum of TEA and ethanol. This stoichiometric relationship indicates that the ethanol was produced by reaction 28. Hence, a part of DEA, DEMA and H_2O produced by reactions 25 and 26 is consumed by reactions 27 and 28 to produce MEA, TEA and ethanol. Therefore, the amount of DEA produced by reaction 26 is given by the sum of DEA and TEA in the exit gas stream. The amount of DEMA produced by reaction 25 is the sum of DEMA and MEA in the exit gas stream. Similarly, the amount of water produced by reaction 25 is given by the sum of water and ethanol in the exit gas stream.

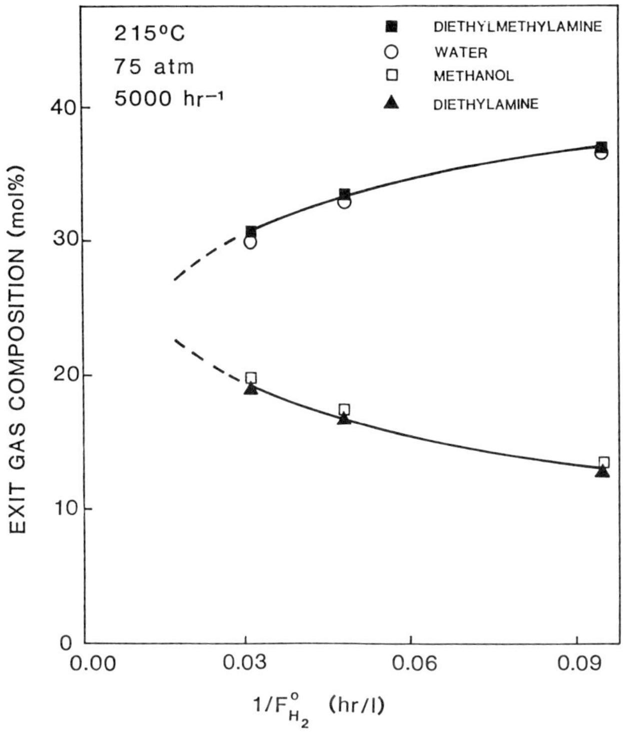

Fig. 2. Diethylmethylamine, diethylamine, methanol and water produced by reactions (25) and (26) as a function of the reciprocal flow rate of hydrogen at 215°C, 75 atm over 2.45 g Cu/ZnO catalyst. The molar feed rate of N,N-diethylformamide was kept at 0.025 x flow rate of hydrogen in mol/hr.

The calculated amounts of DEMA and water produced by reaction 25, DEA produced by reaction 26 and methanol are plotted as a function of the reciprocal flow rate of hydrogen in Fig. 2. The amounts of DEMA and water calculated as due to reaction 25, as well as those of DEA and methanol due to reaction 26 are seen to be in stoichiometric equivalence over the whole range of flow rates, which confirms the validity of the reaction scheme 25-28. The contact time dependence of the product yields in Fig. 2 leads to the following scheme of the primary reactions:

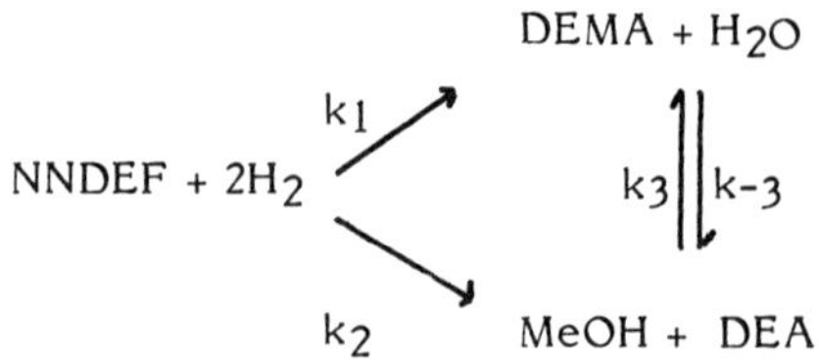

(29

By a kinetic analysis, Vedage et al.[30] showed that the amount of methanol and DEA generated by the primary reaction 26 was 54% and the percentage of NNDEF converted to DEMA and water in the primary reaction 25 was 46%. These initial rates of formamide hydrogenation have mechanistic implications as discussed below.

The mechanism for the hydrogenation of N,N-dimethylformamide over the Cu/ZnO catalyst may be similar to that suggested by Brown and Heim[35] for the reduction of amides by $LiAlH_4$. The reaction scheme, 30, is proposed

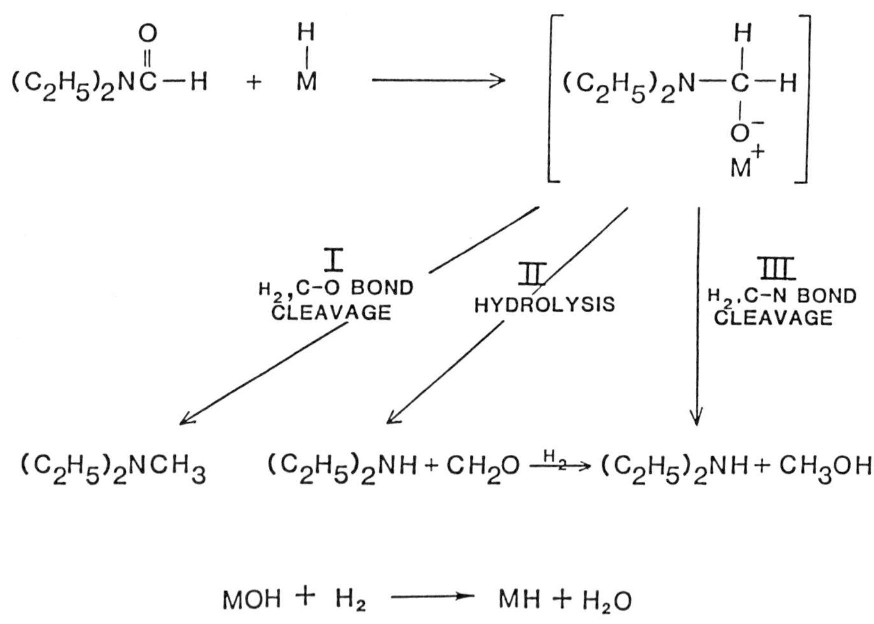

for the hydrogenation of N-N-dimethylformamide over the Cu/ZnO catalyst based, in part, on reference 35. In this reaction sequence. the amide is first attacked by a hydride (MH), probably CuH, to form a tetrahedral intermediate (A). This intermediate can undergo:

Path I: Hydrogenolytic C-O bond cleavage to produce the direct hydrogenation product DMA without the hydrogenolysis of the C-N bond.

PATH II: Hydrolysis to produce DEA and formaldehyde. Formaldehyde can then undergo hydrogenation to produce methanol or couple with DEA to produce DEMA.

Path III: Hydrogenolytic C-N bond cleavage to produce methanol and DEA. Subsequently, these could undergo coupling to produce DEMA via dehydrogenation of methanol to formaldehyde and amine-aldehyde coupling.

Here, Path I corresponds to the primary reaction 25 and Path III to reaction 26. Formaldehyde generated by the hydrolytic Path II can be either rapidly hydrogenated to methanol over the Cu/ZnO catalyst as expected from the earlier found hydrogenation rates of aldehydes 19, or can couple with the amine via reaction scheme 23 above. Thus, Path II can contribute to both reactions 25 and 26. The kinetic analysis above results in 46% probability for reaction 25 and 54% for reaction 26. These numbers indicate that the rates of hydrogenolysis of the C-N bond by combined Paths III and II and of the C-O bond by Path I are comparable.

It is noteworthy that the C-N bonds involving ethyl groups of NNDEF are not severed by hydrogenolysis or by ethylene elimination because neither ethane nor ethylene are among the products. Only hydrolytic cleavage of this C-N bond (Eq. 28) and alkyl transfer (Eq. 27) occur as minor secondary reactions as is seen from the contact time dependence of the product composition in Fig. 1.

CONCLUSIONS

This research has demonstrated that copper-based methanol synthesis catalysts are very active and selective for the synthesis of alkylamines via the reaction of alcohols with amines and methylamines via the reaction of synthesis gas with primary or secondary amines or ammonia. It has been shown that these synthetic reactions proceed by the coupling of a surface aldehydic intermediate such as formyl, pi-bonded formaldehyde, or hydroxycarbene with amine or ammonia. These alkylation reactions proceed selectively at fast rates under mild reaction conditions over Cu/ZnO catalysts, but disproportionation of alkylamines tend to occur over Pd/SiO_2 catalysts.

In the reaction of synthesis gas with alkylamines, the Cu/ZnO catalysts induce monomethylation of the amines to occur. At the same time, the water gas shift reaction readily converts by-product water into carbon dioxide, and methanol, the yield of which is dependent upon the reaction temperature, is produced by a parallel reaction involving the common C_1 intermediate. When the nitrogen-containing reactant is ammonia, it has been demonstrated that elevated ammonia concentrations, e.g. with a H_2CO/NH_3 = 47.9/23.8/28.2 mol % reactant gas stream, methanol synthesis is suppressed and monomethylamine is very selectively produced. Decreasing the ammonia content to particular levels in the H_2/CO = 2.33 synthesis gas provides for the selective synthesis of dimethylamine or trimethylamine. Addition of small amounts of iron to the Cu/ZnO catalyst results in the formation of hydrocarbons and in a very non-selective catalyst for methylamine synthesis.

It has been found that the Cu/ZnO catalyst is a very active catalyst for the hydrogenation of N,N-diethylformamide. However, a variety of products are formed because the hydride intermediate undergoes reaction by (i) hydrogenolytic C-O bond cleavage, (ii) hydrolysis, and (iii) hydrogenolytic C-N bond cleavage, the products of which can undergo subsequent coupling. An analysis of the experimental results excludes the formamide reaction pathways as the route for alkylamine synthesis involving alkylation by alcohols or synthesis gas because secondary alcohols couple very rapidly to amines and the water gas shift, proceeding via a formate intermediate that is common to methanol synthesis, is unaffected by the methylation of amines by synthesis gas.

REFERENCES

1. D.A. Gardner and R.T. Clark, U.S. Patent 4,255,357 (March 10, 1981); assigned to Pennwalt Corp.
2. Y. Ono and H. Ishida, J. Catal., **72**, 121 (1981).
3. L.H. Slaugh, U.S. Patent 4,206,150 (June 3, 1980); assigned to Shell Oil Co.
4. L.H. Slaugh and G.W. Schoenthal, Ger. Otten. 2,844,984 (April 19 1979) and U.S. Patent 4,229,379 Oct. 21, 1980); assigned to Shell Internationale Research Maatschappij, B.V.
5. L.H. Slaugh, Ger. Offen. 2,749,065 (May 18, 1978); assigned to Shell Internationale Research Maatschappij, B.V.
6. R.G. Herman, in "Catalytic Conversions of Synthesis Gas and Alcohols to Chemicals" (R.G. Herman, ed.), Plenum, New York, pp. 433-462 (1984).
7. A. Baiker and J. Kijenski, Catal. Rev., **27**, 653 (1985).
8. V.A. Nekrasova and N.I. Shuikin, Russ. Chem. Rev., **34**, 843 (1965).
9. R.G. Herman, K. Klier, G.W. Simmons, B.P. Finn, J.B. Bulko, and T.P. Kobylinski, J. Catal., **56**, 407 (1979).
10. J.B. Bulko, R.G. Herman, K. Klier, and G.W. Simmons, J. Phys. Chem., **83**, 3118 (1979).
11. M.L. Poutsma, L.F. Elek, P.A. Ibarbia, A.P. Risch, and J.A. Rabo, J. Catal., **52**, 157 (1978).
12. W.A. Dietz, J. Gas Chromatogr., **5**, 68 (1967).
13. R.G. Herman, P. Pendleton, J.B. Bulko, in "Advances in Materials Characterization" (D.R. Rossington, R.A. Condrate, and R.L. Snyder, ed.), Plenum Press, New York, pp. 109-132 (1983).
14. K. Klier, R.G. Herman and G.A. Vedage, U.S. Patent 4,480,131 (Oct. 30, 1984); assigned to Lehigh University.
15. G.A. Vedage, R.G. Herman and K. Klier, "Selective Production of Alkyl Amines from Alcohols and Ammonia, Alcohols and Primary Amines, and Alcohols and Secondary Amines", to be published.
16. H.A. Benesi and B.H.C. Winquist, Adv. Catal., **27**, 97 (1978).
17. W.W. Kaeding, U.S. Patent 4,082,805 (April 4, 1978); assigned to Mobil Oil Corp.

18. J. vonBraun, G. Plessing, and F. Zobel, Ber., **56**, 1988 (1923).
19. G.A. Vedage and K. Klier, J. Catal., **77**, 558 (1982).
20. W. Palczewska, Adv. Catal., **24**, 245 (1975).
21. D.C. Best, U.S. Patent 4,123,462 (Oct. 31, 1978); assigned to Union Carbide Corp.
22. K. Adam and E. Haarer, Fr. Patent 1,483,299 (June 2, 1967); assigned to BASF.
23. A. Baiker and W. Richarz, Helv. Chim. Acta, **61**, 1169 (1978).
24. A. Baiker, Ind. Eng. Chem., Prod. Res. Dev., **20**, 615 (1981).
25. E.J. Schwoegler and H. Adkins, J. Am. Chem. Soc., **61**, 3499 (1939).
26. Y. Omote, K. Iwase and J. Nakamura, Kogyo Kagatu Zusshi, **70** (9), 1508 (1967).
27. G.A. Kliger, L.F. Lazutina, R.A. Friedman, Y.B. Kryukow, A.N. Bashkirov, Y.S. Snagovskii, and R.M. Smirnova, Kinet. Katal., **16**, 660 (1975).
28. A. Baiker and W. Richarz, Preprints, 5th Canad. Symp. Catal., 289 (1977).
29. A. Baiker, W. Caprez and W.L. Hoistein, Ind. Eng. Chem., Prod. Res. Dev., **22**, 217 (1983).
30. G.A. Vedage, R.G. Herman and K. Klier, J. Catal., **95**, 423 (1985).
31. K. Klier, V. Chatikavanij, R.G. Herman, and G.W. Simmons, J. Catal., **74**, 343 (1982).
32. G.A. Vedage, R. Pitchai, R.G. Herman, and K. Klier, Proc. 8th Intern. Congr. Catal., Vol. II, Verlag Chemie, Weinheim, 47 (1984).
33. M.J. Chen and H.M. Feder, in "Catalysis of Organic Reactions", ed. by W.R. Moser, Marcel Dekker, New York, 273 (1981).
34. J.A. Sibilia, J.M. Dominguez, R.G. Herman, and K. Klier, Preprints, Div. Fuel Chem., ACS, 29(5), 261 (1984).
35. H.C. Brown and P. Heim, J. Org. Chem., **38**, 912 (1973).

Heterogeneous Catalysts

10

Relationship Between Catalyst Characterization and Performance

Robert J. Farrauto, Melvin C. Hobson and Nancy L. Brungard

Research and Development
Engelhard Corporation
Edison, New Jersey

ABSTRACT

Characterization of heterogeneous catalysts has as its goal the quantitative measure of those physical and chemical properties of the catalyst assumed to be responsible for performance in a given reaction. Definitive relationships between catalyst properties and performance are elusive due to the heterogeneity of the catalyst surface. Furthermore, mass transfer effects often mask the advantage of an optimized chemical property of the catalyst. This paper will, therefore, concentrate on the practical measurement and significance of some of the critical parameters influencing performance. Most notable will be the nature of the active catalytic components and their location within the support for slurry phase and fixed bed catalysts.

I. INTRODUCTION

Modern instrumentation has proved valuable in defining those chemical interactions on a heterogeneous catalyst responsible for a given performance. Extensive work with single crystals, or crystals with controlled imperfections, using sophisticated surface characterization tools such as Auger, ESCA, X-ray absorption edge spectroscopy, etc., are providing a basis for developing models to aid in designing new and improved catalysts. Although progress is being made, the translation from these fundamental mechanistic studies to commercially useful catalysts will remain a major challenge for years to come due to the complexity of the field of heterogeneous catalysis. There are, however, chemical and physical properties of catalysts easily measured in the laboratory that often can be correlated with performance and it is these which are the subject of this paper.[1-10]

A. General Description of a Catalyst and Catalysis Process

1. Catalysts

Most commercial heterogeneous catalysts are composed of active catalytic components supported or blended with refractory oxides and carbons, while others are bulk metals. With few exceptions they all have pore structures through which reactant and product molecules must pass in order to reach the majority of active sites. The optimization of catalyst preparations involves maximizing the accessible active sites by efficiently dispersing them on a high surface area carrier, but with pore structures sufficiently large to minimize diffusional resistances. A delicate balance of these chemical and physical properties must be achieved to produce a satisfactory catalyst. Molecules vary in size and chemical structure, so different catalysts are necessary for different reaction processes. In this respect the optimum catalyst is specific for each reaction and operating condition. The process of preparing catalysts has been reviewed elsewhere[8] and thus will not be discussed further.

B. Fundamental Steps in Catalytic Processes

In a catalytic process, therefore, reactants must interact with the "active site" on the catalyst. The term "active site" is thought to be that site (or sites) on the catalyst surface in which chemisorption and reaction occur. At least one molecule involved in the catalytic process must (1) diffuse to the surface of the catalyst particle (2) diffuse through the pore network to the active sites (3) chemisorb, (4) form an activated complex, (5) convert to the product while retaining its chemisorbed state, (6) desorb from the active site (7) diffuse through the pore structure and finally (8) diffuse into the bulk fluid. Steps (2) and (7) are controlled by the molecular size and pore structure of the catalyst while steps (3), (4), (5) and (6) depend on the chemical compatibility of reactant and product molecules (or atoms) with the active sites. The mass transfer steps (1) and (8) depend on reactant/product concentrations as well as catalyst amount and particle size (i.e., geometric area of catalyst). One of these steps will control the overall reaction rate. It is essential to understand which is controlling so that the catalyst or conditions of operation can be varied to obtain optimum performance. It is the objective of this paper to describe those physical and chemical properties of a catalyst which influence rection rates. A further objective is to provide some criteria for determining rate controlling steps during catalyst evaluation in both slurry and fixed bed processes.

1. Bulk Mass Transfer Control

Imagine a solid catalyst particle slurried in a fluid containing reactant molecules. The reactant molecules are assumed completely miscible in the fluid. The molecules must diffuse to the surface of the catalyst particle in order to chemisorb and subsequently react. There are physical parameters of the reaction matrix that influence the rate of diffusion and ultimately the extent of reaction in a given time frame. Increasing the amount of catalyst or reducing the particle size both lead to an increase in reactive surface and thus increase rate of conversion. When a reaction is limited by bulk mass transfer the concentration of reactant at the surface of the catalyst is close to zero because consumption at the surface is fast relative to rate of replenishment. Increasing the concentration of the reactant(s) in the bulk fluid will increase the driving force for diffusion and increase the rate of reaction. Other variables such as temperature and stirring rate have only a minor influence on reaction rate; however, stirring strongly influences gas/liquid transport phenomena such as dissolution of H_2 in the reactant solvent.

For the case of a fixed bed reactor increasing the linear velocity of the fluid, with a comparable increase in catalyst volume to maintain a constant space velocity, will increase conversion whenever bulk mass transfer is limiting. An increase in linear velocity enhances turbulence and decreases the stagnant layer of fluid surrounding the catalyst particle.

2. Pore Diffusion Control

Access to the active catalytic centers by reactant molelcules is usually through the internal pore structure of the carrier. Similarly the products must pass through pores to exit the catalyst. Naturally, the physical size of these microscopic channels and the physical geometry of the molecules influence the time required for the "chemical conversion journey". The condition leading to pore duffusion control exists when the rate of bulk mass transfer (1) and (8) and fundamental chemical steps (3)-(6) are fast relative to either step (2) or (7).

Pore diffusion resistance can be minimized by decreasing the particle size of the catalyst thus decreasing the "journey" path of molecules, by using wider pore carriers, and by depositing the active catalytic components nearer the edge of the carrier. One must recognize that by decreasing the carrier particle size the reaction rate may increase due to minimization of pore diffusion and/or by increasing the mass-transfer area. Hence, the structure of the pore and location of the active catalytic component are more definitive factors for pore diffusion control.

One further note of importance is that increasing conversion rates may not always be favorable if selectivity is sacrificed. Conditions can exist in which limiting diffusion of one or more reactants may be advantageous for a particular selectivity.[12]

3. Chemical Reaction Control

Fundamental steps (3) - (6) require specific chemical interaction of reactants and product molecules with active sites and thus rates are influenced by those factors common to chemical reactions. Thus increasing the amount of catalyst, the percentage of active components on the carrier, the process temperature, the concentration of reactants, ets., will usually favor the reaction rate. Of course, exceptions do exist; for example, when product desorption is rate limiting, any change that increases its concentration could decrease the reaction rate.

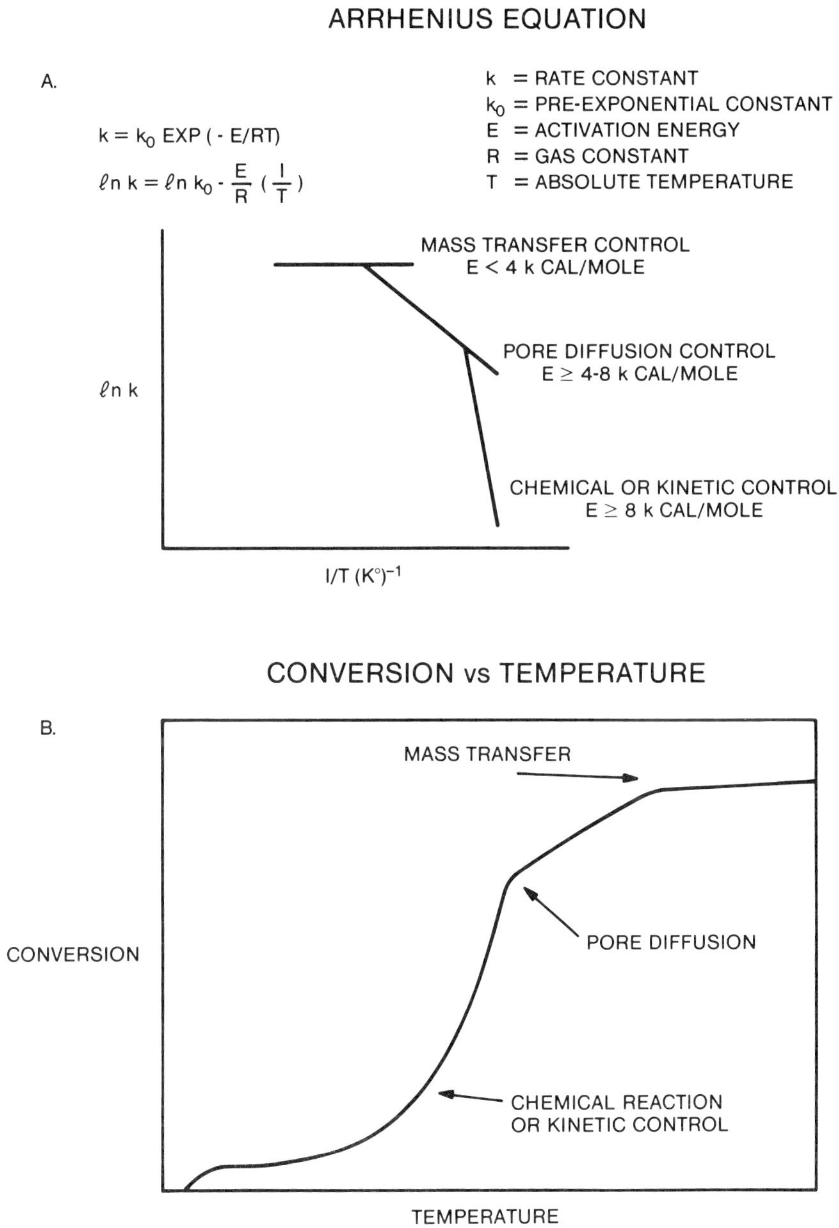

Fig. 1. Rate controlling mechanisms in a catalytic reaction. A, identification of the controlling mechanism from an Arrhenius plot. B, conversion of reactant as a function of temperature showing rate controlling regimes.

C. Summary of Fundamental Process Steps

We have briefly described physical and chemical properties of catalysts that can control the rate of a reaction. It should be evident that considerable planning is required to optimize the properties of the catalyst and te conditions for its use. A feedback network of testing, simulating actual operating conditions and catalyst characterization, is essential for efficient use of catalysts.

The most generally applicable test for rate limiting phenomena is the Arrhenius Plot from which activation energies are determined. An idealized plot is shown in Fig. 1A and a conversion versus temperature profile is shown in 1B. The calculation of the activation energy, E, from the slope of the arrhenius Plot allows an assessment of the rate limiting step. From Fig. 1B it is evident that reactions controlled by chemical kinetics are most sensitive to temperature while mass transfer controlled reactions are the least.

Finally, one may diagramatically represent the three modes of reaction control by Fig. 2. The bulk mass transfer control case (A) clearly shows the surface reactant concentration to be zero, while the kinetically controlled situation (B) shows uniform concentration throughout the catalyst particle. Pore diffusion (C) shows a gradient in reactant concentration within the pellet.

Fig. 2. Concentration gradients in a catalytic system for different rate controlling mechanisms. A, diffusion of reactant to outer catalyst surface is shown relative to reaction at the active site. B, diffusion to the active site is fast relative to the rate of reaction at the active site. C, reaction at the active site is fast relative to diffusion of the reactant in the pores of the catalyst particle.

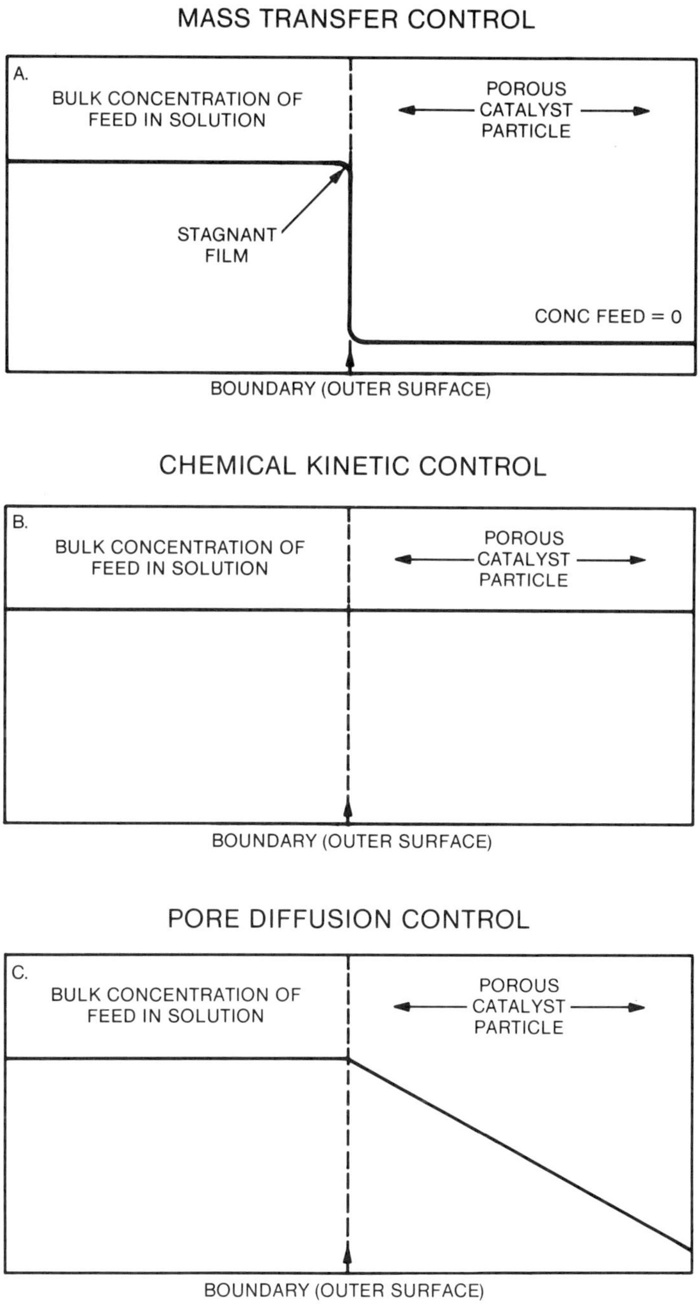
MASS TRANSFER CONTROL
A.
BULK CONCENTRATION OF FEED IN SOLUTION
POROUS CATALYST PARTICLE
STAGNANT FILM
CONC FEED = 0
BOUNDARY (OUTER SURFACE)
CHEMICAL KINETIC CONTROL
B.
BULK CONCENTRATION OF FEED IN SOLUTION
POROUS CATALYST PARTICLE
BOUNDARY (OUTER SURFACE)
PORE DIFFUSION CONTROL
C.
BULK CONCENTRATION OF FEED IN SOLUTION
POROUS CATALYST PARTICLE
BOUNDARY (OUTER SURFACE)

Figure 2

II. CHARACTERIZATION METHODS

The characterization methods most commonly used in the development and production of supported metal catalysts are:

1. Nitrogen adsorption and mercury penetration for surface areas and pore size distributions.
2. Chemisorption of H_2 or CO for metal surface area and dispersion.
3. X-ray diffraction for phase identification and crystallite sizes of metals and compounds.
4. Electron microscopy for morphology, location of catalytic species within a support and crystallite size.
5. Thermal analysis for phase identification, oxidation state and thermal stability.
6. Electron spectroscopy for chemical analysis (ESCA) for oxidation state and location of catalytic species.

A brief description of these methods is presented below.

A. Gas Adsorption

The most commonly used procedure for determining the internal surface area of a porous material, with surface areas greater than 1 or 2 m^2/g and up to about 1200 m^2/g, is based on the adsorption and condensation of nitrogen at liquid nitrogen temperature.[10] It is this surface on which the active catalytic components are dispersed. The partial pressure of nitrogen above the sample is gradually increased and N_2 molecules physically adsorb onto the surface. Monolayer coverage is approached in the linear region above the first steep portion of isotherm shown in Fig. 3A. Each adsorbed molecule occupies an area of the surface comparable to its cross sectional area ($\sim$16.2 A^2). The number of N_2 molecules adsorbed at monolayer coverage are measured by noting the pressure change in a static volume system, from which one can calculate the internal surface area. In practice, as the measured pressure relative to the normal vapor pressure (P/Po) increases, multilayer adsorption begins and at high relative N_2 partial pressures condensation of liquid N_2 in the pores occurs. The volume of

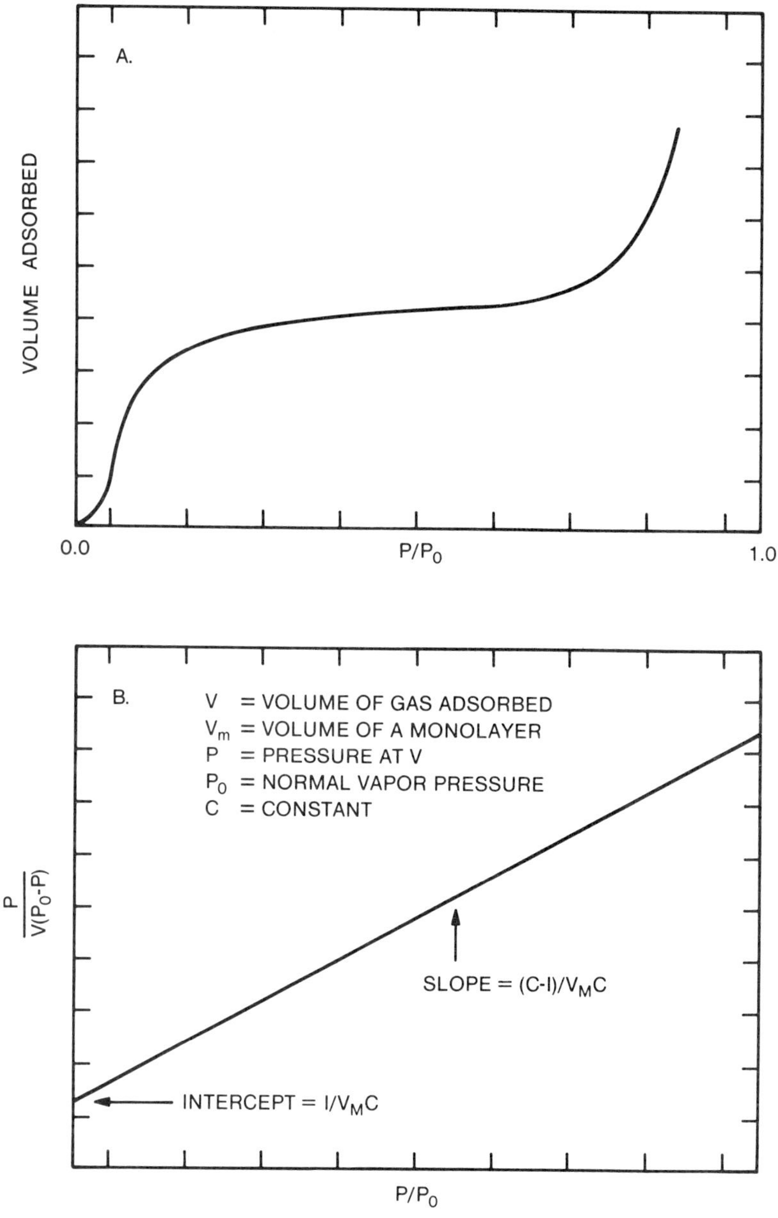

Fig. 3. Measurement of Surface Area by the BET Gas Adsorption Method. A, a typical adsorption isotherm with a relatively flat curve in the region of monolayer adsorption. B, plot of the linear form of the BET equation between p/po = 0.05 to 0.3 used to calculate the monolayer coverage, V_m.

nitrogen adsorbed at monolayer coverage, Vm, is obtained by plotting the linear form of the BET equation, Fig. 3B, where the slope and intercept are used to solve for Vm.

At higher relative pressures, from 0.25 to 1.0, the isotherm data may be used with the Kelvin equation to calculate pore size diameters between 30 and 600Å.

B. Mercury Intrusion

The penetration of mercury into the pores of a material is inversely proportional to the applied pressure, provided the mercury does not wet the surface. By measuring the volume of mercury intruded into the catalyst as a function of applied pressure, a quantitative pore size distribution measurement can be obtained.[11]

Typical pore size distribution data is shown in Fig. 4 where the integral penetration of mercury into the pores is plotted as a function of applied pressure. The calculated pore diameters in angstroms are shown across the top. The integral curve clearly shows a bimodal pore distribution with mean pore diameters at 9000 and 50Å. The latter is at the lower limit of the technique. A nitrogen desorption isotherm is required to obtain an accurate measure in the region below 100Å.

C. Chemisorption

Selective chemisorption is used to measure the accessible metal surface area of a supported catalyst by noting the amount of gas adsorbed per unit weight of catalyst.[8] Stoichiometry of the chemisorption process must be known in order to estimate available catalytic surface area. One assumes as a rule of thumb that the metal surface area is proportional to the number of active sites and thus reaction rate; however, many other factors can override the effect of metal area on rate. Hydrogen and carbon monoxide are most commonly used as selective adsorbates for many supported metals, although other gases have found limited applications. It is always assumed that the gas adsorbs only onto the metal site and not the carrier.

The measurements are carried out usually in a static vacuum system similar to that used for BET surface area measurements, or a pulsed flowing

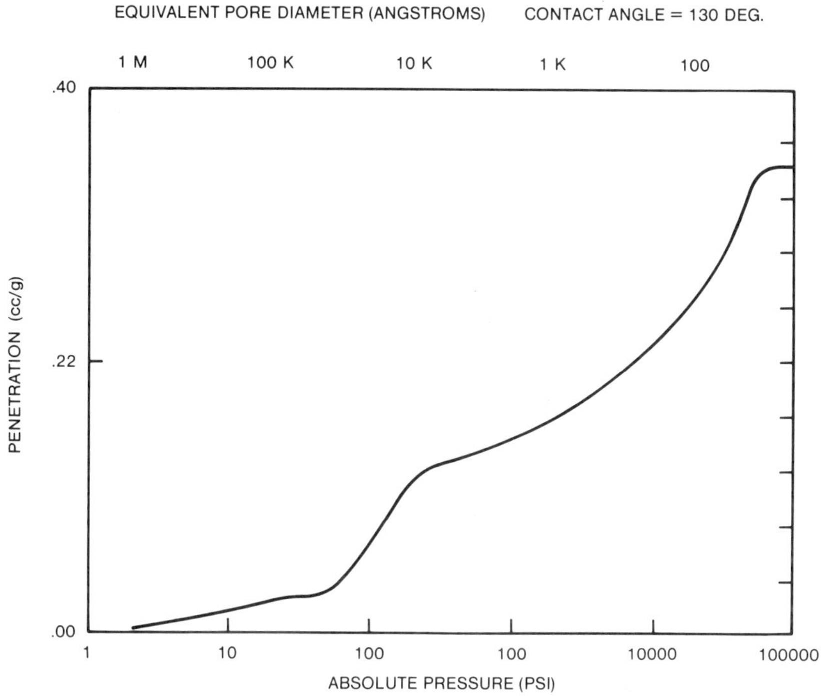

Fig. 4. Typical Pore Size Distribution Curve by the Mercury Intrusion Method.

gas system. Carbon monoxide is used almost exclusively in the pulsed method since it equilibrates rapidly with the metal surface. For the static system, when there is no further adsorption with increasing pressure, one assumes the catalytic surface is saturated with a monolayer of adsorbate. Noting the amount of gas adsorbed and knowing its stoichiometry with the surface site, the number of catalytic sites is determined. Approximating a value for the cross sectional area of the catalytic component based on an assumed geometry, its surface area and dispersion may be calculated. The number of surface species measured is assumed proportional to active sites, but there are many cases in the literature that show no relationship between metal surface area and reaction rates. In such cases, one evokes the concept of crystallite size effects controlling the activity or selectivity of a given reaction.

D. Analysis by X-Ray Diffraction (XRD)

Provided a material is sufficiently crystalline to diffract x-rays and present in an amount greater than about 1%, XRD can be used for qualitative and quantitative analyses. Many catalytic materials, particularly support materials, tend to be amorphous to x-rays due to their small crystallite sizes. The diffraction pattern will exhibit broad, featureless peaks which transform to a sharp, crystalline pattern under sintering conditions leading to substantial growth in crystallite size.

Fig. 5 shows the x-ray diffraction patterns of two Al_2O_3 structures used as carriers, γ-Al_2O_3 and δ-Al_2O_3. The former 5A, is the high surface area, lower temperature structure, while the latter 5B, is produced at high temperatures and has a lower surface area. Deactivation of the catalyst results from this loss of surface area since high catalyst dispersion cannot be maintained. Furthermore, sintering can occlude the active catalytic species.

E. Crystallite Size by XRD

The larger the crystals of a given component the sharper the peaks on the x-ray diffraction pattern for each crystal plane. Thus, the breadth of the peak can be related to the crystal size[8] which subsequently relates to the catalytic components available to reactants, i.e., catalytic area. The breadth at 1/2 peak height of an x-ray diffraction line is inversely proportional to the size of the crystallites. Fig. 5C shows the diffraction pattern of platinum crystallites obtained as a residue by dissolving the γ-Al_2O_3 carrier. Calculations from line broadening data indicate an average crystallite size of 100Å. The platinum appears to have sintered during use since crystallite size is usually about 25Å for "as prepared" samples of this type catalyst.

F. Electron Microscopy

For reactions which are pore diffusion controlled, it is advisable to locate the active catalytic species close to the fluid/solid interface in order to decrease the diffusion path of reactants and products. This applies to all

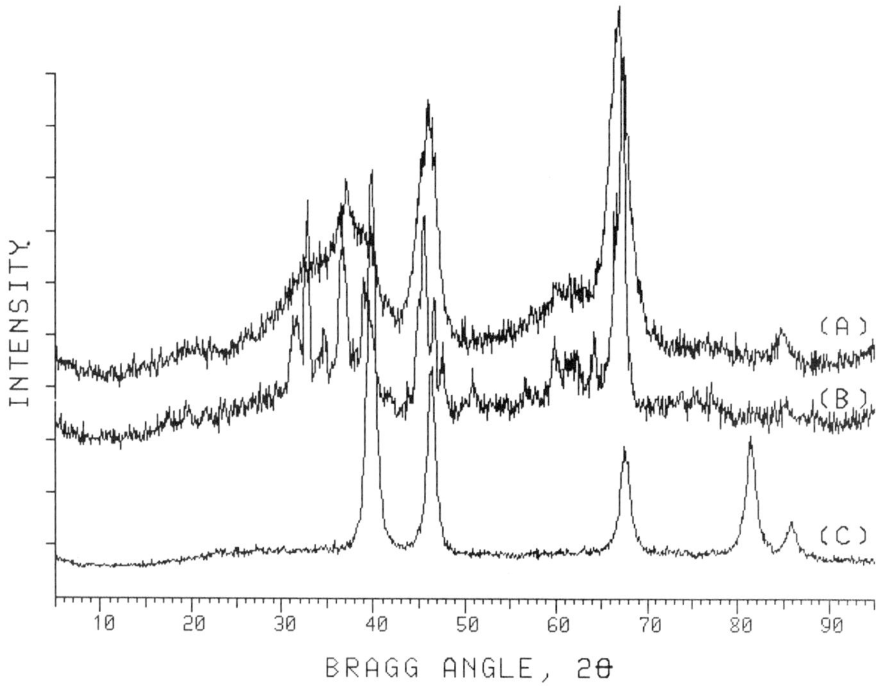

Fig. 5. X-Ray diffraction patterns of sintering of aluminum oxide and platinum supported on γ-Al_2O_3. Pattern A is amorphous γ-Al_2O_3; B is δ-Al_2O_3, a transition phase in the sintering of γ-Al_2O_3 to crystalline α-Al_2O_3. C is the diffraction pattern of platinum crystallites 100Å in diameter remaining after dissolving the α-Al_2O_3 carrier.

forms of catalysts from powders to catalyzed monolith washcoats. Uniform distributions of catalytic species are preferred for reactions controlled by chemical kinetics free from diffusional limitations since catalytic surface area can be maximized when more carrier area is available.

If a process feed contains known catalyst poisons such as Pb compounds present in the exhaust of automobile emissions, it may be wise to locate the active species more deeply into the washcoat to avoid undesirable reaction between the poison and active catalytic component (i.e., Pt, Pd or

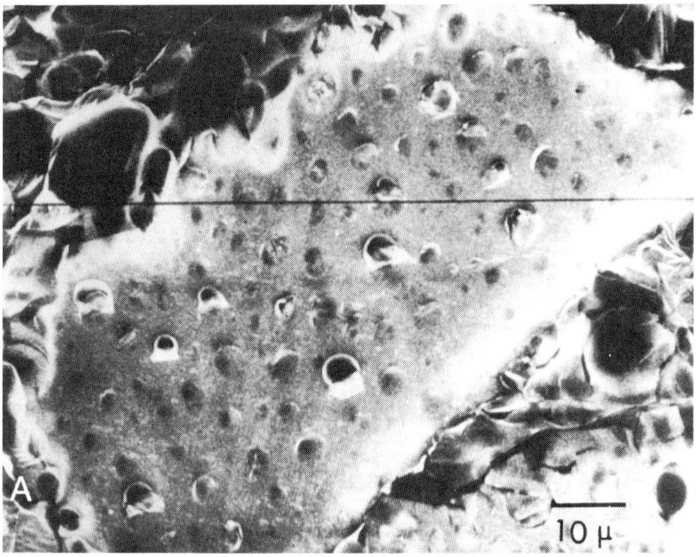

Fig. 6. An edge coated palladium on carbon catalyst. A is a photomicrograph of an irregularly shaped particle of wood carbon. B is a palladium map showing the palladium concentrated on or near the surface of the particle. C is a palladium line scan showing the distribution of the palladium at the edge of the carbon.

Rh). In this way the poisons deposit on the outside periphery of the catalyst before reaching the active sites. A similar situation exists for those cases where abrasion or surface erosion is expected. Obviously, if all the active catalytic component is located at the fluid/carrier interface, it is likely to be lost in an abrasive environment in addition to having poor active surface area due to inadequate dispersion on the carrier.

Both electron microprobe and scanning electron microscope have proven to be very useful in determining catalytic species-location in the carrier particles.[7-9] Morphology and metal location of a typical carbon supported metal catalyst is shown in the SEM photomicrographs, Fig. 6. The carbon particles are potted in epoxy resin and then cut, ground and polished to provide a cross-sectional surface to be examined. From the morphology

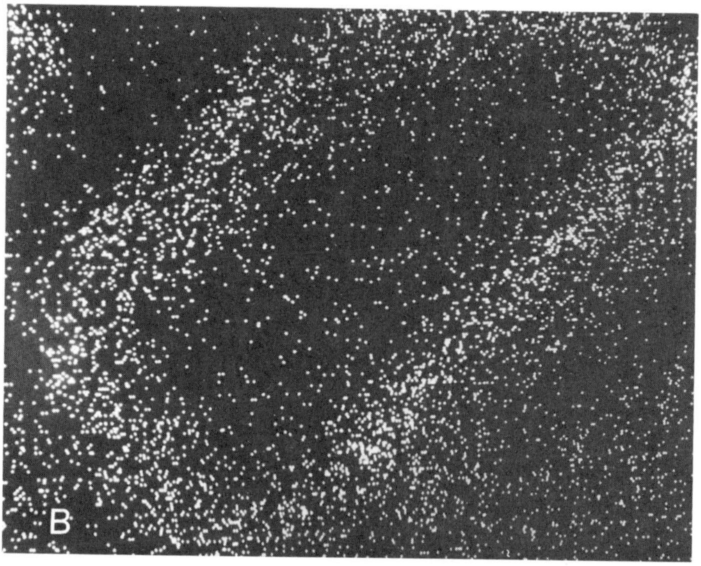

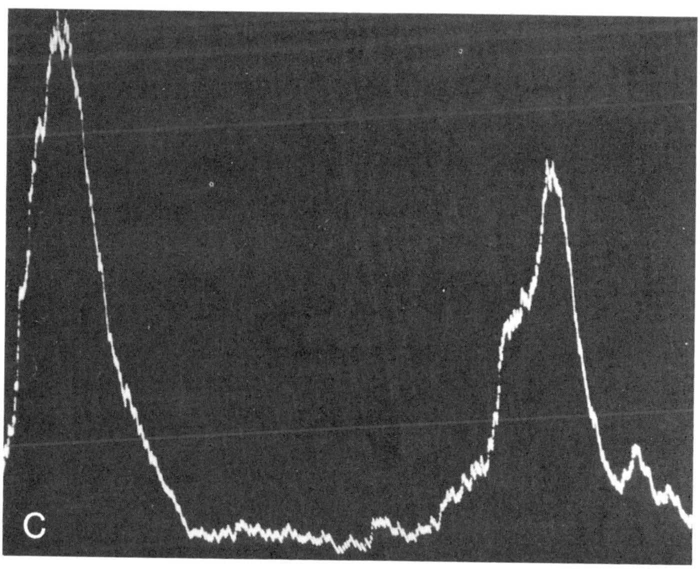

Fig. 6. Continued

shown in Fig. 6, the particles are a wood-based carbon with pores typical of this type. Other types of carbon, coal, peat or coconut, can be identified by their morphology. The palladium x-ray maps (Fig. 6B) clearly outline the edge coating of palladium on the carbon particle. The coating penetrates no more than 15 microns into the interior of the carbon particle which is approximately 50 microns in diameter.

The metal location in a particle can also be determined by running a line scan to detect the x-rays characteristic of palladium. As the electron beam moves across the particle, the detector signal is integrated and displayed on a cathode ray tube as intensity versus electron beam position. The resulting signal is shown in Fig. 6C.

Transmission electron microscopy allows direct observation of catalytic crystallites and their morphology by observing the transmission of electrons through a thin slice of material. Dense crystallites of metal or metal oxides prevent transmission, and thus appear as dark spots on a photomicrograph. By image analysis a size distribution and average crystallite size can be calculated. Electron diffraction can identify the structure and thus the composition of the dark spots. By assuming a shape for the crystallites, the dispersion ratio of surface atoms to total atoms in the crystallite can be calculated. It should be understood that this technique measures only a small fraction of the catalyst and, hence, obtaining data representative of the entire sample is quite difficult. Usually many different areas must be analyzed for proper statistics to be obtained.

G. Thermal Analysis

A very convenient method to approximate the oxidation state of a component within a compound is to reduce or oxidize a sample in a controlled environment and measure the weight change by microbalance techniques such as thermal gravimetric analysis.[7-9] The catalyst weight (0.1 to 0.5g) is equilibrated at a particular temperature in a stream of inert gas. A reactive gas such as H_2 for a reduction, or oxygen for an oxidation, is introduced to the inert gas stream and the associated weight changes monitored. The weight change associated with reduction or oxidation can give information regarding the valence state of the catalytic component.

Comparison with bulk chemical analysis can allow the determination of the extent of oxidized species present.

An alternate technique based on the same principle but called temperature programmed reduction (TPR), measures the H_2 or O_2 consumed in the reaction rather than noting weight change. From the reaction stoichiometry the approximate oxidation state can be deduced.

H. Surface Techniques

The tools available for surface composition characterization are Electron Spectroscopy for Chemical Analysis (ESCA), Auger Spectroscopy (AES), Ion Scattering Spectroscopy (ISS) and Secondary Ion Mass Spectroscopy (SIMS).[7-9] ESCA is more widely used than the others for studying the surface composition and oxidation states of industrial catalysts and will be illustrated with an example in the next section.

The acronym ESCA refers to the technique of bombarding the surface with x-ray photons that produce the emission of the characteristic electrons measured as a function of electron energy. Because of the low energy of the characteristic electrons, the depth to which the analysis is made is only about 20Å. The composition of this thin layer as a function of depth can be obtained by sputtering away layers of the surface and analyzing the underlying surfaces. A number of important catalytic properties have been studied by this technique including oxidation state of the active species, interaction of a metal with an oxide carrier, changes in oxidation state on activation of the catalyst, and deactivation of the catalyst by poisons.

III. RELATIONSHIPS BETWEEN CATALYST PERFORMANCE AND PROPERTIES

Correlations of physical properties of a catalyst with performance in a chemical process provides a rational basis for development and optimization of the catalyst. In this section examples of such correlations leading to product improvement will illustrate the power of traditional and modern characterization methods.

A. Correlations Between Metal Surface Area and Activity

Heterogeneous catalytic reaction rates are proportional to the number of active sites provided the system is free of external or internal diffusional effects. The major difficulty one encounters in developing correlations is that "active sites" are infrequently measured. It was mentioned above that chemisorption, XRD and electron microscopy techniques are utilized in measuring crystallite size of catalytic components; however, these methods do not necessarily measure the precise active sites in the catalytic material. For example, a supported catalytic metal may adsorb H_2 or CO in a chemisorption measurement; however, the active sites for a specific reaction may be the oxide, a compound formed by reaction of the metal species with a component of the carrier, or a surface compound produced by reaction of the catalyst with a feed component. Consequently, the major question that must be asked is, "does the measurement truly measure active sites?" The answer is that it usually does not, and thus any measurement of metal or metal oxide surface area is not as meaningful as actual acitivity/selectivity measurements. Nevertheless, we can learn a great deal about the relationship, or lack of, between the reaction rate and the surface area of the catalytic component. These measurements are also important for quality control since consistency in the measurement gives one confidence that the catalyst being produced day-to-day is the same.

A smooth relationship between CO chemisorption and conversion for a metal catalyzed ozone decomposition reaction is illustrated in Fig. 7. In this case conversion is a linear function of metal area. As conversion approaches the limiting value of 100%, the slope of the curve approaches zero. Zero slope defines a maximum metal surface area requirement at process conditions.

Optimization of a precious metal catalyst must consider also the economics of catalyst manufacture. This means limiting the metal loading without sacrificing performance. Metal loading and its relationship to metal surface area and dispersion is shown in Fig. 8. Intuitively, these relationships hold for all supported metal catalysts, but as stated above, often other chemical phenomena influence reaction rates and selectivities.

The significance of the metal surface area measurement to the catalysis industry is apparent. A limiting metal surface area value can be

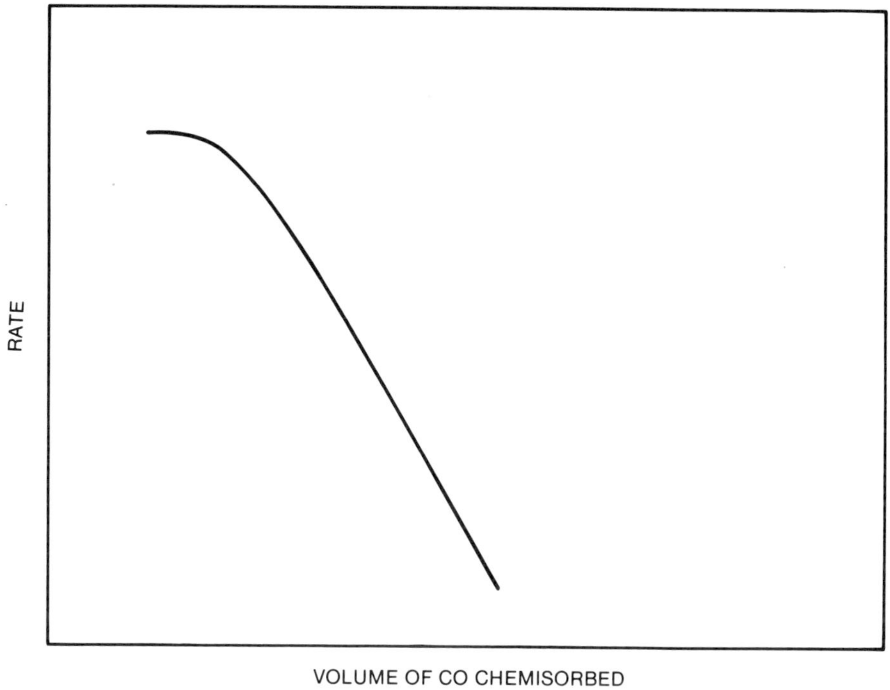

Fig. 7. Effect of Metal Area on Percent Decomposition.

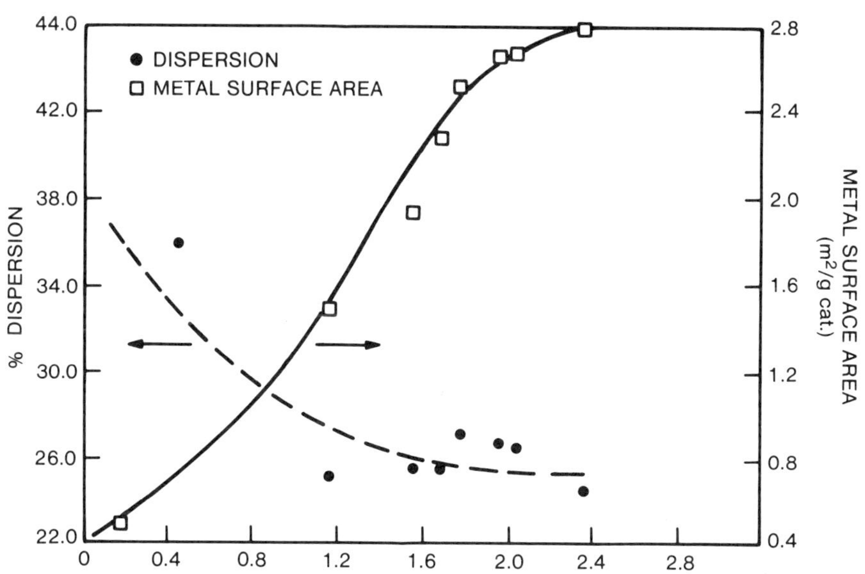

Fig. 8. Relationship Between Metal Dispersion and Precious Metal Content.

determined beyond which no improvement in activity is observed (i.e., the reaction becomes controlled by non-chemical phenomena such as diffusion). An optimum catalyst loading can then be determined which affords the catalyst both acceptable activity and cost-effectiveness.

B. Contaminant Effects on Activity

The specific nature of catalyst poisoning is useful not only for its elimination, but also in the determination of the overall rate controlling mechanism of the reaction. A poison can deactivate a catalyst in two basic ways: 1. physical or chemical interaction with the active catalyst species rendering it no longer active, or 2. physical deposition or chemical interaction with the carrier, blocking reactant species access to the catalyst's active sites. According to the rate controlling mechanism of the reaction and the nature of interaction a contaminant has with the catalyst, the resultant decrease in activity can be negligible or significant.

The analysis of poisoning in industrial catalytic processing can be quite complex. For example, more than one contaminant may be present in a deactivated catalyst; however, not all behave as poisons. Fig. 9 demonstrates this behavior for the deactivated ozone decomposition catalyst mentioned above. Both phosphorous and sulfur were found using XRF and LECO methods, respectively. No strong correlation between contaminant concentration and metal surface area was found for sulfur having a correlation coefficient = 0.714, Fig. 9B. Phosphorous concentration, however, does correlate with metal surface area, with a correlation coefficient = 0.961, Fig. 9A. This suggests that phosphorous is interacting with the active metal surface and poisoning activity much more than sulfur.

Since catalysts are commonly deactivated by poisons, research into poisoning behavior is essential for industrial processes. Once contaminants are identified, their elimination is often not a trivial or inexpensive matter. Thus, identifying only those important for performance has obvious economic benefits.

Characterization of contaminant behavior can also yield valuable clues as to the rate-controlling mechanism of reaction. A reaction which is kinetically controlled will not deactivate due to pore blockage, while a

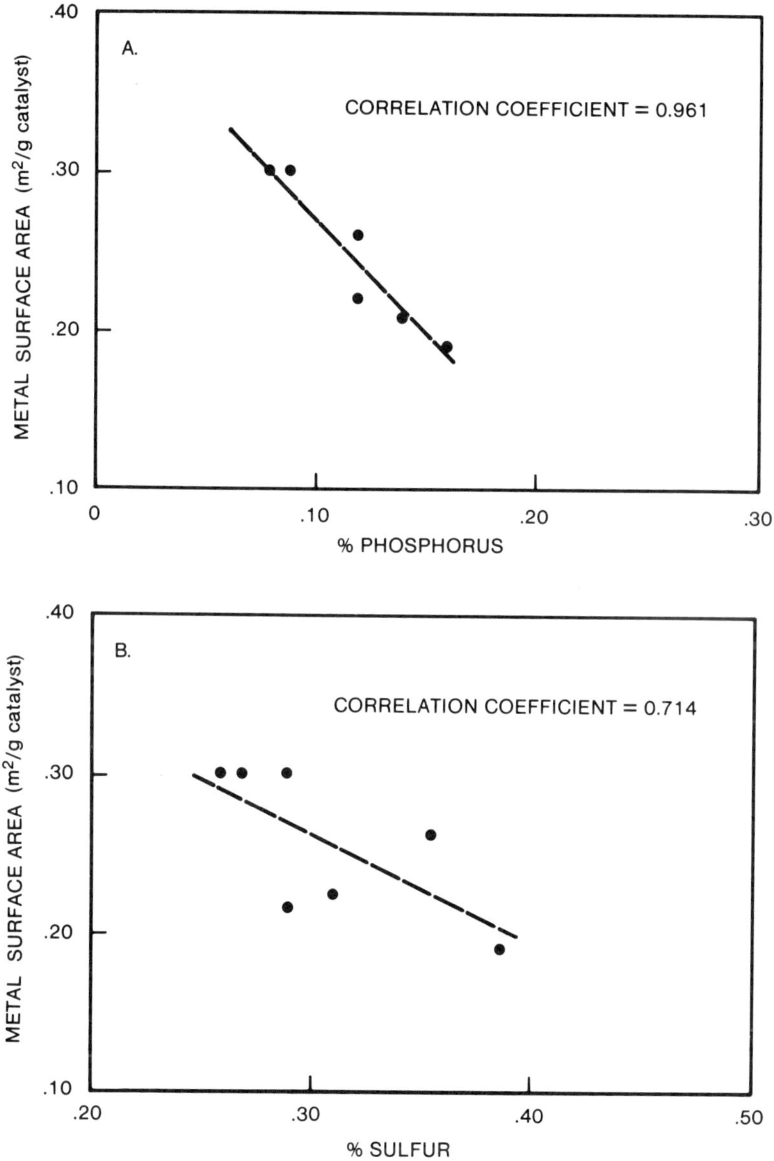

Fig. 9. Metal Surface Area as a Function of Percentage Contamination of Phosphorous and Sulfur.

reaction limited by pore diffusion will. Thus, by characterizing the composition, location, and deactivation consequences of a poison, the rate controlling mechanism of a reaction may be determined.

C. Effect of Catalytic Component Location

Diffusion limitations within the catalyst particle are common-place in industrial applications and must be considered in the formulation of the catalyst. Not only must the catalyst have an optimum metal or metal oxide surface area, but this area must be available to the reactants within the residence time dictated by flow rate. Two methods can be used to optimize location of the catalytic sites within the carrier material. First is to deposit the catalytic species in a thin layer near the carrier surface. Secondly, a carrier with large pores may be used to decrease diffusional resistance for reactants and products. The preferred method will be a function of the application in which both technical and economic factors must be considered.

Determination of catalytic species-distribution is normally done with a scanning electron microscope or electron microprobe. The x-ray fluorescence from a cross section of the catalyst particle provides a semi-quantitative, or at least qualitative, measure of catalytic species location. An example of edge coated palladium-on-carbon is shown in Fig. 6.

An example of the effect of metal location on catalytic activity is shown in Fig. 10. The rate of hydrogenation of nitrobenzene to aniline catalyzed by palladium-on-carbon is plotted as a function of a number of different carbon carriers using two different preparation methods. One method produced edge coated catalyst particles and shows higher activity, while the other distributed the metal much more deeply into the interior of the carbon particles. The carbons, A, B, C, D,...K, have varying physical properties; i.e., surface area, pore volume and particle size distribution. All of these properties, as well as metal surface area, can affect metal surface availability and, consequently, rate of reaction. The scatter in the data shown in Fig. 10 is a complex function of these properties, assuming constant conditions of catalyst preparation. The error bands for a 95% conficence level have a small overlap area, but only one sample from each group lies within the overlap.

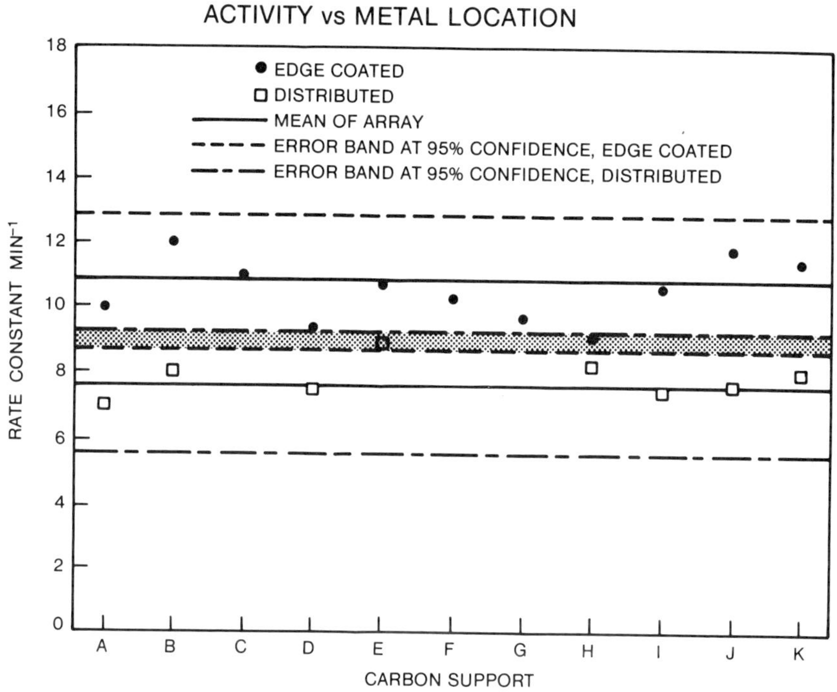

Fig. 10. Rate of Nitrobenzene Hydrogenation to Aniline as a Function of Carbon Support. Statistical Comparison of Edge Coated and Distributed Metal.

This example clearly demonstrates that metal location may be more important than metal dispersion when pore diffusion effects are present. Optimization of a catalyst for a catalytic process of this type requires a delicate balance between metal location, pore structure and metal dispersion leading to a mximum in available metal surface area.

The results of Fig. 11 nicely illustrate the relationship between hydrogenation activity, depth of Ru penetration in an Al_2O_3 particle, and metal surface area. The rate clearly increases with decreasing metal penetration, indicative of a pore diffusion-controlled reaction. The Ru metal surface area, as measured by CO chemisorption, inversely relates to reaction rate since dispersion increases with depth of penetration (Fig. 12).

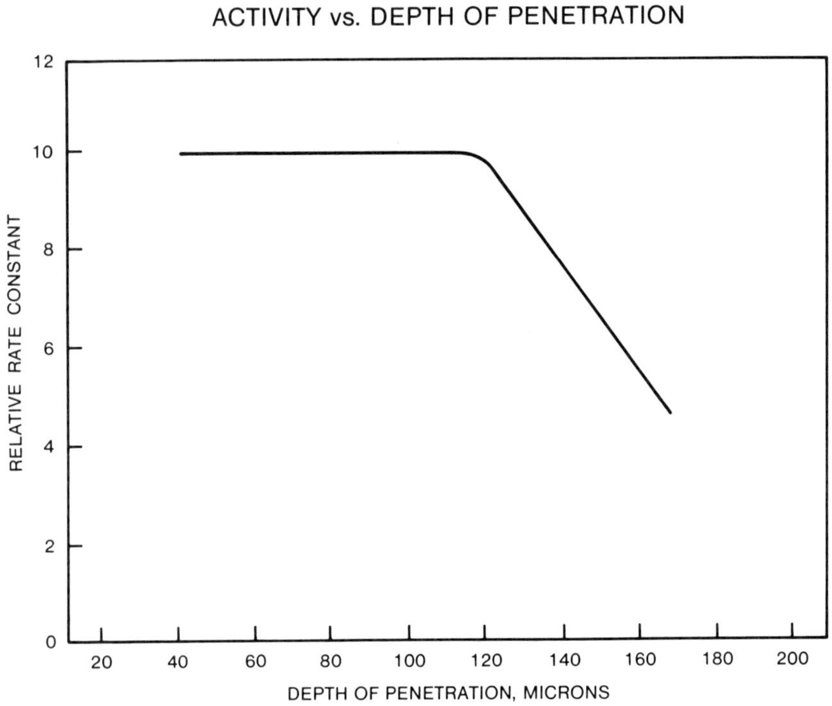

Fig. 11. Relative Rate Constant for MDA Hydrogenation as a Function of Depth of Metal Penetration for 1% Ru/Al_2O_3 Reduced in Hydrogen.

That is the Ru is dispersed better over a larger carrier surface as the penetration increases. Since the reaction rate decreases with metal depth due to increasing pore diffusion resistance, the relationship between metal area and rate is meaningless.

The reaction becomes insensitive to penetration depth below 100 microns, indicating that it is no longer controlled by pore diffusion. This depth, therefore, represents the transition between pore diffusion and kinetic control. One might expect activity to improve with increasing metal area below this transition point; thus catalyst optimization for this process requires balancing Ru depth and metal dispersion.

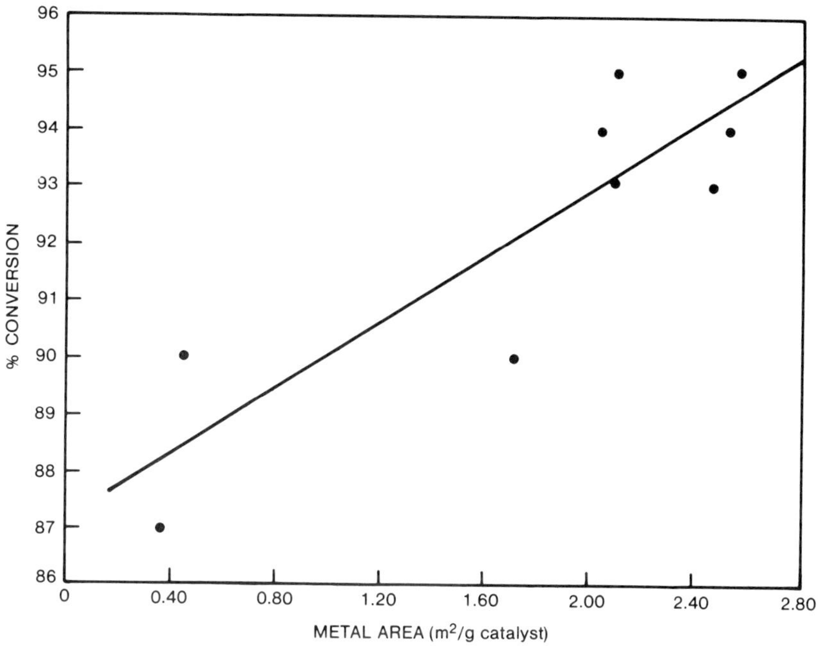

Fig. 12. Correlation of Reaction Rate with Metal Area for Catalyst with Increasing Depths of Metal Penetration.

D. Mixed Rate Controlling Processes

The apparent breakdown in correlations between catalytic activity and physical properties of a catalyst may signal a transition in the rate controlling mechanism of the catalytic reaction. For example, a platinum-on-carbon catalyst, A, was evaluated for activity using nitrobenzene reduction and compared to a similar catalyst, B. Catalyst A had been prepared by an edge coating method, and B by a uniform distribution method. The carbon carrier for both catalysts had a surface area of 1300 sq. m/g and pore volume of 0.82 cm^3/g. Carbon monoxide chemisorption was measured on both, Table I. The activity and chemisorption for A, Ko = 3.90, with a metal dispersion of 2.6%, was comparable to B, Ko = 3.97, with a much

TABLE I

CO Chemisorption in cm^3 (STP)/g Catalyst and Nitrobenzene Activity, K_o, for an Edge Coated Platinum Catalyst (A) Compared to Catalyst B with a Uniform Distribution of Platinum

Catalyst	Volume CO Adsorbed (cc/g cat)	% Dispersion	K_o
A	.03	2.6	3.90
B	.54	47	3.97

greater metal dispersion of 47%. It was surprising to find nearly identical activity with a 2.6% dispersion for catalyst A, although edge coated, compared to catalyst B with 47% cispersion and a uniform metal distribution.

In order to resove this discrepancy, average metal crystallite size measurements were determined on catalyst A using two additional techniques, x-ray line broadening, and transmission electron microscopy, Table II. The calculated crystallite size, 500Å diameter, from carbon monoxide chemisorption, is in clear disagreement with those calculated from x-ray and microscopy data, 51Å and 47Å diameters, respectively. The dispersion calculated from the XRD and TEM measurements is about 30%. However, in the transmission photomicrographs, Fig. 13, the metal crystallites can be seen agglomerated at the carrier edge.

TABLE II

Crystallite Size Measurements for Catalyst A

	CO	X-Ray(Line Broadening)	TEM
Crystallite Size (Avg. Dia. Å)	500[(1)]	51	47

(1) Calculated assuming spherical geometry and a completely accessible surface.

Fig. 13. Transmission electron microscope (TEM) photomicrograph of a platinum-on-carbon catalyst showing large aggomerates of platinum crystallites on the perimeter of a carbon particle.

The apparent anomalous results are, therefore, resolved. The carbon monoxide chemisorption is correct as a measure of the available metal surface. The average metal crystallite size of 51Å measured by x-ray line broadening has been confirmed by an estimate of 47Å average crystallite size from transmission electron microscopy photomicrographs, although the crystllites are agglomerated into much larger clusters. Since carbon monoxide chemisorption measurements indicate an average crystallite size of 500Å, only the external surface of agglomerates must be available to chemisorbing species.

Resolution of the metal dispersion problem provides a plausible explanation for the poor correlation between metal dispersion and nitrobenzene activity data. The nearly equal rates for the two catalysts is probably accidental. The rate observed for catalyst A is high since its metal location compensates for poor dispersion. The observed rate for catalyst B is not as high as expected from its dispersion due to the deeper metal penetration. Thus, the overall reaction is influenced by both metal location and dispersion.

E. Importance of Oxidation State

The chemical state of the surface active sites is a critical factor in the activity of the catalyst. Hydrogenations usually need the active component in the metallic state. Oxidations require an oxide surface, such as, palladium oxide or vanadium pentoxide, in which the active site is V^{4+} in equilibrium with V^{5+} at the surface. Any change in the chemical state of the active site results in deactivation of the catalyst. This mode of deactivation may be caused by reaction with the environment, carrier, or impurities or decomposition of the active species.

As one might expect, a very large and diverse number of experimental techniques are used to identify and measure the chemical state of catalytic component. These techniques include many types of spectroscopy, thermal analysis and x-ray diffraction. The list is long and proper characterization often requires the application of more than one method.

Methane oxidation by a palladium oxide-on-alumina catalyst is an example of a system that exhibits catastrophic deactivation caused by decomposition of the active oxide to the metal. Oxidation is an exothermic

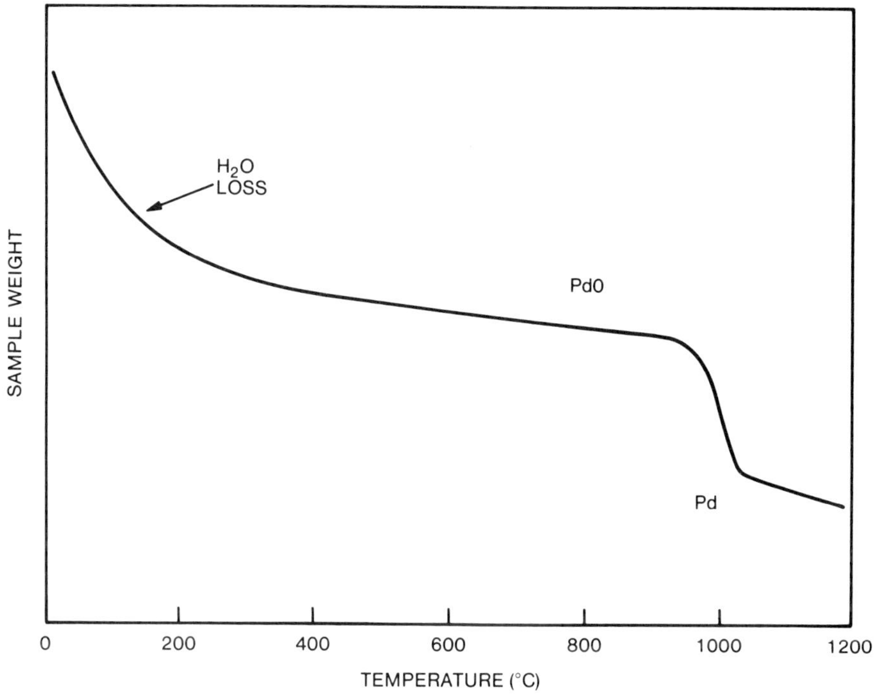

Fig. 14. Decomposition Temperature of a PdO/Al_2O_3 Catalyst in air.

reaction. Under adverse reactor conditions heat released is not rapidly dissipated and the temperature of the catalyst rises exponentially until decomposition of the oxide occurs. The oxide decomposes to the metal and the catalyst deactivates. A thermogravimetric study of this catalyst as a function of temperature is shown in Fig. 14. The resultant palladium metal is quite mobile on an alumina surface and sinters rapidly. The active surface area decreases by an order of magnitude. Thus reoxidation does not produce the well dispersed oxide and consequently irreversible deactivation occurs.

This palladium oxide-on-alumina catalyst has provided an example of irreversible catalytic deactivation caused by a decompositional change in surface composition followed by rapid sintering. The thermogravimetric analysis established a stringent temperature criterion for process control.

As already noted, supported palladium is a good hydrogenation catalyst in its metallic state as well as a good oxidation catalyst in its oxide state. Temperature programmed reduction data on a typical unreduced palladium-on-carbon demonstrated that palladium oxide in this type of catalyst reduces at subambient temperatures. Yet, in commercial practice this catalyst has been found inactive for some hydrogenation reactions unless it had been prereduced. Usually, this inactivity occurs with stirred autoclave type processes where reactions are carried out with both temperature and hydrogen pressure above ambient. Palladium oxide should reduce to metallic palladium under process conditions.

A study using electron spectroscopy for chemical analysis (ESCA) of a 5% palladium-on-carbon catalyst in the unreduced and reduced state has shown that significant differences in surface composition do exist dependent on pretreatment conditions (Fig. 15). Fig. 15A is the spectrum of the "as prepared", unreduced catalyst showing the palladium 3d 3/2 and palladium 3d 5/2 peaks at 342.1 and 336.8 eV, respectively. The spectrum is characteristic of palladium oxide. This catalyst was reduced in hydrogen and passivated by exposure to oxygen at partial pressures ranging from 0.02 to 0.2 atmospheres (Fig. 15B). The palladium 3d spectrum shown in this figure resulted from the most severe passivation treatment, exposure to air for twenty hours. The peak positions in Fig. 15B are characteristic of metallic palladium with the palladium 3d 3/2 and palladium 3d 5/2 located at 340.5 and 335.3 eV, respectively. The only evidence for reoxidation of the palladium is some tailing at the high energy wings of the two peaks. The ESCA spectra of other samples passivated under less severe conditions were identical to Fig. 15B.

A chemisorbed layer of oxygen that retards further oxidation of the palladium crystallites would explain the ESCA data in Fig. 15B. The "inactivity" observation reported in some commercial application is probably a long, unnoticed or unacceptable, induction period caused by a slow, diffusion-controlled reduction of the oxide (15A) to the metal (15B). The length of the induction period will be a complex function of solvent properties, hydrogen pressure, stirring speed, particle size and dispersion of the metal oxide. Adjustment of these process parameters to reduce or

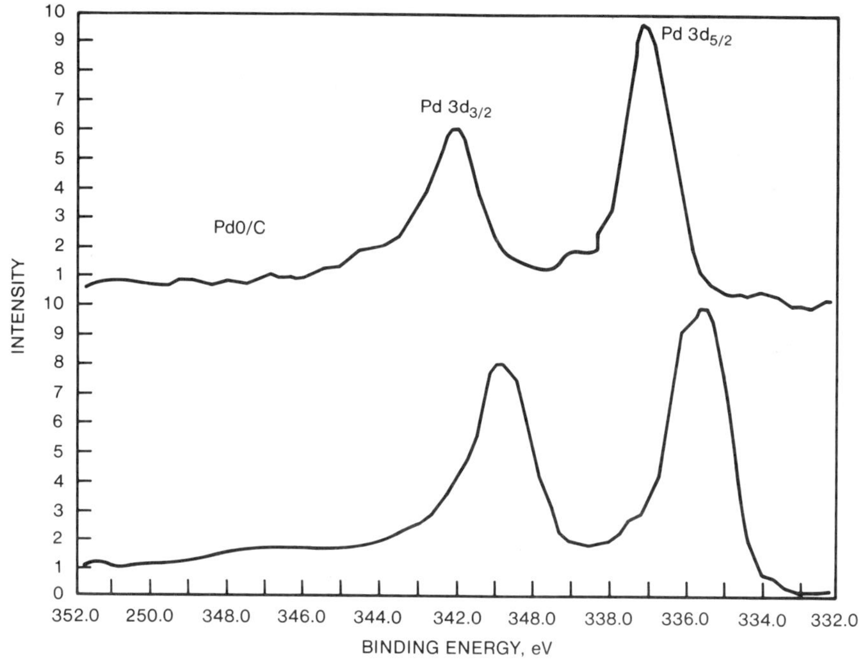

Fig. 15. Palladium 3d ESCA Spectra of a 5% Pd/C catalyst. A. Spectrum of unreduced catalyst; B. Spectrum of the catalyst reduced in hydrogen at 200°C for 2 hours and then exposed to ambient air for 20 hours.

eliminate a long induction period may be impossible, or at least impractical. The viable alternative is prereduction and passivation of the catalyst. A monolayer of chemisorbed oxygen is much more easily and rapidly removed from the surface to expose the metal then reduction of oxide crystallites to metal crystallites. Although the nature of the problem might have been deduced from plant observations, the evidence was all anecdotal. The ESCA study demonstrated that the prereduced palladium catalyst did not reoxidize to its unreduced state, and that the "inactivity" problem was real and related to the chemical state of the catalyst.

IV. CONCLUDING REMARKS

We have attempted to illustrate examples of when catalytic surface properties, metal location and poisoning may be important in assessing the efficiency of a catalyst. It should be understood that commercial catalytic processes require that the catalyst have activity, selectivity and life, all of which are strongly dependent on the "active sites". Fundamental research should be continued to more precisely understand the nature of these sites. Once achieved our ability to predict performance from basic principles will be greatly advanced.

REFERENCES

1. ASTM Standards on Catalysis, 2nd Ed., American Society for Testing and Materials, Philadelphia, (1984).
2. G.C. Bond and G. Webb, editors, **Catalysis**, Vol. **6**, Royal Soc. Chem., London, 1984, and previous volumes in this annual review.
3. F. Delannay, **Characterization of Heterogeneous Catalysts**, 1st Edition, Marcel Dekker, Inc., New York, N.Y. (1984).
4. D.D. Eley, H. Pines, and P.B. Weise, editors, **Advances in Catalysis**, Vol. **33**, Academic Press, Inc., New York, N.Y. (1984), and previous volumes in this annual series.
5. H. Heinemann and J.J. Carberry, editors, Catalysis Reviews - Science and Engineering, Vol. **26**, Marcel Dekker, Inc., New York, N.Y. (1984), and previous volumes in this quarterly.
6. C.N. Satterfield, **Heterogeneous Catalysis in Practice**, 1st Edition, McGraw Hill Book Co., New York, N.Y. (1980).
7. R.J. Farrauto and M.C. Hobson, Catalyst Characterization, in **Encyclopedia of Physical Science and Technology** (M. Yelles, ed.), Academic Press, in press.
8. R.L. Moss Preparation and Characterization of Supported Metal Catalysts, in **Experimental Methods in Catalytic Research**, Vol. II (R.B. Anderson and P.T. Dawson, ed.), Academic Press, New York, N.Y. (1976).

9. J.M. Thomas and R.M. Lambert, editors, **Characterization of Catalysts**, J. Wiley and Sons, New York, N.Y. (1980).

10. A. Clark, **The Theory of Adsorption and Catalysis**, Academic Press, New York, N.Y. (1970).

11. R.L. Bond, editor, **Porous Carbon Solids**, Academic Press, New York, N.Y. (1967).

12. P.N. Rylander, **Catalytic Hydrogenation in Organic Synthesis**, Academic Press, Inc., New York, N.Y. (1979).

11

On the Wizardry of Heterogeneous Palladium Catalysis

Wilhelm F. Maier

Department of Chemistry
University of California, Berkeley and the
Center for Advanced Materials at the
Lawrence Berkeley Laboratory
Berkeley, California

ABSTRACT

Among the unique properties of Palladium catalysts is their sensitivity towards modification procedures which result in improved special selectivities. The modifications appear to be connected to the ability of Palladium in the presence of hydrogen to undergo drastic morphological changes which are affected by additives and modification conditions. These changes,, as characterized by TEM studies of Pd-catalysts on various supports, represent changes in the surface structure of the catalyst. A pronounced surface structure sensitivity is detected for various hydrogenation and hydrogenolysis reactions. The combination of facile morphological changes and the sensitivity of the selectivity of catalytic reactions to surface structures suggests a direct link which may explain part of the action of a wide variety of catalyst modification procedures and special reaction conditions.

I. INTRODUCTION

In heterogeneous catalysis Pd differs from other group VIII elements mainly in its unique selectivity properties obtainable after catalyst modifications.[2] Famous examples of such modified catalysts are the Lindlar catalyst for *cis*-selective hydrogenation of alkynes,[1] and the Rosenmund catalyst for the selective reductions of alkanoyl chlorides to aldehydes. The Lindlar catalyst is Pd on calcium carbonate which is treated with lead-acetate[1] and the Rosenmund catalyst is a Pd on barium sulfate catalyst used in the presence of quinoline and sulfur.[2] However, there are many other modification procedures recommended in the literature for these reactions, which range from quinoline with and without sulfur, manganous chloride, sodium acetate, N,N-dimethylaniline, thiourea, thiophene, dibenzothiophene, lead acetate, to ethyldiisopropyl amine.[3] It is not likely that such a broad range of "additives or modifiers" is directly related to the catalytic reaction itself. Although little is known about the true action of such additives, modifiers are commonly used (and therefore apparently found effective) in heterogeneous reactions based on palladium and nickel but much less common in the catalysis based on other transition metals. Palladium in particular seems magically connected to selectivity in hydrogenation reactions. When we conducted a recent literature search with CAS-on line based on the keywords "selective hydrogenation" 338 references were based on palladium, 215 on nickel, and 99 on platinum. Among the Pd references a wide variety of recipes for the selective hydrogenation of carbon dioxide[4] of cyclopentadiene,[5] of polyenes and alkynes,[6] of aldehydes,[7] of carbon monoxide,[8] and of dienes[9] was detected; to mentioned just the most frequent substrates.

Palladium is a unique element in many respects. Starting with its electronic structure, it is the only transition metal with a d^{10} atomic ground state configuration. It is the least dense and lowest melting of the Platinum group of metals. Palladium is also unique in its interactions with elemental hydrogen. It has a high diffusivity (twice that of nickel[10]) for hydrogen even at low temperatures,[10] which led to its use as a membrane for hydrogen purification. Palladium also effectively chemisorbs hydrogen into its bulk structure, a process which can lead to a considerable lattice expansion.[11] Palladium can absorb up to 900 times its volume of hydrogen at room temperature, a property that has led to the application of palladium as one of the basic materials for hydrogen storage devices.[12] This affinity to elemental hydrogen is also responsible for palladium specific problems in the characterization of surface area and dispersion of palladium catalysts by hydrogen adsorption techniques, where special care must be taken to differentiate among surface and bulk absorption (hydride phase). This hydride phase forms readily under normal hydrogen pressure and room temperature; it is a non-stoichiometric material of the same lattice type as palladium itself, namely fcc.[13] Among the group VIII transition metals only nickel and palladium form hydride phases.[10] The relation of this hydride form to the catalytic properties of palladium has not been fully established.[13]

II. RESULTS AND DISCUSSION

If the hydride phases of palladium are related to the catalytic properties, a special activity for reactions like hydrogenations of double or triple bonds would have to be expected, especially with bulk·metal particles. However, Bond reported a reduction in the ability to hydrogenate acetylene and a complete inhibition of the hydrogenation of ethylene with palladium catalysts upon prolonged exposure to hydrogen.[14] Such "poisoning" effect of palladium hydrides for olefin hydrogenations was confirmed by later studies.[15] These studies, however, are rather isolated and do not allow a direct comparison of the hydrogenation activity of different group VIII metals in their bulk form.

The hydrogenation rate of bulk Pd foil was therefore compared with that of Pt, Rh, and Ni.[16] The rate of hydrogenation of 2-hexyne was used as reference reaction. Indeed, as can be seen in Table 1, palladium is more than an order of magnitude more active for hydrogenation reactions than rhodium and platinum in their bulk form. Furthermore, while there was never any problem to obtain reaction on the Pd foil, the Pt and Rh had to be freshly oxidized and reduced in order to get any activity. A similar turnover frequency as with the palladium foil was obtained with the nickel foil indicating familiarity of bulk nickel and palladium catalysts under

TABLE 1

Turnover Frequency of the Hydrogenation of 1-Hexyne at Room Temperature and Normal Hydrogen Pressure with the Metal Foils Indicated

	Pd	Pt	Rh	Ni	Pd-β-hydride	
TURNOVER FREQUENCY	3	.05	.1	1.6	3.5	[mol./site and second]

hydrogenation conditions. To confirm the presence of the bulk β-hydride form under our reaction conditions, the experiment was repeated with a palladium foil that has been pretreated for 12 h under a hydrogen pressure of 500 psi, conditions sufficient to cause quantitative conversion to the β-hydride.[12] Slow pressure release to one atmosphere was followed by immediate addition of 2-hexyne, whose hydrogenation proceeded essentially at the same rate (T_f 3.5 mol/site x sec) as the regular foil. Apparently the β-hydride form of palladium is highly active for the hydrogenation of alkynes and no indication of poisoning of the alkyne hydrogenation was obtained.

That such unique catalytic properties of the bulk forms of palladium and probably also nickel had not been recognized before is probably due to the fact that the majority of catalysis studies in the past have been performed on highly dispersed metals. In such catalysts, where many or most of the metal atoms are surface atoms any special activity of the hydride form of Pd or Ni could not be recognized.

With such evidence about the involvement of the hydride forms in the heterogeneous catalysis of Pd at normal or higher pressures of hydrogen the connection to selectivity and to the action of additives and modifiers mentioned above had to be established. Aside from direct interference with the catalytic reaction, which seems unlikely due to the unrelated nature of the modifiers, the mode of action may be to change or modify the catalyst particle structure (sintering?) as indicated in earlier studies.[3]

The sintering of supported metal catalysts has been reviewed.[17] The effect of support on the properties of catalysts has not been established satisfactorily. Metal support interaction (MSI) as an important factor in catalyst preparation and stability has been recognized and is an area of active studies.[18] Pd-particles on TiO_2 supports were found to sinter more effectively than on alumina support.[19] Chen and Ruckenstein monitored the sinter behavior of Pd crystallites supported on thin film alumina upon heating to 800°C by electron microscopy and observed a steady increase in particle size from 40 to 164 A.[20] Intensive sintering of Pd on charcoal was observed by Brownlie et al. in 1969.[21] In 1971 Pope et al. reported excessive sintering of the Pd-particles on a charcoal support after attempts to reduce or outgas the catalyst. The catalyst was found to be very sensitive to treatments; even the mildest surface cleaning (exposure to H_2 at RT, followed by evacuation at 100°C) induces sintering.[22] All reports indicate an unusually labile behavior of Pd particles in the presence of hydrogen. Since there is no evidence for excessive sintering of Pt or Rh in the presence of hydrogen the ease of sintering of Pd is most likely, due to the hydride phase of Pd.

However, aside from the thermal effect there may be an effect of solvent, modifier, or substrate on the sintering of Pd(hydride?)-particles. Little attention has been paid to the effect of substrate interaction on the morphological changes of catalysts. The complete rearrangement of Pd black to metallic palladium under the influence of triethyl amine at 150°C indicating strong substrate effects on the morphological rearrangement of the palldium particles has been reported in 1983.[3] These initial observations were confirmed in a study about the effect of pretreatment conditions on the nature of Pd-catalysts in the Rosenmund reduction.[23] Here the active metal surface area of a 10% Pd/C catalyst (determined by hydrogen/oxygen titration) was found to drop from 9.5 to 2 m^2/g upon simple reflux in xylene under hydrogen and to 0.5 m^2/g upon reflux in xylene under hydrogen in the presence of quinoline and sulfur indicating an accelerating effect of the modifiers quinoline and sulfur on the rearrangement process. This change was accompanied by a change from unselective hydrogenation in the case of the untreated catalyst to a highly selective catalyst in the case of both xylene treated catalysts, with and without quinoline. Examination of these catalysts by transmission electron microscopy confirmed major structural changes in the catalyst morphology. While the fresh catalyst exhibited a homogeneous, rather amorphous dispersion of the Pd on the amorphous carbon with efficient coverage of the entire carbon surface, the catalyst after xylene reflux showed large islands of uncovered carbon together with islands of palladium indicating excessive migration of Pd over areas of 1000 and more A. Three types of palladium particles were distinguishable, thin palladium particles on the carbon support, concentrated agglomerations of palladium crystallites, and isolated palladium particles on otherwise uncovered carbon. However, not only did palladium change its particle size and shape under pretreatment conditions, it did so even under mild rection conditions. The active metal surface area was found to change from 9.5 to 3.5 m^2/g after a 20 h exposure to heptanoyl chloride in cyclopentane as solvent at room temperature under hydrogen, reaction conditions that resulted in the unselective formation of heptanal and heptanol. This rearranged catalyst, when reused, exhibited the same selectivity as the pretreated catalysts confirming that the structural change of the palladium particles were directly related to the selectivity observed.[23]

In general it might be concluded that sintering of Pd in the presence of hydrogen leads to a gradual increase of particle size and the extent of this process may be directly related to the metal support interaction as well as the interaction of additives and modifiers. However, when the effect of substrate interaction with a Pd/alumina catalyst was investigated we obtained results which contrast such a preliminary conclusion. Figure 1 shows representative micrographs of a Pd/alumina catalyst before and after treatments with hydrogen as well as with hydrogen in xylene for 2 hr at 140°C. Several important features can be noted.

The fresh catalyst is not a well dispersed Pd/alumina catalyst. It has large Pd-particles (ϕ 1500 A) which, upon closer inspection, are composed of rather small Pd-crystallites. Such particles have just recently been recognized by Heller et al. as the structure of transparent metal (20 - 50 A crystallites packed in 800 A ϕ spheres).[24] Our finding of Pd particles of a similar morphology on completely differently prepared catalysts indicate that such agglomerations of small crystallite size in these catalysts is an intriguing, yet unexplored property. Treatment with hydrogen at 140°C for only 2 hr results in a decrease of the particle size to about 800 A. Treatment with xylene, however, reduced the aggregate size to about 300 A

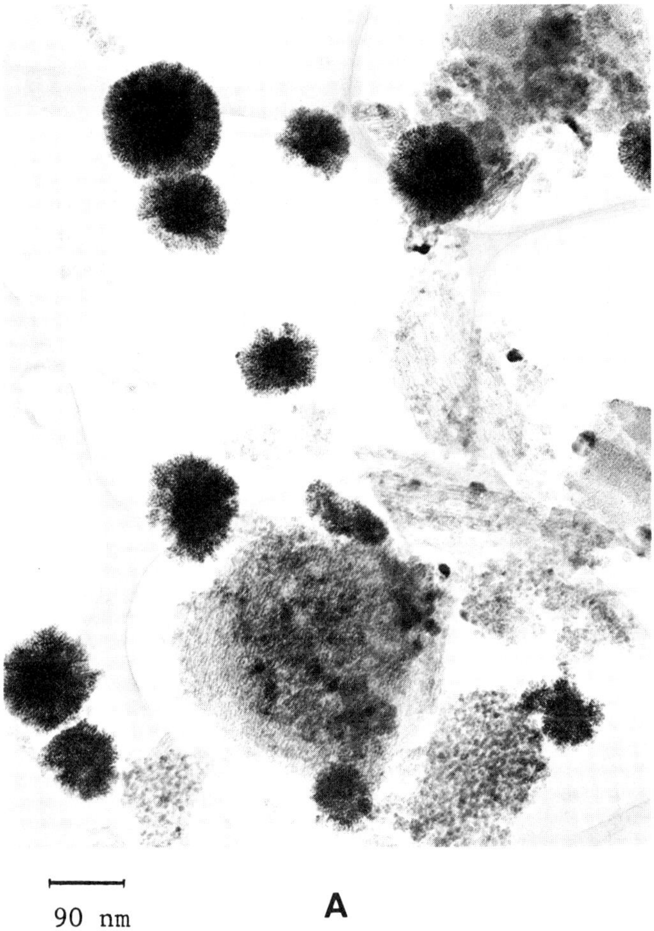

Fig. 1. TEM micrographs of palladium aggregates on alumina support.

(A) fresh Pd/alumina catalyst showing Pd-aggregates

(B) catalyst (A) after reflux in xylene with quinoline and sulfur under hydrogen for 4 h

(C) catalyst (A) after reflux in xylene under hydrogen for 4 h. higher magnification

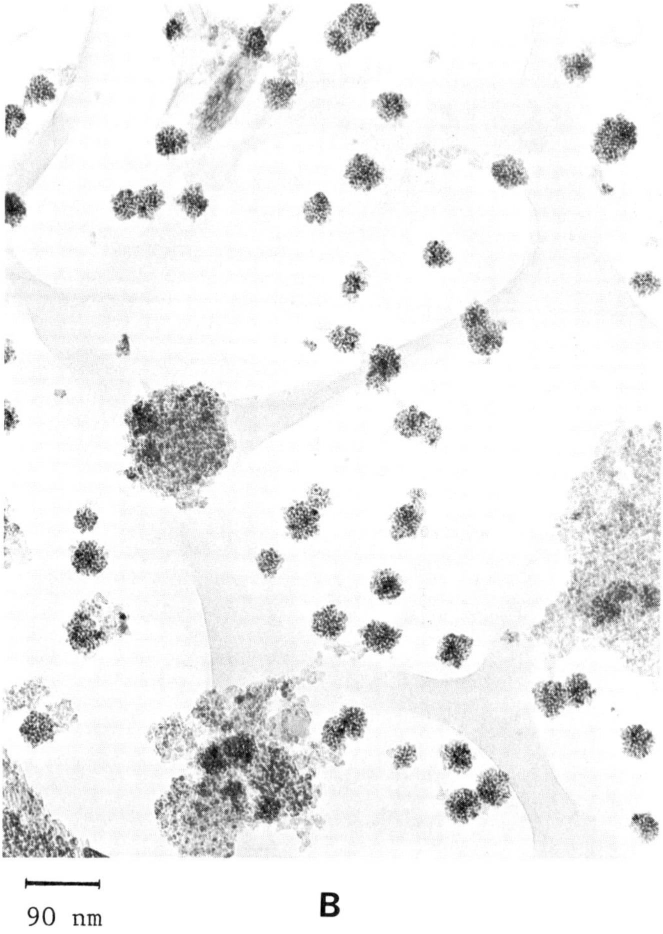

Fig. 1. Continued

indicating an effective redispersion of the Pd particles in the presence of the organic solvent. However, as can be seen in Fig. 1c, the crystallite size does not change under these treatments. The controlled preparation of catalysts with such uniform crystallite size might be a useful way for the formation of better defined catalysts, an aspect which requires further investigations.

Fig. 1. Continued

The complete lack of analogy of the particle morphology observed on our Pd/alumina catalyst with that of the alumina film supported Pd[20] is an indication that there is more to particle morphology in supported catalysts than just surface area and metal support interaction. However, granted that there is a large variety in particle morphology on supported palladium catalysts, how does a change in morphology affect catalysis?

Careful inspection of single crystal surfaces as well as metal particles reveals, that there are few fundamentally different metal atoms present on metal surfaces (see Scheme 1). These are terrace atoms characterized by 9 nearest neighbor atoms on (111) surfaces and 8 nearest neighbor atoms on (100) surfaces; step atoms with 7 and kink atoms with 6 nearest neighbors. The surfaces of real catalysts, known to consist of a range of metal particles, and solid metal, composed of a certain grain size distribution, cannot be characterized as readily. However, in a first order approximation cubo-octahedral shapes may be assumed for these particles, as identified as the common shape of group VIII metal particles by Yacaman.[25] Close inspection of the cubo-octahedral particles reveals, that the surface atoms can be readily identified as kink type atoms (6 nearest neighbors), step type atoms (7 nearest neighbors), and terrace type atoms (8 and 9 nearest neighbors).[26] With increasing particle size the relative concentration of terrace atoms will increase and the number of kink and step atoms will

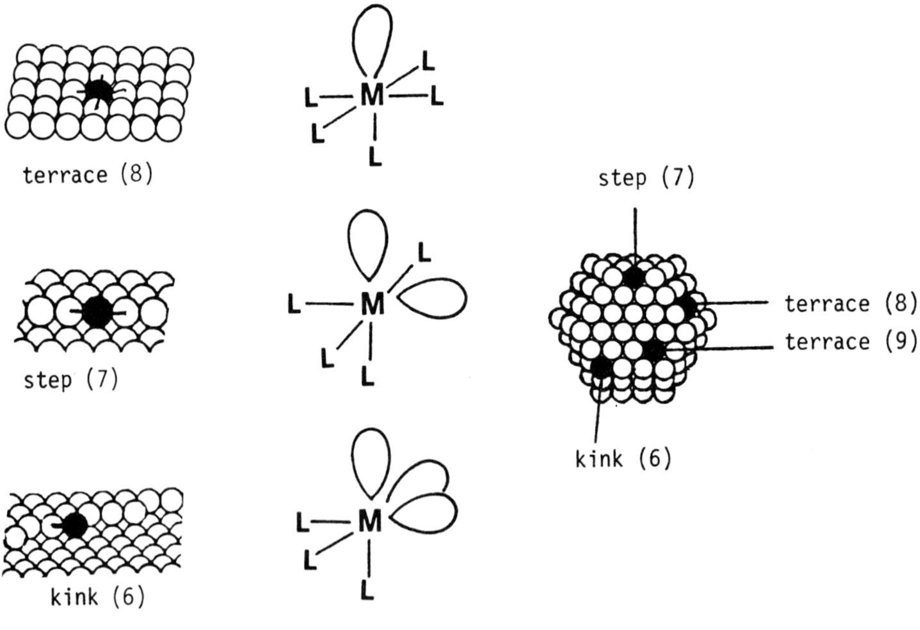

Scheme 1. Representation of principal surface and particle sites and their relation to transition metal complex fragments.

decrease, while with decreasing particle size kink type surface atoms will dominate. This qualitative trend should be general, independent of the actual shape of the metal particles and represents the justification for structure sensitivity studies in which the sensitivity of rates of catalytic reactions towards a wide range of dispersion is determined. The frontier orbitals of these sites are analogous to the ML_5, ML_4, and ML_3 transition metal fragments (see Scheme 1), which implies that they should exhibit distinctly different reactivity. However, simple dispersion studies cannot distinguish between pure surface structure effects, particle size effects (the wavefunction of metal particles especially in the high dispersion range changes from the rather simple MO picture of di or triatomic metal clusters to the complex band structure of the bulk metal in larger particles) and the differences in the steric environment of a kink or step atom on a two dimensional metal surface and the kink atom on a spherical particle, all of which may affect the interaction with organic substrates.

However, before such detailed problems can be addressed, we have to determine whether there is <u>any</u> effect of morphological or surface structural changes on the selectivity of heterogeneous reactions on Pd. Hydrogenation reactions, as all C-H, O-H, and H-H bond making and breaking reactions, have generally been considered surface structure insensitive.[27] However, these conclusions were based on few experiments which often did not cover the complete dispersion range. Contradicting results have been reported. Boitiaux observed a strong sensitivity to Pd dispersion in the hydrogenation of butynes[28] and Carturan observed a similar effect in the semi-hydrogenation of phenylacetylene.[29] Kung et al. have shown in a structure sensitivity study that there are large effects of dispersion on the relative selectivity of hydrogenation of alkynes and cis-olefins.[30] Since hydrogenolysis and hydrogenation reactions represent the most common use of heterogeneous catalysts in organic synthesis we have examined the structure sensitivity of some representative organic reactions under the mild reaction conditions common in organic synthesis.

The Rosenmund Reaction was for a long time the only useful method to convert carboxylic acid derivatives into aldehydes.[2] The reaction is based on supported palladium catalysts and special reaction conditions or pretreatments to prevent the undesired over-reductions. A study on the effect of the pretreatment has revealed drastic losses in active Pd surface.

That these changes in active Pd surface are not due to poisoning or surface carbonation but to morphological changes of the Pd particles was documented by TEM studies (see above).[23] The selective catalyst (after treatment with xylene alone, or in the presence of quinoline and sulfur, see above) showed the buildup of thin single crystal Pd rafts exposing one-dimensional lattice fringes which indicated the presence of stepped surfaces. Since such particles were not present on the fresh, unselective Pd/C catalyst it was concluded that they may represent the active and selective sites for acid chloride hydrogenolysis.[23] As shown in Scheme 2 experiments with single crystals have shown[23] that the low energy terrace surface (111) is not active for the Rosenmund reduction while rapid conversion of heptanoyl chloride to heptanal was observed on a stepped (755) and on a stepped and kinked (976) Pd crystal. It was concluded, that step type surface atoms have to be present for the Rosenmund reaction to occur.[23]

In contrast to the Rosenmund reaction, the other important selective hydrogenation reaction based on "modified" Pd catalysts, the <u>cis</u>-selective hydrogenation of alkynes, reveals a very different surface structure sensitivity. The hydrogenation of hexyne was studied with single crystal experiments and it was found that the low energy terrace, the (111) surface, was most active and selective for <u>cis</u> selective hydrogenations supporting earlier reports by Janko, Palcewska, and Szymerska.[31] The reaction profiles, obtained with a (111) and a (110) single crystal (see Scheme 3),

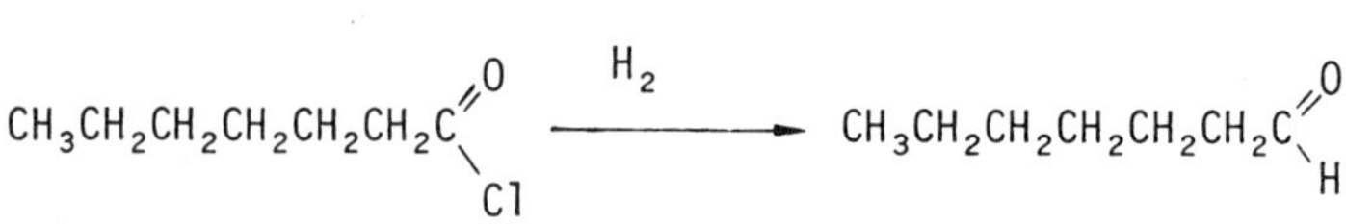

Pd-crystal face	turnover frequency	(mol./site and second)
[111] (terrace)	no reaction	0.0
[755] (stepped)		1.2
[976] (kinked)		6.0

Scheme 2. The activity of different Pd surface structures for the reduction of heptanoyl chloride at room temperature in cyclopentane as solvent.

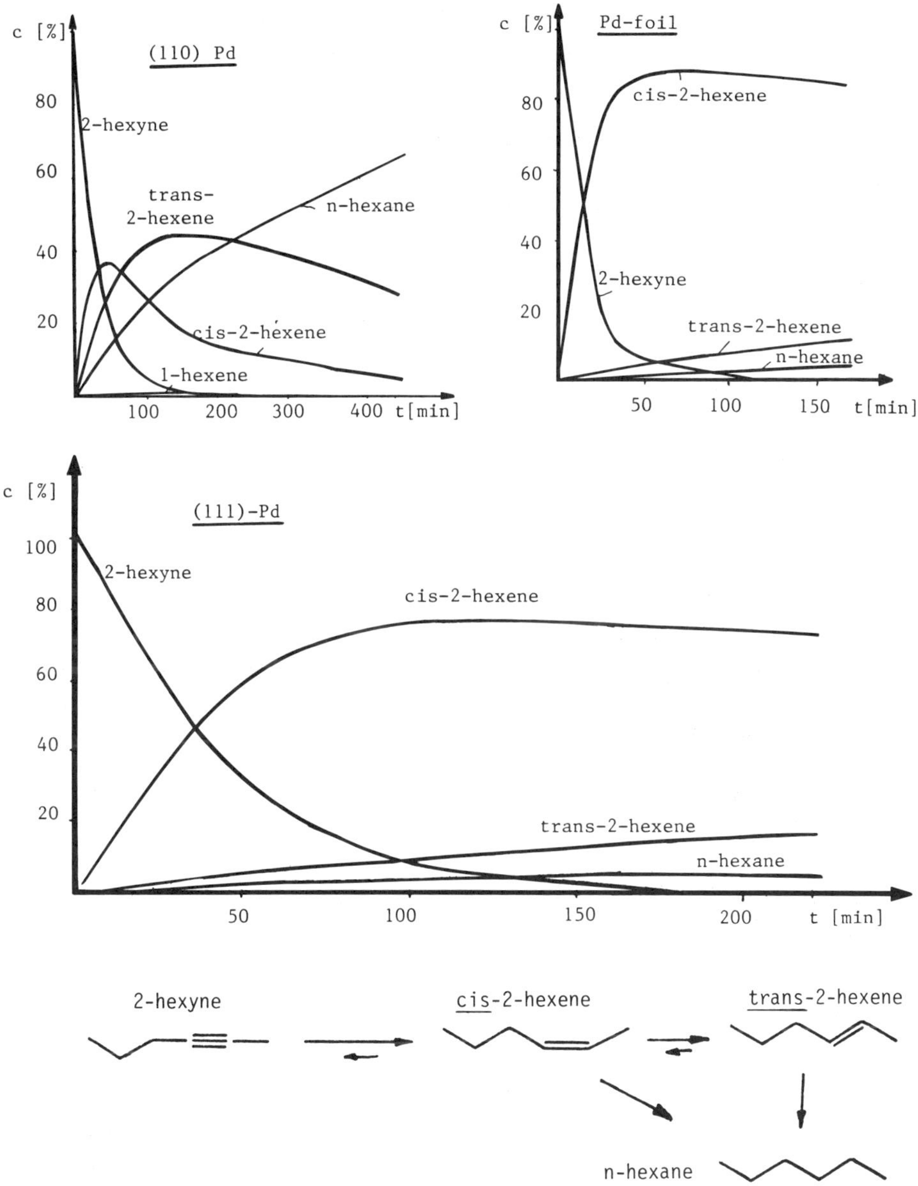

Scheme 3. Reaction profiles for the reduction of 2-hexyne on Pd foil and (111) and (110) Pd single crystals obtained at room temperature in n-heptane as solvent.

document a high sensitivity of the selectivity of this hydrogenation reaction reaction to the surface structure of the Pd.[32] Apparently _cis-trans_-isomerization and over-hydrogenation proceed much more rapidly on stepped or kinked surface sites.[32] The scheme also shows, that the reaction profile of the commercial Pd-foil is quite similar to that of the (111) Pd single crystal. In addition, the study also revealed that the alkyne hydrogenation itself is insensitive to the surface structure or Pd phase (the pregenerated β-hydride phase Pd was as active as untreated Pd-foil), while the _cis-trans_-isomerization and hydrogenation of the olefins was found to be rather sensitive to surface structure and the presence of Pd-hydride phases (no hydrogenation of _cis_-2-hexene was observed on the Pd-β-hydride confirming earlier observations on the "poisoning" effect of Pd hydride phases on the reduction of alkenes[13]). This difference in sensitivity between alkyne and alkene hydrogenations is confirmed by another observation. While the turnover frequency of 2-hexyne was found to vary only from 3-8 on a variety of differently treated Pd-foils, variations from $1.4 - 10^{-3}$ were recorded in the turnover frequency of the hydrogenation of 2-hexenes.[32]

Another indication of the importance of the Pd hydride phases under our reaction conditions were obtained in the single crystal experiments. After a two day hydrogenation experiment at room temperature macroscopic cracks were noted on the surface of the (111) single crystal. These cracks are precisely oriented (see fine lines on the crystal surface in Fig. 2) along the symmetry of the (111) surface of the single crystal providing a macroscopic reflection of the bulk orientation of the crystal. The formation of such cracks can readily be explained by β-hydride formation which, due to the lattice expansion, introduces high strain into the perfect lattice of the single crystal.

The study of allylic hydrogenation reactions also showed a strong structure sensitivity.[33] Table 2 summarizes some of the results achieved in the hydrogenation of allylphenolether. While the bulk Pd foil appears very selective, the dispersed Pd/C, (20% dispersion) shows a selectivity of only 83% with cyclohexanone as the major by-product. Pretreatment of the Pd/C in xylene (exp. 3) in the presence of quinoline and sulfur, a treatment known to decrease the particle size and dispersion and thus believed to increase the

Fig. 2. (111)-Pd single crystal surface showing one of the three sets of parallel cracks that developed upon 2 day exposure to 2-hexyne in n-heptane at 20°C and 1 atm of hydrogen.

relative amount of terrace atoms, resulted in a selectivity of 99%. Even simple annealing of the Pd/C catalyst in a hydrogen atmosphere improved the selectivity from 80 to 95% (not shown). This effect was amplified with a freshly prepared Pd/$CaCO_3$ catalyst whose selectivity of 67% increased to 98% upon annealing in hydrogen at 550°C. Although little effect of the dispersion on the reaction rate was noted, the selectivity was found to be almost quantitative with either very low or very highly dispersed catalysts.[33]

The hydrogenation of 3-bromocyclohexene shows a somewhat different behavior. The reaction rate of this more hindered substrate showed little

TABLE 2

Reactivity and Selectivity of Different Pd Catalysts for the Hydrogenation of Allyphenol (from ref. 33)

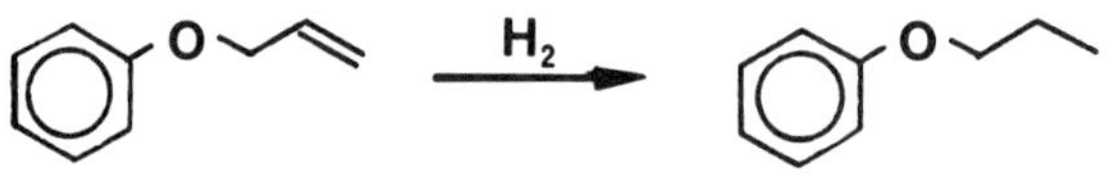

	Catalyst	Selectivity	Turnover frequency
1	Pd-foil	98%	3.5×10^{-1}
2	10% Pd/C	83%	0.9×10^{-1}
3	10% Pd/C[a]	99%	0.8×10^{-1}
4	30% $Pd/CaCO_3$	67%	-
5	30% $Pd/CaCO_3$[b]	98%	-
6	0.3% Pd/Al_2O_3	100%	0.9×10^{-1}

[a]same catalyst as 2, but refluxed in xylene in the presence of quinoline and sulfur for 4 h prior to the reaction

[b]like 4, but heated in H_2 at 550 °C for h prior to the reaction. 5 is about 5 times slower than 4.

sensitivity over a wide range of dispersion. However, as shown in Scheme 4, a two orders of magnitude increase was obtained with the highly dispersed Pd catalyst.[34] The huge difference in turnover frequency between the 92% dispersed catalyst and the 100% dispersed catalyst, which was accompanied by a drop in selectivity, may well be due to atomic dispersion of the latter, which contained only 0.1% of Pd and was prepared from allyl palladium with the intention to achieve maximum dispersity. The selectivity for double bond hydrogenation was almost quantitative on the low dispersed Pd while the highly dispersed Pd catalyst was only 30% selective with n-hexylbromide and cyclohexylbenzene as the major side products.[34]

These examples of recent studies document a high sensitivity of hydrogenation reactions to surface structure and/or particle size. A better control of particle shape and surface structure of heterogeneous catalysis is therefore desirable and can be expected to lead a better selectivity and

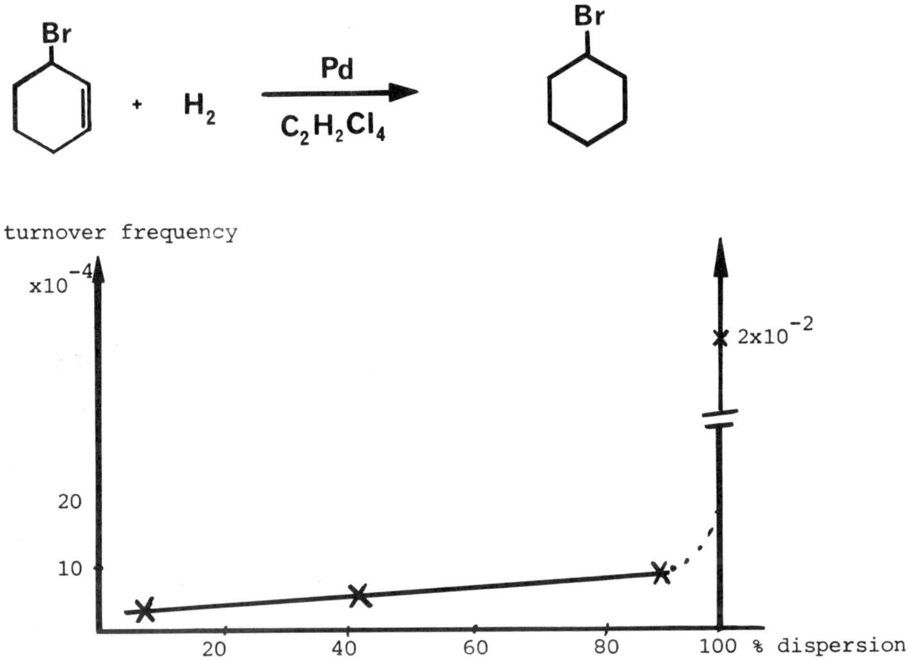

Scheme 4. Surface structure sensitivity of the hydrogenation of 3-bromocyclohexene in refluxing tetrachloroethane.

control of heterogeneous reactions. However, the above studies also indicate an unprecedented liability of the morphology and thus surface structure of Pd catalysts which does not encourage attempts to control the particle size and shape of Pd-catalysts.

Several solutions to such a problem come to mind. One way to stabilize Pd-particles is through the formation of alloys or bimetallic catalysts.[35] A cleaner solution, which is presently being probed in our laboratory, is the use of epitaxial Pd-films as catalysts of uniform surface structure. The picture in Scheme 5 shows the electron diffraction pattern of a Pd on mica and a Pd on tungsten film. The dominating (111) orientation is clearly visible and was found not to change upon treatment under reaction conditions. The reaction profile obtained for the hydrogenation of 2-hexyne on these Pd films (shown below) are almost superimposible to the profile obtained on the (111) single crystal which indicates that epitaxial palladium films might be useful and practical catalysts of uniform surface structure for mechanistic studies as well as for small scale applications as selective catalysts in organic synthesis.

electron diffraction pattern of Pd/mica film (from TEM) showing (111) orientation)

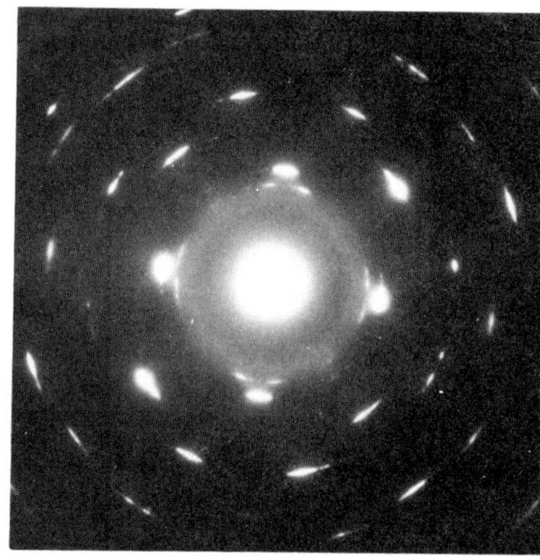

electron diffraction pattern of Pd/tungsten (from TEM) showing (100) orientation

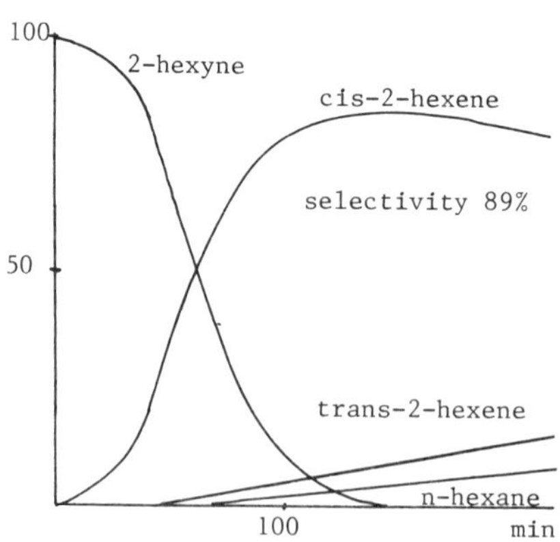

Pd/mica

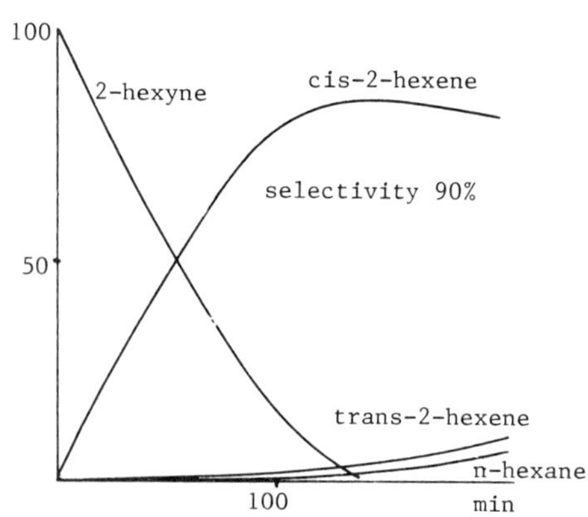

Pd/tungsten foil

Scheme 5. Diffraction pattern and reaction profiles of epitaxial palladium films.

Conclusion

Palladium in the presence of hydrogen is an instable solid which readily undergoes morphological changes and rearrangements. These changes are likely to originate from the β-hydride phase of palladium, which forms readily under normal synthetic organic reaction conditions. It is highly likely that the action of modifiers, poisons, and special treatment procedures, recommended to improve the selectivity of palladium catalyzed reactions, is to change the shape, size, and thus surface structure of the palladium particles to achieve the desired selectivity. Hydrogenation and hydrogenolysis reactions on heterogeneous palladium were found to be sensitive to the surface structure and morphology of the catalyst particles. In order to obtain a better understanding of catalytic processes as well as to obtain a better control over catalytic reactions detailed characterization of the structure of the catalyst is as essential as the development of a more precise understanding of the differences of heterogeneous reactions on different surface sites. This requires the development of new ways to synthesize stable catalyst surfaces in a controlled fashion as well as the application of new analytical techniques which allow to monitor the catalyst as well as the organic reaction of interest.

Acknowledgement

Thanks go to Dr. R.S. Rai from the Center of Advanced Materials who provided the TEM study of the Pd/alumina catalyst and Dr. D.A. Smith from the IBM J.T. Watson Research Center for the TEM studies of the Pd-thin films. Support for this work was obtained in part from the Director, Office of Basic Energy Research, Office of Basic Energy Sciences, Materials Science Division of the U.S. Department of Energy under Contract No. DE-AC03-76SF00098 and the National Science Foundation Grant No. CHE-840093.

References

1. E.N. Marvell and T. Li, Synthesis, 457 (1973).
2. E. Mosettig and R. Mozingo, Org. React., 4, 362 (1948).

3. A.B. McEwen, M.J. Guttieri, W.F. Maier, M.R. Laine and Y. Shvo, J. Org. Chem., **48**, 4436 (1983).

4. E. Ramaroson, R. Kieffer and A. Kienneman, J. Chem. Soc. Chem. Commun., **12**, 645 (1982).

5. L. Cerveny, J. Vopatova and V. Ruzicka, React. Kinet. Catal. Lett., **19**, 223 (1982).

6. J.P. Boitiaux and J. Cosyns, Eur. Pat. **EP 89252** A1.

7. L. Fischer, M. Hausen M. and K. Wember, Ger. Offen DE 3151086 A1.

8. Y. Kikukzono, S. Kagami, S. Naito, T. Onishi and K. Tamaru, Faraday Discuss. Chem. Soc., **72**, 135 (1982).

9. G. Pregaglia, G. Ferrari, A. Andreetta and G. Gregorio, Ital. IT 912648.

10. J. Völkl and G. Alefeld in "Hydrogen in Metals", Topics in Appl. Phys., J. Völkl, G. Alefeld, Eds.; Vol 28, springer Verlag, Berlin 1978.

11. E. Wicke and H. Brodowsky in ref. 10.

12. F.A. Lewis "The Palladium/Hydrogen System" Academic Press, New York, 1967.

13. W. Palczewska, Adv. Catal. **24**, 245 (1975); Z. Paal and P.G. Menon, Catal. Rev.-Sci. Eng. **25**, 229 (1983).

14. G.C. Bond, D.A. Dowden and N. Machenzie Trans. Faraday Soc., **54**, 1537 (1958).

15. R.J. Kokes Catal. Rev. **6**, 1 (1972); M. Kowaka, J. Jap. Inst. Metals, **23**, 655 (1959).

16. J.G. Ulan, W.F. Maier and D.A. Smith J. Am. Chem. Soc. submitted.

17. S.E. Wanke and P.C. Flynn Catal. Rev.-Sci. Eng. **12**, 93 (1975).

18. G.B. McVicker and J.J. Ziemiak, J. Catal. **95**, 473 (1985); T.C. Chang, J.J. Cheng and C.T. Yeh, J. Catal. **96**, 51 (1985).

19. R.T.K. Baker, E.B. Prestridge and G.B. McVicker, J. Catal., **89**, 422 (1984).

20. J.J. Chen and E. Ruckenstein, J. Catal., **69**, 254 (1981).

21. I.C. Brownlie, J.R. Fryer and G. Webb, J. Catal., **14**, 263 (1969).

22. D. Pope, W.L. Smith, M.J. Eastlake and R.L. Moss, J. Catal., **22**, 72 (1971).

23. W.F. Maier, S.J. Chettle, R.S. Rai, and G. Thomas, J. Am. Chem. Soc., in press.

24. A. Heller, D.E. Aspnes, J.D. Porter, T.T. Sheng and R.G. Vadimsky, J. Phys. Chem., **89**, 4444 (1985).

25. M.J. Yacaman, Appl. Surf. Sci., **19**, 348 (1984); Appl. Catal. **13**, 1 (1984); M.J. Yacaman and J.M. Dominguez E., J. Catal., **64**, 213 and 223 (1980).

26. C.B. Lebrilla and W.F. Maier, J. Am. Chem. Soc., **107**, in press.

27. M. Boudart, Adv. Catalysis **20**, 153 (1969); A. Frennet, G. Lienard, L. Degols, and A. Crucq Bull. Soc. Chim. Bel., **88**, 621 (1979); W.H. Moogue and J.R. Katzer, J. Catal. **32** 166 (1974); J.A. Cusumano, G.W. Dembrinski and J.H. Sinfelt, J. Catal. **5**, 471 (1968); T.A. Dorling and R.L. Moos, J. Catal. **5**, 111 (1966); M. Boudart, J. Mol. Catal., **30**, 27 (1985).

28. J.P. Boitiaux, J. Cosyns and S. Vasudevan, Applied Catalysis, **6**, 41 (1983).

29. G. Carturan, G. Facchin, G. Cocco, S. Enzo and G. Navazion, J. Catal., **76**, 405 (1982).

30. H.H. Kung, R.J. Pellet and R.L. Burwell, Jr., J. Am. Chem. Soc., **98**, 5603 (1975).

31. A. Janko, W. Palczewska and I. Szymerska, J. Catal., **61**, 267 (1980).

32. J.B. Ulan, W.F. Maier, and D.A. Smith, J. Am. Chem. Soc., submitted.

33. G. Lau, Ph.D. thesis, U.C. Berkeley, 1986.

34. A. Jhingan and W.F. Maier, submitted.

35. J.H. Sinfelt "Bimetallic Catalysts: Discoveries, Concepts, and Applications" Wiley, New York 1983.

12
Surface Characterized Dispersed Metal Catalysts

R.L. Augustine, R.W. Warner, K.P. Kelly,
M.M. Thompson, K. High, and P. O'Hagan

Department of Chemistry
Seton Hall University
South Orange, NJ

ABSTRACT

The Single Turnover (STO) procedure has been used to characterize the surface of a number of different dispersed metal catalysts. With this technique it is possible to identify and measure the relative densities of five different types of surface sites. There are two types of sites responsible for the direct hydrogenation of alkenes, on one H_2 is reversibly adsorbed while on the other it is irreversibly adsorbed. These are referred to as ^3M_R and ^3M_I sites respectively. There are also sites on which double bond isomerization takes place (^2M_C), those on which a stable half-hydrogenated state or metal alkyl is formed (^2M_S), and those which adsorb H_2 but do not take part in alkene hydrogenation (^1M).

This technique has been used to evaluate those factors in catalyst preparation and pretreatment which influence catalyst activity as well as the nature of the solvent effect and some specific types of metal support interactions. The use of these STO characterized catalysts in other reactions has made it possible to identify the types of sites on the metal surface responsible for mono- and polydeuterium exchange on alkanes and the dehydrogenation and hydrogenolysis of cyclohexane. This later work provides a direct correlation between single crystal studies and reactions

of course, some problems with this approach. In the first place it can only be applied to structure sensitive reactions.[4] Second, it is difficult to imagine that all of the metal particles in a supported catalyst would have the same shape and, further, that they would have the ideal forms used in this approach.

Other work in this area has been associated with attempts to define the reaction characteristics of the different types of surface sites by comparison with homogeneous catalysts having established mechanistic pathways. An early example of this approach[5] was the comparison of the Horiuti-Polanyi[6] olefin hydrogenation mechanism as it would take place on an fcc metal particle with the corresponding reaction using $(\phi_3P)_3$ RhCl[7] as the catalyst. In this case, the reactive homogeneous intermediate, **1**,[7] is readily extrapolated into an fcc arrangement (Fig. 1). If this represents a corner atom the two H's and the double bond will be on vacant surface orbitals and transfer of the H's to the alkene gives the alkane directly as in

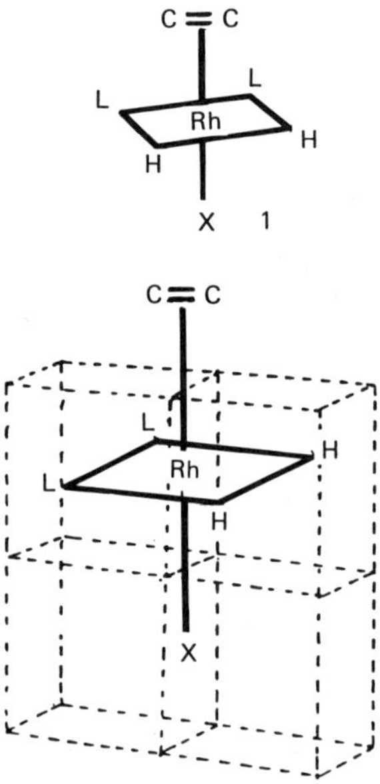

Fig. 1 Octahedral complex analogy to fcc crystal orientation.

the case of **1**.[7] Fig. 2 illustrates the type of reaction sequence proposed for a surface atom such as one on an edge which can adsorb one H and the double bond. In this case, saturation is accomplished only by the sequential addition of the adsorbed hydrogen to give the half-hydrogenated state or metal alkyl, **2**, which can then reverse to give an isomerized olefin or react further only after more H has been adsorbed.[5]

This approach was expanded on by comparing the corner, edge, and face atoms in this fcc arrangement with **1** and other homogeneous catalysts.[8] As shown in Fig. 3, corner atoms have three unoccupied surface orbitals, edges have two and face atoms one. These have been defined as ^{3}M, ^{2}M, and ^{1}M sites respectively,[8] and they correspond to kink, step, and terrace atoms using single crystal terminology.[2]

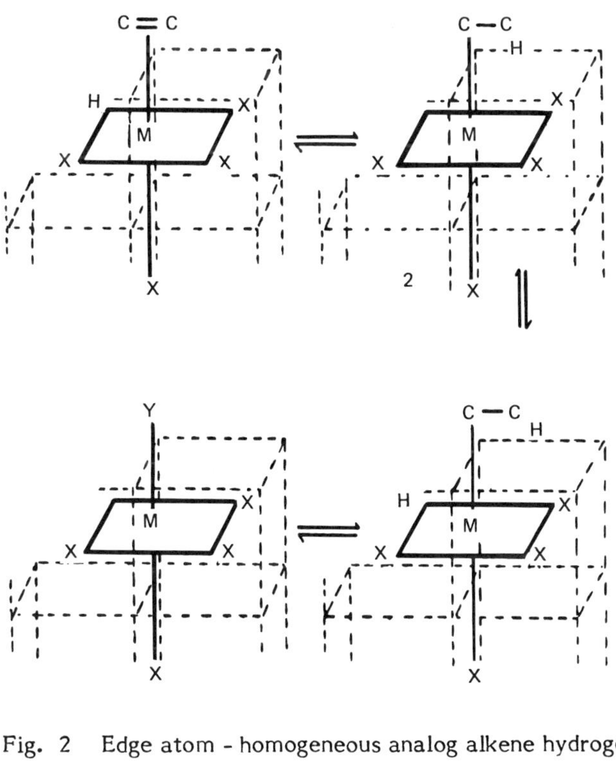

Fig. 2 Edge atom - homogeneous analog alkene hydrogenation mechanism.

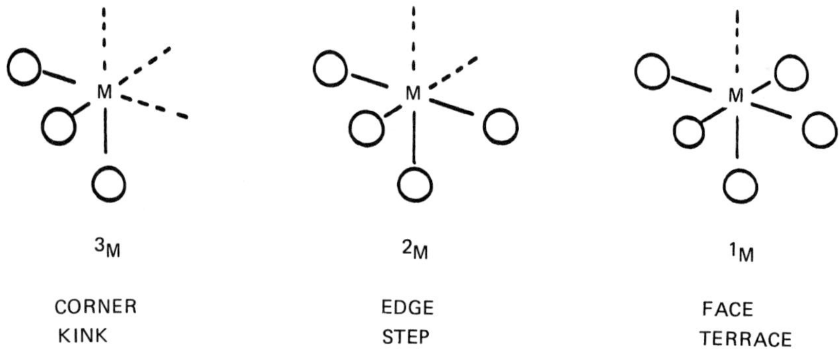

Fig. 3 Surface site descriptions.

II. DISCUSSION

It would seem, then, that with a mechanistic differentiation between the different types of surface sites, one should be able to use a product analysis as a means of determining the relative amounts of these different sites present on the catalyst surface. This would be possible if all of the sites reacted at the same rate but that is highly improbable. It is possible that under standard steady state reaction conditions a small number of very reactive sites could dominate product formation. So, in the absence of any specific site rate factors such an approach cannot be applied to the determination of relative site densities.

A. The Single Turnover (STO) Approach

We have, however, recently developed a procedure in which each surface site of a dispersed metal catalyst reacts only once, thus permitting a direct correlation between product distribution and reactive site densities.[9,10] This Single Turnover (STO) approach is illustrated in Fig. 4. The catalyst is placed in a flow reactor under a stream of purified, O_2-free helium. It is then exposed to a pulse of H_2 of sufficient size to completely saturate the metal surface with the excess H_2 removed in the carrier gas stream. In this way the only H_2 available for further reaction is that adsorbed on the catalyst surface. A pulse of 1-butene, again large enough to react with all

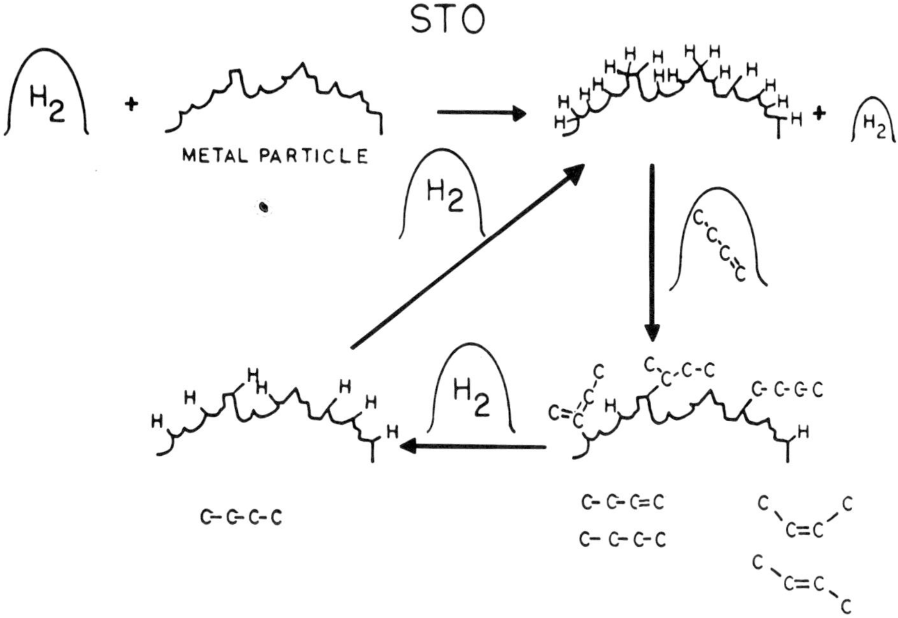

Fig. 4 Single Turnover (STO) reaction sequence.

of the H_2 on the catalyst, is then introduced into the reactor and the products formed analyzed by on-stream gas chromatography which separates the unreacted 1-butene, butane, and the cis and trans 2-butenes. Some of the alkene remains on the surface as a stable half-hydrogenated state or metal alkyl[10] and is removed, as butane, by a second H_2 pulse to complete the turnover cycle. Successive turnovers are accomplished by a repitition of this cycle: successive pulses of H_2 - ene - H_2, H_2 - ene - H_2, etc. With proper timing of the second H_2 pulse the gas chromatogram shown in Fig. 5a is obtained. This contains all of the product formation data needed for active site determinations and is typical of that observed with all of our Pt/CPG (Controlled Pore Glass) catalysts.

It has been established[10] that for Pt catalysts, at least, each active site reacts only once during each turnover. Further, experimental data is available which shows that under the conditions used for this reaction (room temperature and atmospheric pressure) H_2 migration over the catalyst

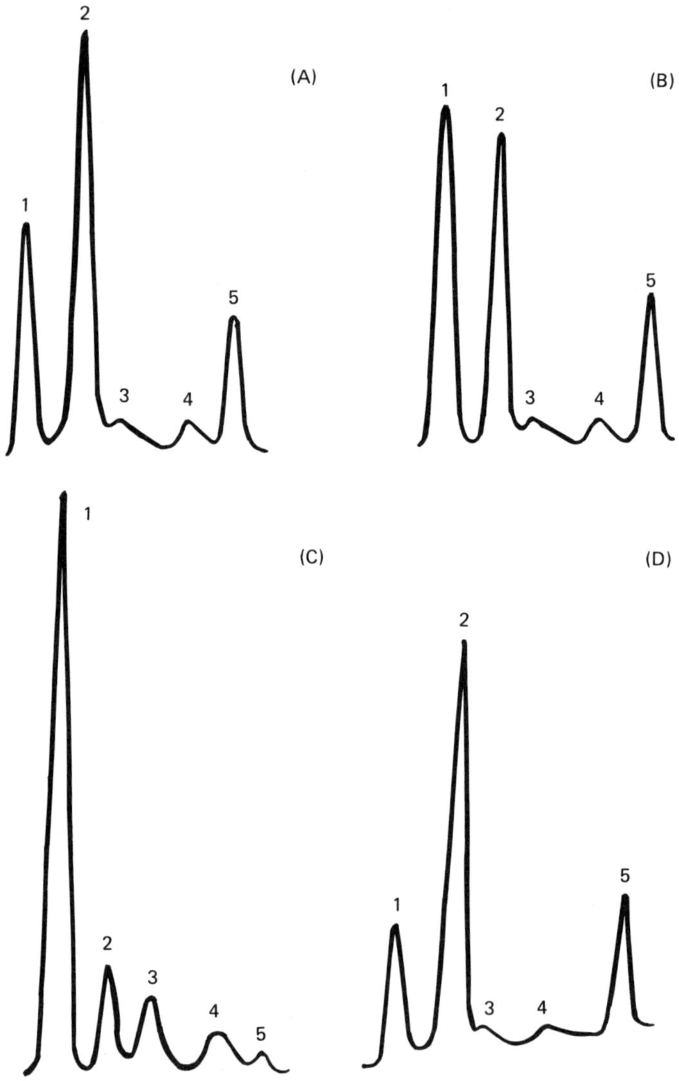

Fig. 5 STO chromatograms of a Pt/CPG catalyst: 1=1-butene; 2=butane; 3=<u>cis</u> 2-butene; 4=<u>trans</u> 2-butene; 5=butane from second H_2 pulse. (a) Normal STO chromatogram; (b) After 30 min. sweepoff between H_2 and 1-butene pulses; (c) After heating in He at 250°C for 30 min. and then cooling to room temperature; (d) After heating as in (c) and then exposure to air at room temperature for 33 days.

surface does not take place.[11] Thus, the amount of butane formed by reaction of the 1-butene pulse with the H_2-saturated catalyst (Fig. 5a, peak 2) is a direct measure of the number of sites on which direct alkene saturation to alkane takes place. Each molecule of butane is formed on an individual saturation site. The number of isomerization sites is measured by the amount of cis and trans 2-butene formed (Fig. 5a, peaks 3 and 4) and the number of sites on which stable half-hydrogenated state or metal alkyl are formed is determined by the amount of butane formed by the second H_2 pulse (Fig. 5a, peak 5). It was found that if the time between the first H_2 pulse and the 1-butene pulse were lengthened some of the H_2 initially adsorbed on the catalyst is swept off in the carrier gas stream. This sweep-off effects only the amount of initial butane formation (Fig. 5b) so the H_2 removed must have been present only on the saturation sites. Thus, there are two saturation sites, one on which the H_2 is more tightly or irreversibly bound, and another with more loosely held reversibly adsorbed H_2.

Fig. 6 shows the relationship between the products formed and the specific sites present on the catalyst surface. In keeping with the Siegel[8] definition and the reaction descriptions given above we have labeled the saturation sites as ^3M_I and ^3M_R for those with irreversibly and reversibly adsorbed H_2, respectively. As depicted in Fig. 3 these ^3M sites have three

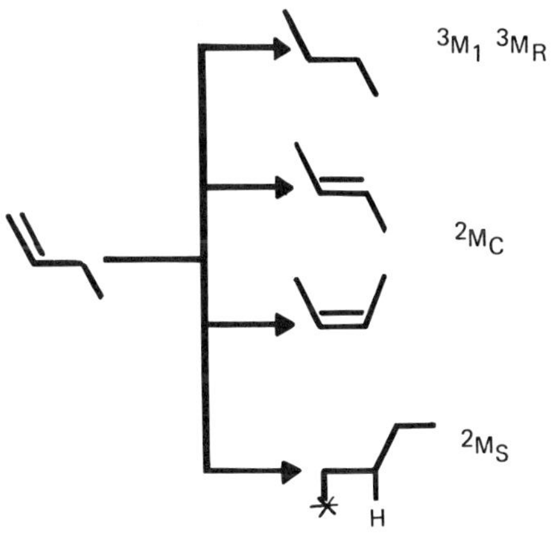

Fig. 6 STO product formation, site description correlation.

vacant surface orbitals so they can adsorb two H's and a double bond and as with **1** (Fig. 1), can give the saturated alkane directly. ^2M sites have two vacant surface orbitals and, thus, can adsorb only one H and the double bond and react as shown in Fig. 2. Those sites on which metal alkyl, **2**, formation reverses to give isomerized olefins are labelled ^2M_C and those on which **2** is stable and reacts with the second H_2 pulse are termed ^2M_S sites.

A fifth type of site, ^1M, has a single vacant surface orbital and can only adsorb either a H or a double bond but not both at the same time. These sites, therefore, do not take part in the alkene hydrogenation reaction. The number of such sites present on the catalyst is determined as shown in eq. 1-3 from a knowledge of the total quantity of H_2 adsorbed (H_T)

$$^3M_I + {^3M_R} = {^3M_T} \tag{1}$$

$$H_T = 2 \times {^3M_T} + {^2M_S} + {^2M_C} + {^1M} \tag{2}$$

$$^1M = H_T - (2 \times {^3M_T} + {^2M_S} + {^2M_C}) \tag{3}$$

and the amounts of each of the reactive sites present. Since each ^3M site adsorbs two H's and the others only one, the total number of H atoms adsorbed is equal to twice the ^3M site density plus the numbers of ^2M and ^1M sites present (eq. 2). Since H_2 chemisorption data is readily obtainable and the STO procedure gives the numbers of the various types of active sites present on a given catalyst the ^1M site density is easily calculated (eq. 3).[10] When the site densities are expressed as moles/mole of metal the dispersion of the catalyst is simply the sum of all the site densities (eq. 4). Table I

$$\text{Dispersion} = {^3M_T} + {^2M_S} + {^2M_C} + {^1M} \tag{4}$$

lists a summation of the different types of sites and the specific reactions used for their determination.

TABLE I

Single Turnover (STO) Site/Reaction Correlation

Site	Reaction
$^3M_1$	Hydrogenation
^3M_R	Hydrogenation
^2M_S	Metalalkyl Formation
^2M_C	Isomerization
^1M	H_2 Adsorption

B. Catalyst Preparation and Pretreatment Parameters

STO investigations have shown that the amounts of these various surface sites present on a catalyst are a function of both the amount of metal present on the supported catalyst and the temperature and method used in the reduction of the supported salt to give the dispersed metal catalyst.[12] While our initial work described this effect using Pt/CPG catalysts subsequent work has shown that changing the reduction temperature varies the site densities of a number of metal-support combinations. Thus, it is possible to prepare series of the same supported metal catalysts having differing site densities by changing the temperature and mode of reduction of the supported salt precurser.

A recent study has utilized the STO approach to determine the nature of the effect which preheating has on the activity and selectivity of a catalyst.[13] It has been known for some time that catalysts which have been heated in H_2 to about 400°C have reduced activity and reaction selectivities different from those observed with the unheated species. The original activity and selectivity are regenerated on exposure of the catalyst to O_2.[14,16] Various explanations such as support reductions,[14] strong H_2 chemisorption,[15] and surface reconstruction[16] have been proposed to explain these observations. Fig. 5 shows the STO chromatograms with the same catalyst treated in different ways. Fig. 5a is the standard STO and Fig. 5c is that obtained after the catalyst was heated in He at 250°C for 30 min. It can be seen that the heated catalyst had a marked decrease in saturation activity (peak 2) and an increase in isomerization capability (peaks 3 and 4). This type of change is typical of all supported Pt catalysts studied in this way. Since no H_2 was used in this thermal deactivation, the reduction of the support[14] or strong H_2 chemisorption[15] cannot be invoked to explain this change. Further, as shown by the STO chromatogram in Fig. 5d, exposure of the heated catalyst to air regenerated the original STO characteristics for this catalyst so whatever effect is operative it cannot involve any extensive surface reorganization. One explanation for these observations[13] is that on heating the surface, adsorbed O_2 is lost resulting in a reorganization of the surface atoms into a less active array[16] but reexposure to O_2 brings about a refaceting of the surface to a more active arrangement.[17] In the case illustrated in Fig. 5 it appears that the original surface structure has been regenerated so the changes occurring on heating could not have been too great. However, while O_2 exposure regenerates the activity of all catalysts studied in this way, the original STO data is not always obtained on O_2 treatment of the heated catalysts[13] so in some cases this heat induced surface reconstruction can be more pronounced than in others.

C. Catalyst-Support Comparisons

While all of the work described thus far has utilized Pt/CPG catalysts the STO procedure can be applied to a variety of other metal-support combinations. Activated carbon supports cannot be used, though, because they absorb the hydrocarbon pulses in the STO sequence so no product analysis can be made. Extension of this technique to Pt supported on SiO_2, Al_2O_3, or TiO_2 was straightforward with STO chromatograms obtained which were quite similar to those found with Pt/CPG (Figs. 5 and 7a).[18] Palladium catalysts, though, were found to behave quite differently in that they give a very large amount of isomerized olefins with a <u>trans</u>/<u>cis</u> ratio of about 2.5-3.[9,19] With Pt catalysts only a small amount of isomerization is observed with a <u>trans</u>/<u>cis</u> ratio of about 1. The STO chromatograms which

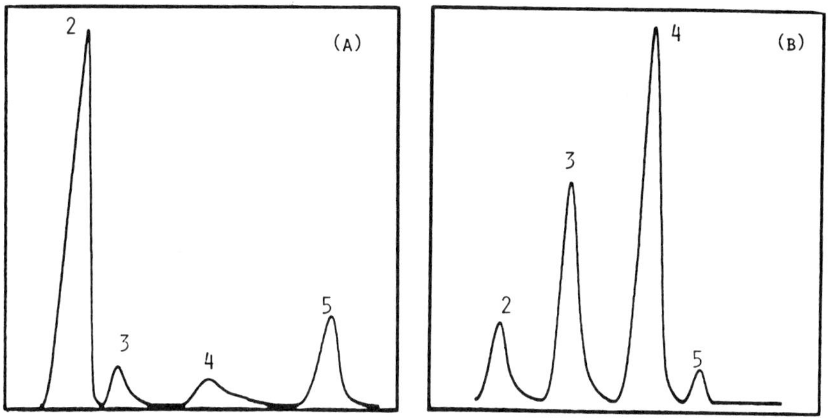

Fig. 7 Computer modified STO chromatograms, 1-butene peak removed: 2=butane; 3=cis 2-butene; 4=trans 2-butene; 5=butane from second H_2 pulse. (a) Pt/CPG; (b) Pd/CPG

have been computer modified to remove the 1-butene peak shown in Fig. 7 for Pt/CPG and Pd/CPG illustrate these differences. With the Pd catalysts the isomerization data is a result of the non-stoichiometric isomerization reactions with the $^{2}M_{C}$ sites reacting more than once during the STO sequence.[9] Because of this there is, at present, no direct correlation between the extent of isomerization and the number of $^{2}M_{C}$ sites on Pd catalysts. The ^{3}M and $^{2}M_{S}$ site densities are, however, measurable and work is underway to provide a method for $^{2}M_{C}$ site determination with supported Pd catalysts.

The STO chromatograms for some supported Rh catalysts are shown in Fig. 8. With these catalysts, as with Pt, the STO reaction is stoichiometric with each site reacting only once during the turnover sequence. There are, however, substantially more isomerization sites on Rh catalysts than there are on Pt. Further, with these Rh species the trans/cis ratio is dependent on the nature of the support used. Preliminary data indicates that with these Rh catalysts the support is influencing the metal particle morphology thus imparting different steric constraints around the isomerization sites.[20]

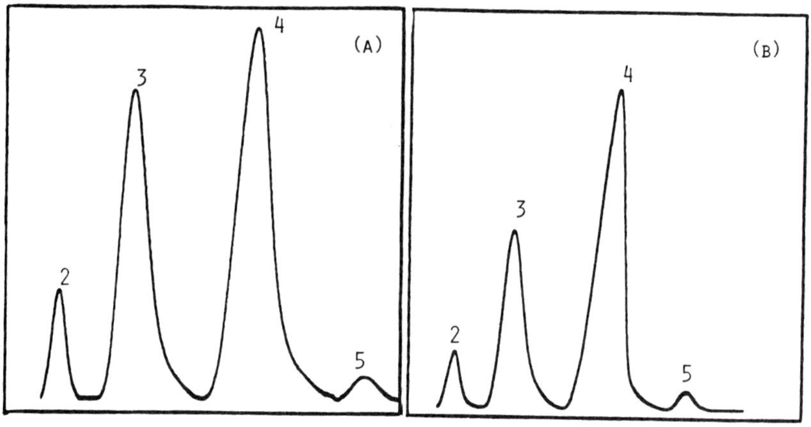

Fig. 8 Computer modified STO chromatograms, 1-butene peak removed: 2=butane; 3=*cis* 2-butene; 4=*trans* 2-butene, 5=butane from second H_2 pulse. (a) Rh/Al_2O_3 (b) Rh/SiO_2

D. Physical Description of ^3M Sites

The above discussion shows that we now have the ability to determine the densities of a variety of surface sites on a number of different types of supported metal catalysts. While the mechanistic correlations presented above present some argument for the physical descriptions of these different types of sites shown in Fig. 3, there has been no direct evidence presented thus far to substantiate these mechanistic and structural assumptions. While we know, for instance, that the STO procedure gives the number of saturation sites on a Pt catalyst, we do not know the nature of these sites. The ^3M description is merely a working hypothesis because we are unable to determine instrumentally the physical nature of the surface of these dispersed metal catalysts. This is not the case, though, with metal single crystals where LEED data can be used to describe exactly the nature of the surface. When a series of Pt single crystals having differing kink atom densities were used for the catalytic conversion of cyclohexane at 250-300°C it was found that the extent of benzene formation was related to the number of kink atoms on the surface. n-Hexane formation was a

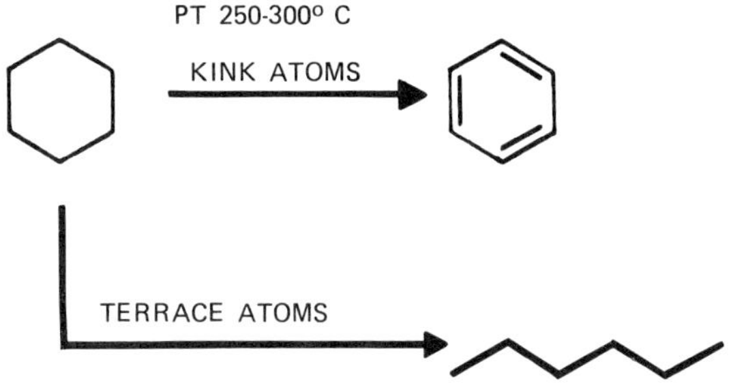

Fig. 9 Cyclohexane transformations.

function of the terrace atom density (Fig. 9). Thus, C-H bond cleavage is a property of kink or corner atoms and C-C bond breaking occurs on terrace or face atoms.[21] It was felt, then, that this reaction could prove to be useful in ascertaining the exact nature of the hydrogenation sites present on a supported Pt catalyst. If these ^{3}M sites are, indeed, corner or kink atoms then cyclohexane dehydrogenations run over a series of Pt catalysts having different ^{3}M site densities would result in the amount of benzene formation increasing with an increase in the number of these hydrogenation sites present. A series of ten STO characterized catalysts having a wide range of ^{3}M site densities were used to promote cyclohexane dehydrogenation at 300°C under a set of standard reaction conditions.[22] As mentioned above, heat has a marked effect on the surface characterization of these supported catalysts so each catalyst used was first heated to 300°C in He for 30 minutes and then cooled to room temperature for an STO evaluation. Fig. 10 is typical of the changes observed; a marked decrease in hydrogenation (peak 2) and an increase in isomerization (peaks 3 and 4). These catalysts were then heated to 300°C and exposed to a pulse of H_2 saturated with cyclohexane and the products analyzed by on-stream gas chromatography. Fig. 11 shows the relationship between the extent of benzene formation over these catalysts and the percent ^{3}M sites present as determined from the STO characterizations of the preheated catalysts (e.g. Fig. 10b). The observed relationship shows that, by analogy with the single crystal results, the ^{3}M or hydrogenation sites on these supported catalysts are, indeed, corner or kink atoms as previously implied using mechanistic considerations

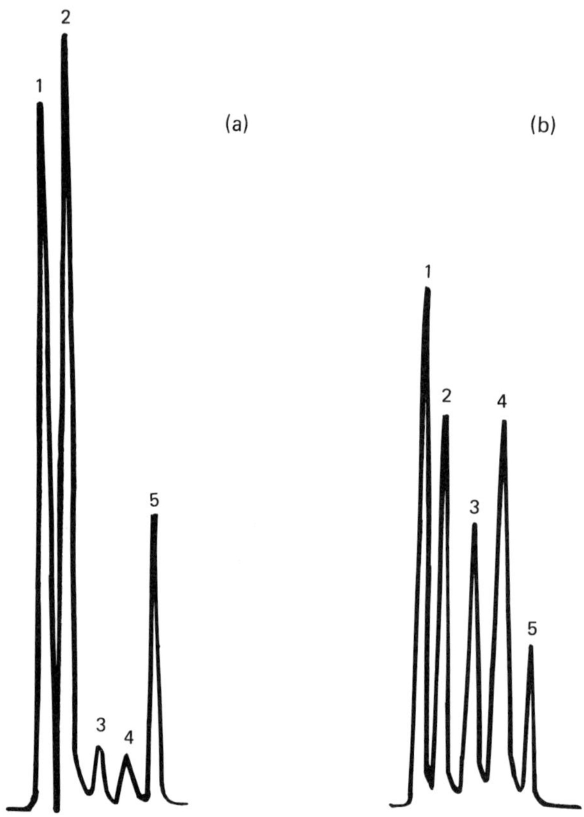

Fig. 10 STO chromatograms of a Pt/CPG catalyst: 1=1-butene; 2=butane; 3=<u>cis</u> 2-butene; 4=<u>trans</u> 2-butene; 5=butane from second H_2 pulse. (a) Normal STO chromatogram; (b) After heating in He at 300°C for 30 min. and cooling to room temperature.

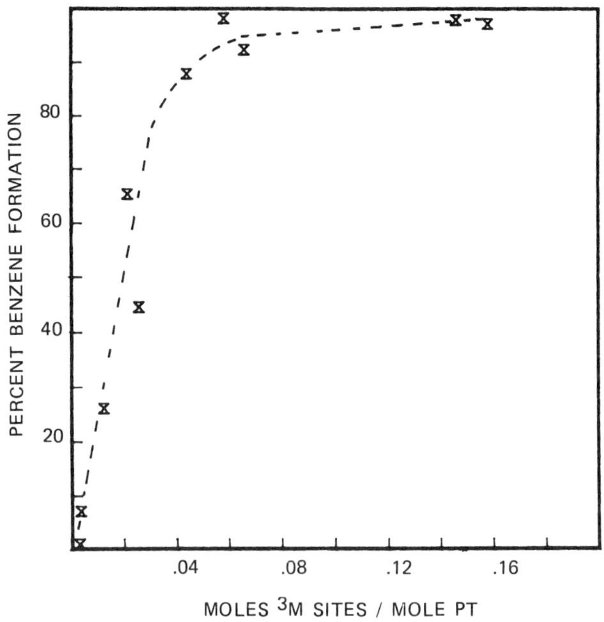

Fig. 11 Relationship between the extent of benzene formation and number of ${}^3M$ sites on Pt/CPG catalysts.

and homogeneous analogies (Figs.1-3). A similar correlation between C-C bond cleavage products and ${}^1M$ site densities was not obtained because these heated catalysts exhibit a marked decrease in H_2 chemisorption ability[13,16] so ${}^1M$ site determination was not possible.

E. Solvent Effect

With this ability to measure the densities of various sites present on the catalyst surface it was felt that the STO procedure should be applicable to the determination of the solvent/catalyst interaction which occurs in solution phase catalytic reactions. Previous work concerned with the solvent effect in olefin hydrogenations led to the conclusion that in alcohol there was a competition between alcohol molecules and H_2 for adsorption on the catalyst surface.[23] To test this a Pt/CPG catalyst having the STO chromatogram shown in Fig. 12a was exposed to methanol vapor introduced

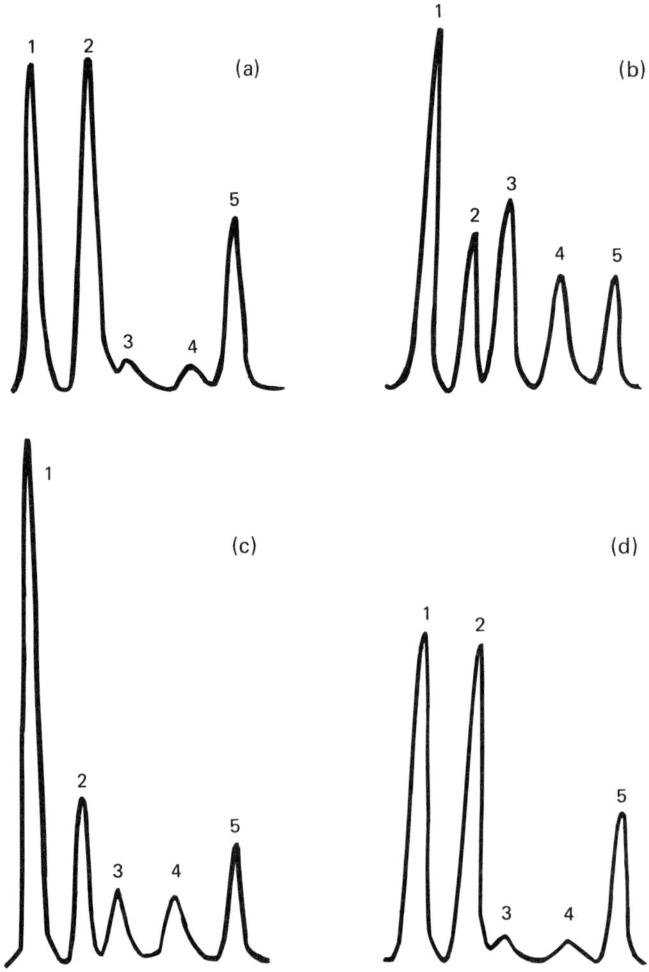

Fig. 12. STO chromatograms of a Pt/CPG catalyst: 1-1-butene; 2-butane; 3=*cis* 2-butene; 4=*trans* 2-butene; 5=butane from second H_2 pulse. (a) Normal STO chromatogram; (b) After exposure of catalyst to methanol vapor; (c) After exposure of catalyst to THF vapor; (d) After exposure of catalyst to pentane vapor.

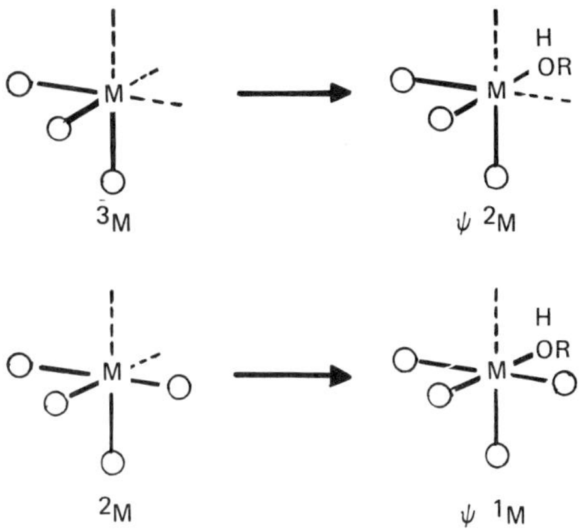

Fig. 13 Mode of alcohol interaction with surface sites.

into the carrier gas stream and then used for additional STO characterization.[24] The results, shown in Fig. 12b, show a decrease in the number of saturation sites and an increase in isomerization capability. Other solvents were also investigated. The results obtained with THF are shown in Fig. 12c, again showing considerable modification in the reaction characteristics of the catalyst. With pentane, a solvent not expected to have any competitive adsorption capability, an STO chromatogram (Fig. 12d) identical to that of the non-solvent treated catalyst (Fig. 12a) was obtained. Similar results were obtained with supported Pd catalysts.[24]

An explanation of these results[24] (Fig. 13) is that the alcohol, or other solvent, adsorbs on the catalyst surface occupying one of the vacant orbitals available at each site. In this way hydrogenation sites, ^3M, are converted into isomerization sites, ^2M, accounting for the decrease in saturation and increase in isomerization observed.

With the successful use of solvent vapor doped catalysts in this solvent effect study it was felt that a similar approach might be used in the study of solution phase reactions involving an STO procedure run on a solvent modified catalyst.[25] To test this, methylcyclopentene (**3**) was deuterated (eq. 5) over both Pt/C and Pd/C in three ways: in an EtOD solution; using a

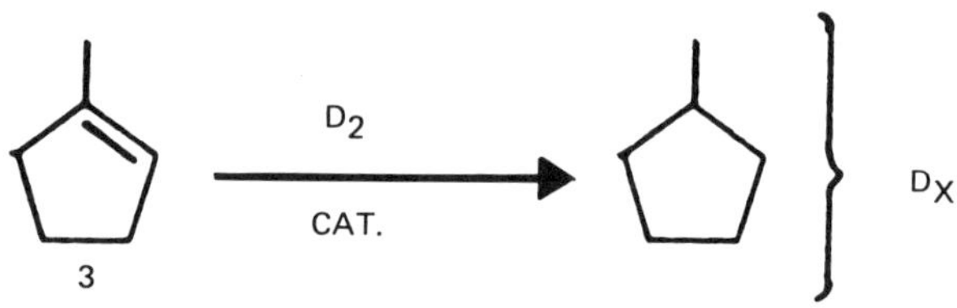

(5)

standard STO vapor phase procedure (D_2 pulse, **3** pulse, D_2 pulse); and with an STO sequence using a catalyst which had been exposed to EtOD vapor before the reaction sequence. The deuterium distributions for the products obtained from the Pt and Pd catalyzed reactions using each of these procedures are shown in Figs. 14 and 15 respectively. With both catalysts the solution phase deuteration data is quite different from that obtained using the normal STO deuteration procedure but very much like that

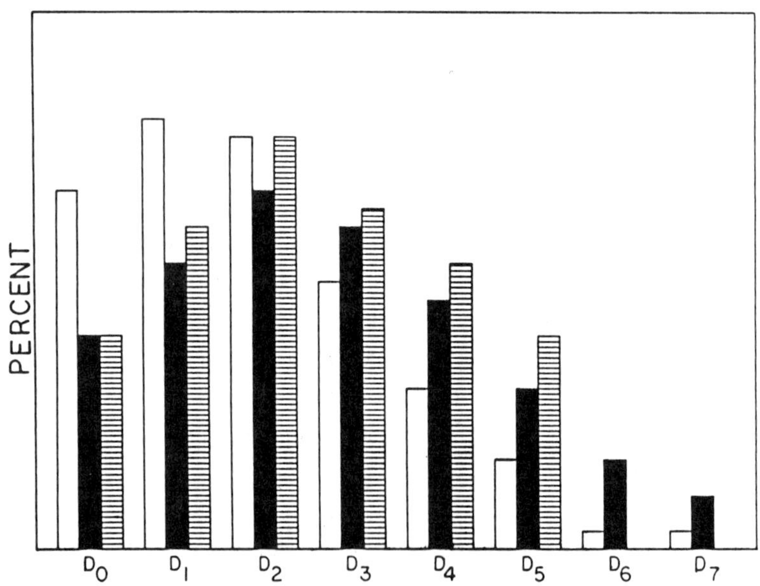

Fig. 14 Deuterium distribution in product from the deuteration of methylcyclopentene over Pt/C. Open bars-STO reaction; solid bars - EtOD saturated catalyst in STO procedure; Hatched bars - EtOD solution phase reaction.

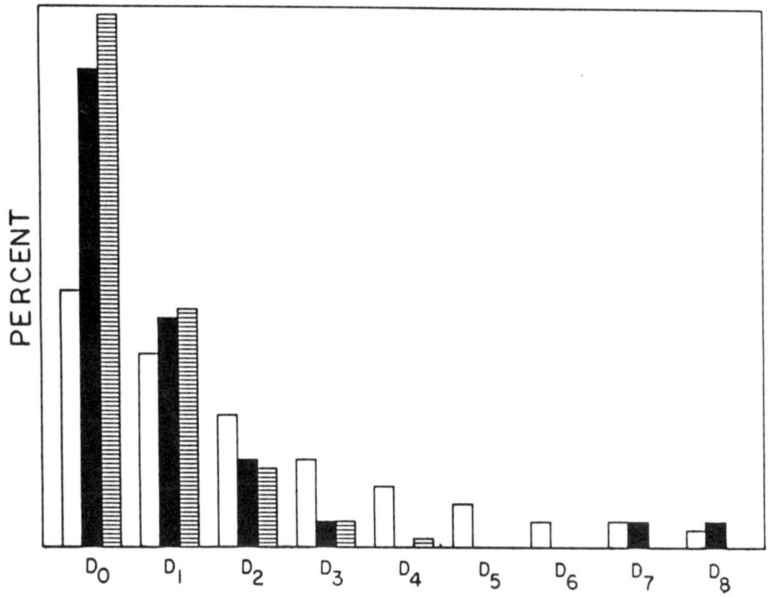

Fig. 15 Deuterium distribution in product from the deuteration of methylcyclopentene over Pd/C. Open bars STO reaction; solid bars - EtOD saturated catalyst in STO procedure; Hatched bars - EtOD solution phase reaction.

obtained using the ethanol modified catalyst in the STO sequence. Thus, this STO procedure is applicable not only to studies of vapor phase processes but it can also be modified for the gas phase investigation of reactions normally run in solutions.

F. Deuterium Exchange Sites

The catalytic H/D exchange on alkanes and cycloalkanes has been extensively studied as a means of determining the nature of the interaction between hydrocarbons and catalyst surfaces.[26] The results obtained were not always interpretable on the basis of a single type of metal/hydrocarbon interaction and a number of times it was proposed that at least two different types of sites had to be present on the catalyst surface to account for the observed exchange patterns.[27-29] To test this the deuterium exchange on cyclopentane (4) (eq. 6) was run over five Pt/CPG catalysts

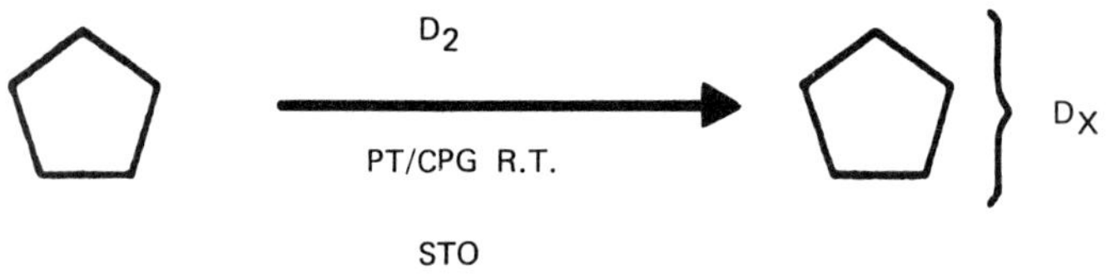

(6)

having a wide range of site densities.[30] A plot of the extent of monodeuterium exchange with respect to the number of ^1M sites present is

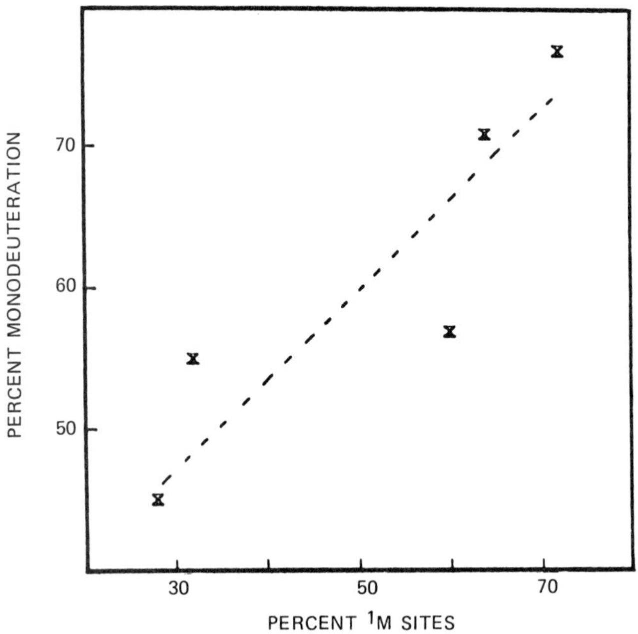

Fig. 16 Relationship between the amount of monodeuterium exchange product and the amount of ^1M sites on a series of Pt/CPG catalysts.

given in Fig. 16 while the relationship between polydeuteration and ^{3}M plus ^{2}M site densities is shown in Fig. 17. These results show that monodeuterium exchange occurs on the ^{1}M or face sites while polyexchange takes place on the more coordinatively unsaturated ^{3}M and ^{2}M sites. These conclusions are understandable in view of the mechanism for exchange shown in Fig. 18.[26] Monoexchange takes place on ^{1}M sites by way of reactions 8 and -8. Further exchange does not take place because the site does not have sufficient coordinative unsaturation for reaction 9 to occur. When this further unsaturation is present as with ^{3}M and ^{2}M sites polyexchange takes place via reactions 9 and -9.

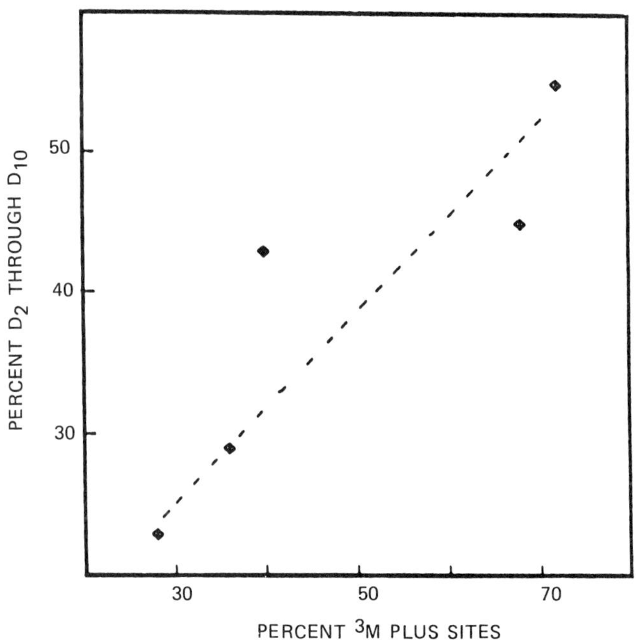

Fig. 17 Relationship between the amount of polydeuterium exchange products and the amount of ^{3}M and ^{2}M sites on a series of Pt/CPG catalysts.

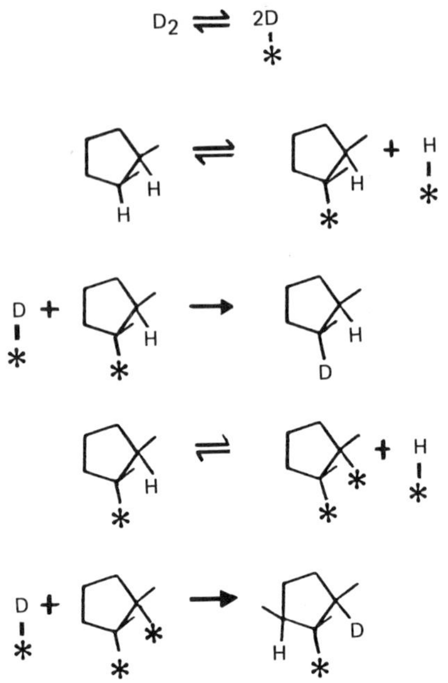

(7)

(8)

(-8)

(9)

(-9)

Fig. 18 Deuterium exchange mechanism.

G. Specific Site Rate Determination

All of the reactions presented so far have been investigated using an STO or pulse reaction sequence. While a reasonable amount of information is available from such an approach it might be considered to be somewhat limited and, perhaps, not directly related to standard steady state reaction studies. To extend this work, then, a series of STO characterized Pt/CPG catalysts were used in a classic steady state hydrogenation study. In order to interpret these results, though, it is necessary to redefine the reaction characteristics of the various surface sites as listed in Table II. The ^3M sites promote hydrogenation under both steady state and STO conditions and the ^2M_C sites promote isomerization in both cases. The difference is in the ^2M_S sites, those on which a stable half-hydrogenated state is formed in the STO sequence. In a steady state hydrogenation H_2 is available to react with this species so these ^2M_S sites become hydrogenation sites.

TABLE II

Site/Reaction Correlation for STO and Steady State Hydrogenations

Site	STO Reaction	Steady State Reaction
^3M_I	Hydrogenation	Hydrogenation
^3M_R	Hydrogenation	Hydrogenation
^2M_S	Metalalkyl Formation	Hydrogenation
^2M_C	Isomerization	Isomerization
^1M	None	None

The reaction used in this steady state reaction study was the isomerization and saturation of (+) apopinene (**5**), a reaction used by Smith[31] as a means of characterizing dispersed metal catalysts. In this reaction (Fig. 19) the ratio of isomerization to addition, k_i/k_a, is obtained by the analysis of the amount of apopinane (**7**) formed and the loss of optical activity resulting from the conversion of **5** to its mirror image **6** by double bond isomerization and by the saturation of both **5** and **6** to **7**. The k_i/k_a ratios for this reaction were obtained by Smith[32] for four STO characterized Pt/CPG catalysts and the results and listed in Table III along with the various site densities for these catalysts.[12]

Since it would not be expected that the various isomerization and hydrogenation sites would all react at the same rate a direct correlation between the ratio of isomerization sites/hydrogenation sites and k_i/k_a is not possible. If however, the relative rate factors A, B, C, and D are introduced as in eq. 10 such a correlation should be feasible. A computerized iterative

$$\frac{K_I}{K_A} = \frac{\text{Isomerization}}{\text{Hydrogenation}} = \frac{A \times {}^2M_C}{B \times {}^3M_1 + C \times {}^3M_R + D \times {}^2M_S} \quad (10)$$

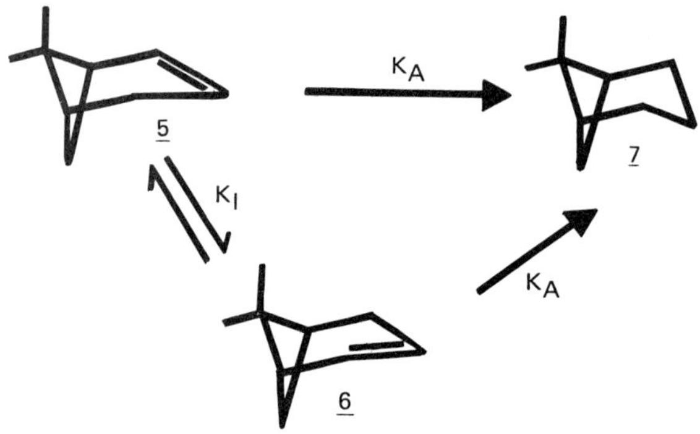

Fig. 19 Saturation and isomerization of (+) apopinene (**5**).

TABLE III

Site Densities and k_i/k_a Data for Pt/CPG Catalyst

Catalyst[a]	Site Densities[b]					k_i/k_a	
	^3M_I	^3M_R	^2M_S	^2M_C	^1M	Exp.[c]	Calc.[d]
III-250	0.040	0.009	0.020	0.005	0.038	0.127	0.130
IV-400	0.003	0.028	0.010	0.009	0.102	0.147	0.150
II-400	0.038	0.026	0.017	0.025	0.137	0.220	0.225
IV-250	0.049	0.022	0.018	0.039	0.062	0.313	0.321

[a]Ref. 13

[b]moles site/mole Pt

[c]Ref. 32

[d]Calculated using eq. 10 and values found for A, B, C, and D.

approach was used to find values for A, B, C, and D which would be applied to the various site densities as in eq. 10 and fit the experimental k_i/k_a results to within ± 3% for all four catalysts. The values found were:

$$A = 1.9$$
$$B = 1.2$$
$$C = 5.4$$
$$D = 1.0$$

The k_i/k_a ratios calculated using eq. 10 for each of the Pt/CPG catalysts are also listed in Table III. Fig.20 shows the relationship between the experimental k_i/k_a values and those calculated as described above.[32]

These relative rate factors for the hydrogenation sites are consistent with the site descriptions presented earlier. The most active sites are those having the more loosely held H_2 (^3M_R). They are about four times more

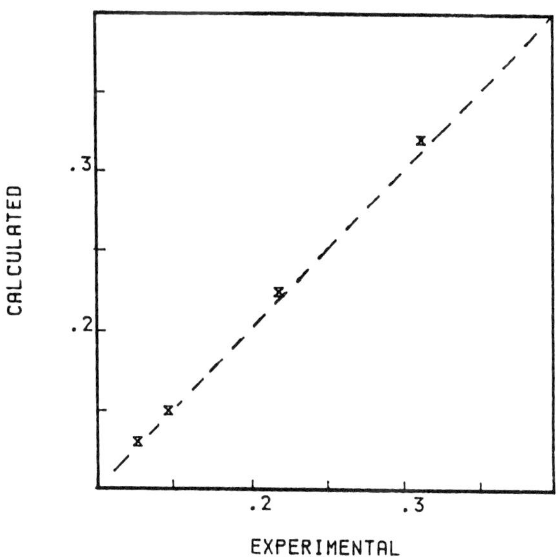

Fig. 20 Relationship between the calculated and experimental values of the k_i/k_a ratios for a series of Pt/CPG catalysts.

active than are those sites on which the H_2 is more tightly bound (^3M_I). Those sites (^2M_S) on which the alkane is formed by successive two step additions of H atoms are the least active.

III. CONCLUSIONS

Single turnover (STO) catalyst characterization studies provide data from which the amounts of five different types of sites present on the surface of a dispersed metal catalyst can be determined. This procedure can be used for both Pt and Rh catalysts on a variety of supports and is, at present, partially effective for the characterization of supported Pd catalysts. The STO approach can be used to evaluate the effect which catalyst preparation and pretreatment factors have on active site densities as well as to emulate solution phase conditions, so the more simple vapor phase reaction conditions can be used for the study of solution phase reactions.

STO characterized dispersed metal catalysts have been used to correlate STO site descriptions with those provided by single crystal studies. These catalysts can also be used to study the reaction characteristics of the different types of surface sites present and with the appropriate computerized assessments can provide data concerning the relative rates at which these different types of sites react.

ACKNOWLEDGEMENTS

The initial phase of this research was supported by grants from the Ameron Corporation and the Procter and Gamble Company. The National Science Foundation provided a grant (PRM-8112908) for the purchase of a mass spectrometer. Present research is supported by Grant DE-FG02-84ER 45120 from the U.S. Department of Energy.

REFERENCES

1. G.D. Halsey, Adv. Catal., **4**, 259 (1952).
2. B. Lang, R.W. Joyner, and G.A. Somorjai, Surf. Sci., **30** 440, 454 (1972).

3 R. VanHardeveld and F. Hartog, Surf., Sci., **15**, 189 (1969).
4. M. Boudart, Adv. Catal., **20**, 153 (1969).
5. R.L. Augustine and J.F. VanPeppen, Ann. NY Acad. Sci., **172**, 244 (1970).
6. I. Horiuti and M. Polanyi, Trans. Faraday Soc., **30**, 1164 (1934).
7. J.D. Osborn, F.H. Jardine, J.F. Young, and G. Wilkinson, J. Chem. Soc. (A), 1711 (1966).
8. S. Siegel, J. Outlaw, Jr., and N. Garti, J. Catal., **52**, 102 (1978).
9. R.L. Augustine and R.W. Warner, J. Org. Chem., **43**, 2614 (1981).
10. R.L. Augustine and R.W. Warner, J. Catal., **80**, 358 (1983).
11. S. Tsuchiya, M. Nakamura, and N. Yoshioka, Bull. Chem. Soc. Japan, **51**, 981 (1978).
12. R.L. Augustine, K.P. Kelly, and Y.-M. Lay, Appl. Catal., **19**, 87 (1985).
13. R.L. Augustine and K.P. Kelly, J. Chem. Soc., Faraday 1, In Press.
14. G.J. DenOtter and F.M. Dautzenberg, J. Catal., **53**, 116 (1978); F.M. Dautzenberg and H.B.M. Walters, J. Catal., **56**, 26 (1978).
15. P.G. Menon and G.F. Froment, J. Catal., **59**, 138 (1979).
16. R. Burch and L.C. Garla, J. Catal., **73**, 20 (1982).
17. G.A. Somorjai, J. Catal., **27**, 453 (1972).
18. S. Chandra, P.J. O'Hagan, and R.L. Augustine, Unpublished Results.
19. D.R. Baum and R.L. Augustine, Unpublished Results.
20. K.G. High and R.L. Augustine, Unpublished Results.
21. R.K. Herz, W.D. Gillespie, E.E. Petersen, and G.A. Somorjai, J. Catal., **67**, 371 (1981).
22. R.L. Augustine and M.M. Thompson, J. Catal., Manuscript submitted for publication.
23. R.L. Augustine, V. Nocito, J.F. VanPeppen, R. Warner, and F. Yaghmaie, in **Catalysis in Organic Synthesis** (W.H. Jones, ed), Academic Press, 1980, p. 173.
24. R.L. Augustine, R.W. Warner, and M.J. Melnick, J. Org. Chem., **49** 4853 (1984).
25. R.L. Augustine and M.E. Lenczyk, J. Catal., **97**, 296 (1986)
26. R.L. Burwell, Jr., Acc. Chem. Res., **2**, 289 (1969); Catal. Rev., **7**, 25 (1972).

27. J.R. Anderson and C. Kemball, Proc. R. Soc. London, Ser. A., **223**, 361 (1954).

28. H.C. Rowlinson, R.L. Burwell, Jr., and R.H. Tuxworth, J. Phys. Chem., **59**, 225 (1955).

29. Y. Inoue, J.M. Herrmann, H. Schmidt, R.L. Burwell, Jr., J.B. Butt, and J.B. Cohen, J. Catal., **53**, 401 (1978).

30. R.L. Augustine and R. Wesdyk, Langmuir, **1**, 262 (1985).

31. G.V. Smith, O. Zahraa, A. Molnar, M.M. Khan, B. Rihter, and W.E. Brower, J. Catal., **83**, 238 (1983).

32. A manuscript describing the joint study with G.V. Smith on Pt/CPG catalysts is in preparation.

13

Acid-Promotion in Dehydrogenation Process in Diarylamines from Primary Arylamines and Alicyclic Ketones

Russell E. Malz, Jr. and Harold Greenfield

Uniroyal Chemical Company
Elm Street
Naugatuck, CT

ABSTRACT

Diarylamines have been prepared in high yield by the in situ catalytic dehydrogenation of the product of the reductive alkylation of an alicylic ketone and a primary aryl amine in the presence of a hydrogen acceptor. This process is greatly accelerated by the addition of a small amount of acid. Palladium appears to be the catalyst of choice, an aromatic nitro compound the preferred hydrogen acceptor, and sulfuric acid the best promoter.

I. INTRODUCTION

Hydrogen transfer reactions do not have wide application when compared to general heterogeneous catalytic hydrogenation. Rylander[1] has said they are useful when one is zoned against using hydrogen. He does acknowledge five instances which show selectivity advantages.

Uniroyal's interest in this type of chemistry came about because we were hoping to eliminate a side-product in the production of phenyl alpha-naphthylamine by utilizing hydrogen-transfer in place of an alcohol and amine condensation reaction. We successfully developed a process which involved promoting a known process[2] with the addition of small amounts of acid.[3]

In general we are referring to hydrogen transfer reactions under mild conditions (less than 200°C), and which are catalyzed by a heterogeneous catalyst. There are three types of hydrogen transfer reactions:

1. A disproportionation reaction in which the same molecule serves as both donor and acceptor. An example is the conversion of cyclohexene to benzene and cyclohexane. We will discuss this chemistry from a mechanistic viewpoint.
2. The reactions in which hydrogen transfer is used in place of conventional hydrogenation. Here again we will be concerned with mechanistic considerations, and interesting selectivity.
3. The reactions which involve the title of our work. In simplest form it is represented by Equation 1.

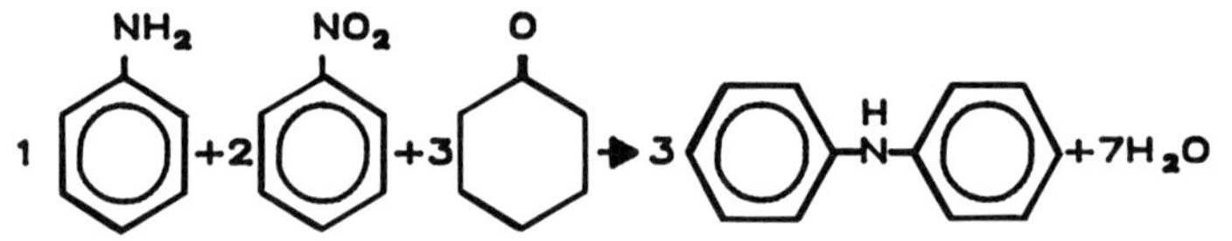

(1)

In this reaction the donor is cyclohexanone and the acceptor is nitrobenzene. The nitrobenzene is converted to aniline while the acceptor is converted to an aromatic system incorporated into the product. We ran these reactions with a stoichiometry based on two moles of nitrobenzene and four moles of aniline to five moles of cyclohexanone. Cyclohexanone is the limiting reagent. The reaction can be conveniently followed by the formation of water, except in those cases where a side reaction produces water. We will discuss this point later.

These reactions were run in various aromatic hydrocarbons at one atmosphere under vigorous reflux. The aromatic hydrocarbons permitted a range of temperatures and smoother reactions with less violent bumping than using the reagents neat.

Vigorous reflux is an absolute necessity. This was shown by Linstead and Michaelis[4] who found that boiling tetralin (207°C) is dehydrogenated over a palladium catalyst, but there is little reaction under tranquil conditions at 200°C, and by Braude[5] who pointed out that the metal-catalyzed hydrogen transfer hydrogenation of nitro compounds will come to a standstill after 10 to 20 percent completion if the reaction mixture is stirred mechanically below its boiling point.

II. MECHANISTIC CONSIDERATIONS

Braude et al.[5-9] discuss hydrogen transfer in a classic series of papers. Most of their discussion centers around heterogeneously catalyzed disproportionations (transfer of hydrogen between identical donors and acceptors) and transfer of hydrogen between unlike donors and acceptors.

In their paper dealing with cyclohexene as a hydrogen donor for the reduction of ethylenic and acetylenic bonds,[7] they feel their results are best interpreted by the co-adsorption of the donor and acceptor on the catalyst surface. Cyclohexene readily disproportionates, as shown in Equation 2.

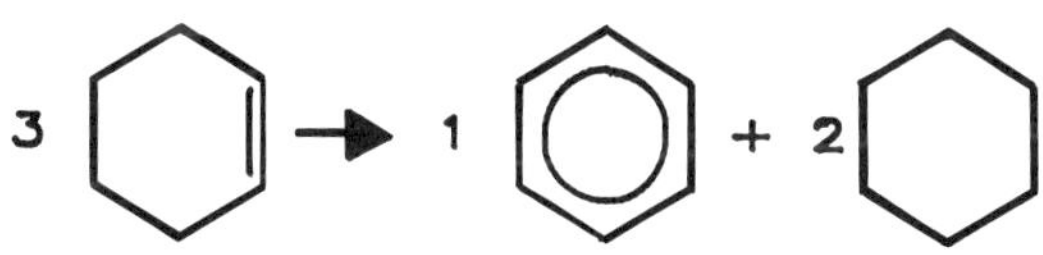

(2)

This reaction was first described by Zelinski and Parlov[9] subsequent to the work of Wieland,[10] who was the first to show a metal-catalyzed hydrogen-transfer under mild conditions. He studied the disproportionation of dihydronapthalenes to naphthalene and tetalin. Weiland considered palladium a reagent rather than a catalyst. He proposed a distinct two step mechanism. The first step involved a dehydrogenation of one molecule with the formation of chemisorbed hydrogen. The second step involved the hydrogenation of another molecule. He postulated this proceeded through a palladium hydride intermediate as shown in Equations 3 and 4.

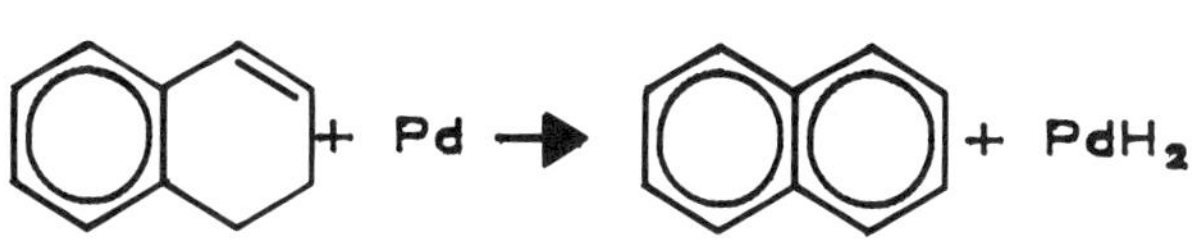

(3)

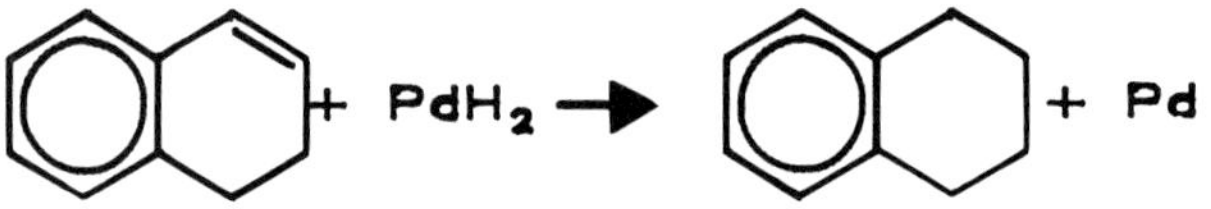 (4)

Braude et al[6] argues that the disproportionation is a linked process with dehydrogenation and hydrogenation taking place simultaneously on the catalyst surface. He points out:

1. that with cyclohexene in THF the reaction proceeds stoichiometrically even under conditions of rapid boiling which would favor the removal of absorbed hydrogen.
2. they were not able to detect any hydrogen evolution. They also argue that the mechanism could be termolecular since they were not able to detect any cyclohexadiene and such a reaction is not kinetically impossible. Brieger[11] points out that later studies were able to detect cyclohexadiene and the data best fit second-order kinetics.

Chemisorbed hydrogen is necessary before disproportionation can take place.[8,12] Braude[7] argues that this absorbed hydrogen may serve as a bridge between non-adjacent absorbed cyclohexenes to reduce the degree of proximity required for the reaction. Brieger[11] proposes because of the limited amount of experimental studies directed toward the determination of this mechanism, that we should turn toward the classical catalytic hydrogenation studies.

The Horiuti and Polyani[13] mechanism could account for many of the results found in catalytic transfer-hydrogenation. This mechanism, is shown in Equations 5 through 8.

$$H_2 + 2* \rightleftharpoons 2\ \underset{*}{\overset{\displaystyle H}{|}} \qquad (5)$$

$$R_2C = CR_2 + 2* \rightleftharpoons \underset{*}{\overset{\displaystyle R_2C}{|}} - \underset{*}{\overset{\displaystyle CR_2}{|}} \qquad (6)$$

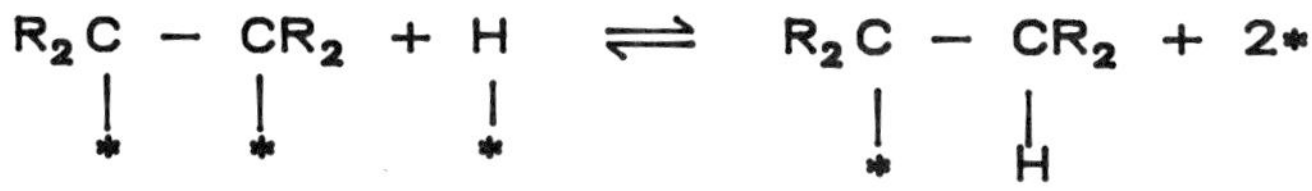

(7)

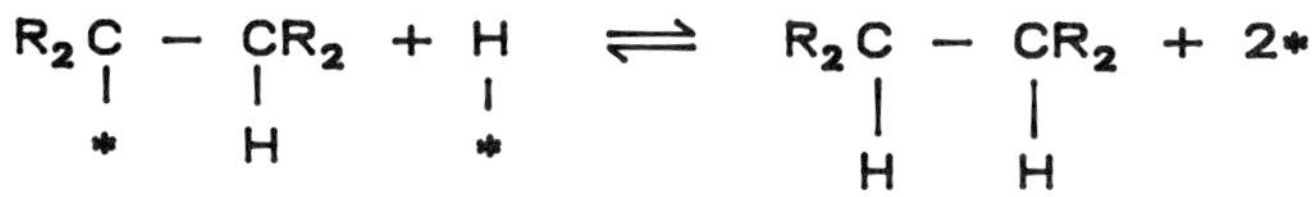

(8)

This has the absorption of both hydrogen and the acceptor as reversible processes. If we now recognize that palladium is capable of forming pi complexes as illustrated in Equation 9, we can arrive at the mechanism of Brieger[11] shown in Equations 10-13.

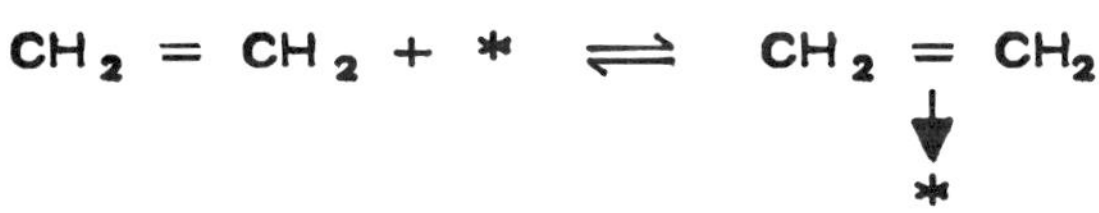

(9)

Step 1

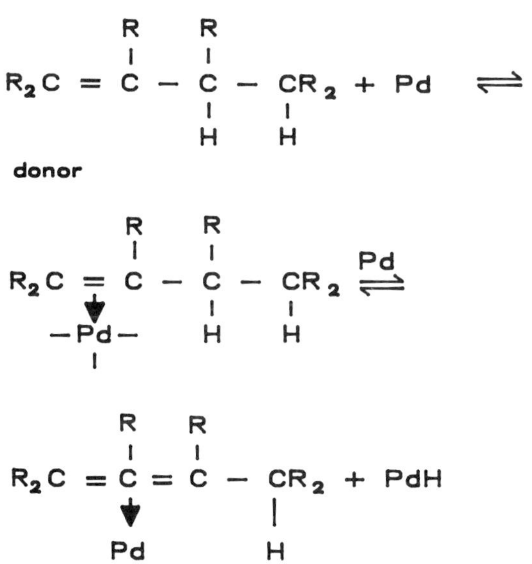

(10)

Step 2

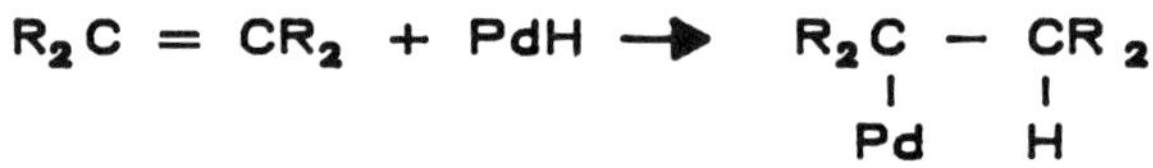

(11)

acceptor

Step 3

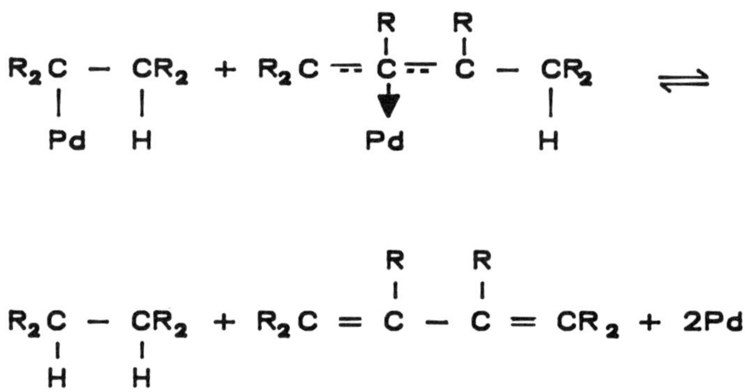

(12)

Alternatively

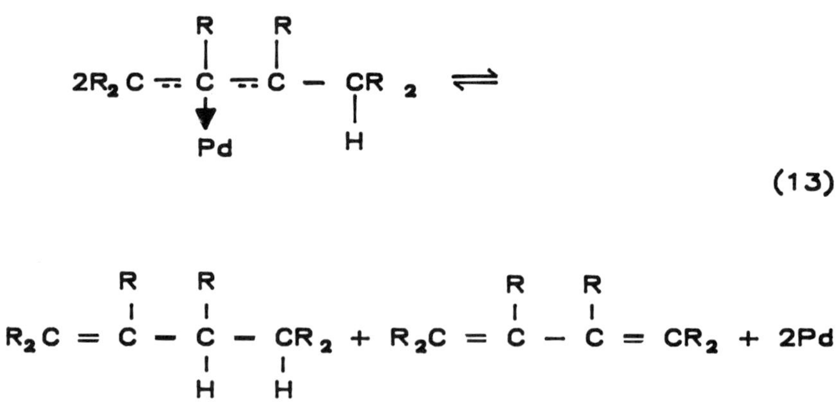

(13)

III. REACTIONS WITH NITRO-COMPOUNDS AS HYDROGEN ACCEPTORS

Nitro-compounds serve as convenient hydrogen acceptors. Braude et al[5] prepared a series of primary amines from aliphatic and aromatic nitro-compounds using a palladium catalyst and cyclohexene as a hydrogen donor.

The results can be generally summed up by saying the yields often compare favorably with those obtained by other methods. However, there are certain cases where entirely unexpected results are obtined. These offer unique preparative possibilities as well as interesting insight into the mechanism of these reductions. For example consider the results obtained with the three dinitrobenzenes shown in Table I.

This sensitivity to structural variation in hydrogen transfer reactions is in contrast to direct reduction with hydrogen with either platinum or palladium catalysts where the meta as well as the ortho and para gives predominantly the diamine. These differences lend support to the postulate that this hydrogen transfer is a step-wise process. It could be suggested these differences are due to poisoning effects of the specific compounds or their intermediates. However, neither the ortho or meta compound inhibit the disproportionation of cyclohexene.[5]

TABLE I

Nitro Groups as Hydrogen Acceptors[a]

Starting Material	Product	Yield (%)
o-dinitrobenzene	o-nitroaniline	<2
m-dinitrobenzene	m-nitroaniline	80
p-dinitrobenzene	p-nitroaniline	<2
2,4-dinitrotoluene	2,4-diaminotoluene	75
1-t-butyl-2,4-dinitrobenzene	1-t-butyl-2,4-diaminobenzene	95

a Each reaction was run at vigorous refulx using palladium black to starting material ratio of them 10 mg/g to 25 mg/g in different solvents with boiling points between 66 to 81°C and for between 5 and 40 hours.

Hydroxylamines and nitroso compounds have been postulated as intermediates in the reduction of nitro compounds. When hydrogen transfer using cyclohexene as donor and nitrosobenzene as acceptor was tried intractable products were obtained. When the same reaction was attempted using the hydroxylamine as acceptor, aniline was cleanly formed.

Braude[2] argues that this lends additional credence to the view that this is a linked process of donor-acceptor rather than separate donor-dehydrogenation, acceptor-hydrogenation steps. He also aruges that if one accepts the first intermediates as the addition of two hydrogen atoms to give the hydroxylamine N-oxide or the isomeric dihydroxylamine (nitroso-hydrate), followed by hydrogenolysis of the N-O bonds, then he can explain this difference in specificity. This involves the nitro group lying flat on the catalyst surface alongside a hydrogen donor molecule. In the process of converting NO_2 to $N(OH)_2$, the nitrogen atom goes from a planar to a tetrahedral arrangement.

The conventional cases proceed smoothly. If there is a second functional group present with high affinity for the catalyst, it will inhibit the transfer reactions because it will hinder the geometric change neccessitated by converting the nitrogen from sp^2 to sp^3. The addition of a large group prevents the simultaneous absorption of the second absorbable group and hydrogen transfer can again take place in the presence of a strongly absorbed group such as amino.

IV. CATALYSTS

Brieger's[11] review of catalytic transfer hydrogenation concludes that palladium is the catalyst of choice. He states "under standard conditions where palladium was effective, neither platinum nor rhodium catalysts work, at least at temperatures below 200°C." However, the literature on this reaction is not adequate to finalize this assertion. Braude et al[5] shows a comparison of catalysts for the transfer hydrogenation of p-nitrotoluene with cyclohexene as hydrogen donor. The data are found in Table II.

Palladium is clearly the catalyst of choice in preference to either platinum or nickel. With Raney nickel we have the added complication that

TABLE II

Comparison of Catalysts for Transfer Hydrogenation of p-Nitrotoluene[a]

Catalyst	Amount mg	Time hr	% Yield of p-toluidine
10% Pd/C	190	7	95
10% Pt/C	350	7	13
Raney Ni	100	22	2
W 7 Raney Ni	100	100	2

a Each reaction used 2.4 g p-nitrotoluene and 15 ml of cyclohexane with the indicated catalyst for the indicated time at reflux.

this catalyst contains absorbed hydrogen and it may be functioning as a hydrogenation rather than a hydrogen transfer catalyst.

Van Verth and Wilder[2] show that in cases where palladium gives a 75% yield of diphenylamine from aniline, nitrobenzene and cyclohexanone, osmium, irdium and platinum give only low yields of product.

V. DEHYDROGENATION PROCESS FOR DIARYLAMINES FROM PRIMARY ARYLAMINES AND ALICYCLIC KETONES

Van Verth and Wilder[2] discuss the dehydrogenation reaction which involves an aryl amine, an aryl nitro compound and cyclohexanone with a 5% palladium-on-carbon catalyst. This reaction can be generally illustrated by Eq. 14. A large number of secondary aryl amines were prepared in this manner.

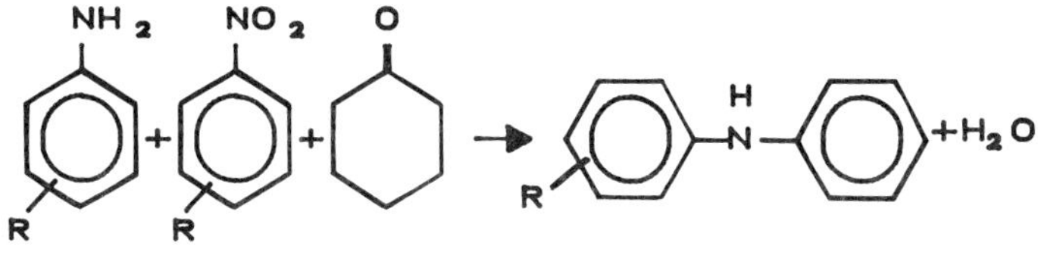

(14)

The yields were reported in almost all cases and varied between six percent with N-methylaniline to quantitative with N-cyclohexyl-N'-phenyl-p-phenylenediamine. Most of the yields are between 50 and the mid seventies.

In some cases, instead of adding amine, the authors started with the nitro compound which they partially reduced before adding the ketone.

A majority of the reactions were run without solvent with the minority using aromatic hydrocarbons. Running the reaction without solvent and repeating with xylene as solvent did not alter the results significantly. The addition of water or propanol was said to be without effect.

In only one case do the authors start with the anil and that is shown in Eqs. 15 and 16. They do not use an aryl nitro compound as hydrogen

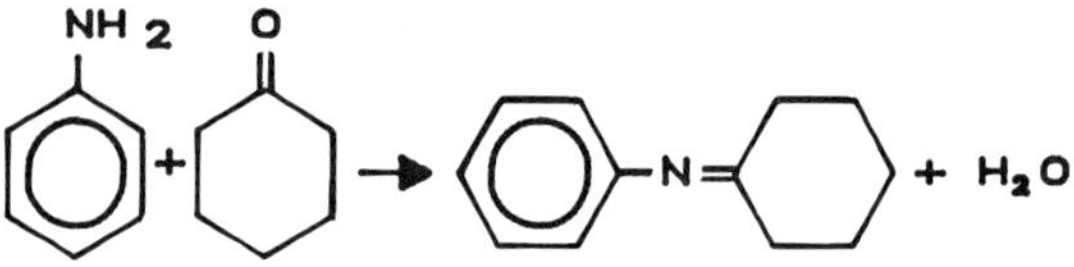

(15)

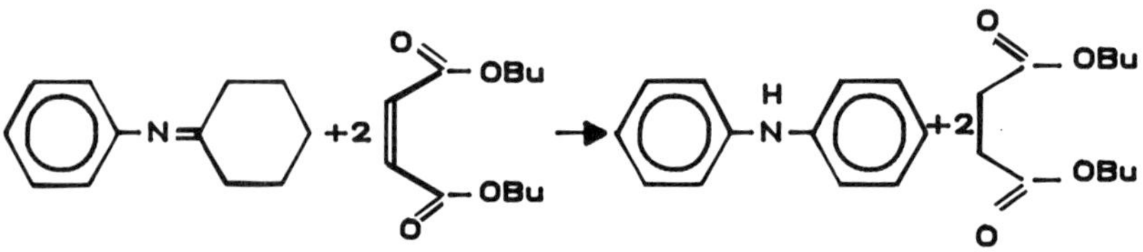

(16)

acceptor but rather a substituted maleimide. Phenyl-alpha-naphthylamine and phenyl-beta-naphthylamine were produced as shown in Eq. 17. No yield

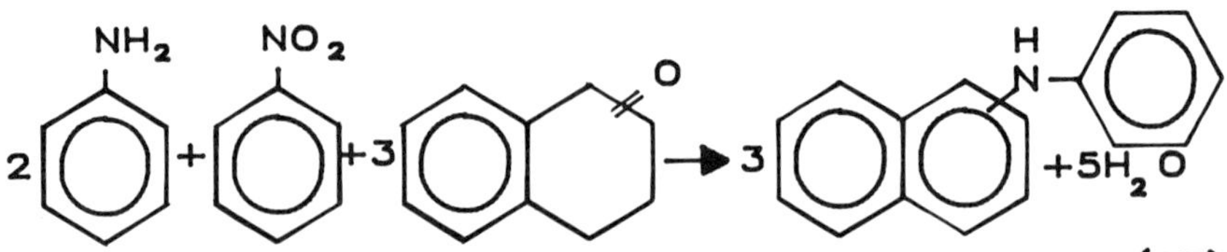

(17)

is reported using the alpha isomer, while a 67% yield is reported with the beta isomer. In almost all cases the yield of water was essentially quantitative.

The catalyst levels used in these reactions were extremely high. For example, with the reaction to produce diphenylamine (DPA) levels of catalyst to aniline varied from a ratio of 0.12 to 0.26 grams of catalyst per gram of aniline.

VI. EFFECT OF ACID ON A DEHYDROGENATION PROCESS

Marschik and Rylander[14] show the effect of a minor amount of acetic acid on the preparation of 4-amino-3-cyanopyridine from 4-imino-3-cyanopiperidine as shown in Eq. 18. The results in Table III show that a maxiumum yield of 80.7% occurred at 1.32% acetic acid. Further addition of acid caused a loss of yield, giving complete suppression of the desired reaction at the 6.6% level.

TABLE III

Preparation of 4-amino-3-cyanopyridine[a]

% Acetic Acid[b]	% Yield
None	51
0.05	56.6
0.66	76.0
1.32	80.7
2.0	77.2
3.3	71.8
6.6	0

a Each reaction was carried out using 15.0 g (0.124 mole) 4-imino-3-cyanopiperidine, 300 ml nitrobenzene, and 5 g 5% Pd/Al_2O_3 at 170°C and 160 mm with stirring for 1 hr while the water formed during the reaction was removed with a nitrogen sweep.

b Based on the weight of piperidine.

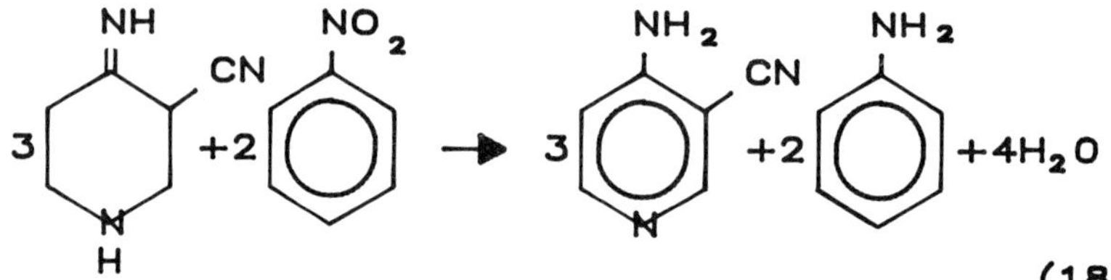

(18)

Van Verth and Wilder[2] mention the use of an acid clay in the formation of phenyl-beta-naphthylamine in 67% yield from beta-tetralone, aniline and nitrobenzene. This is their only mention of an acid-type promoter.

VII. ACID-PROMOTION IN DEHYDROGENATION PROCESS FOR DIARYLAMINES FROM PRIMARY ARYLAMINES AND ALICYCLIC KETONES

The stoichiometry can be calculated from the sum of three consecutive reactions (i.e. equations 19, and 20), which can be summed as Eq. 1. The same products are obtained starting with the ketimine.[2]

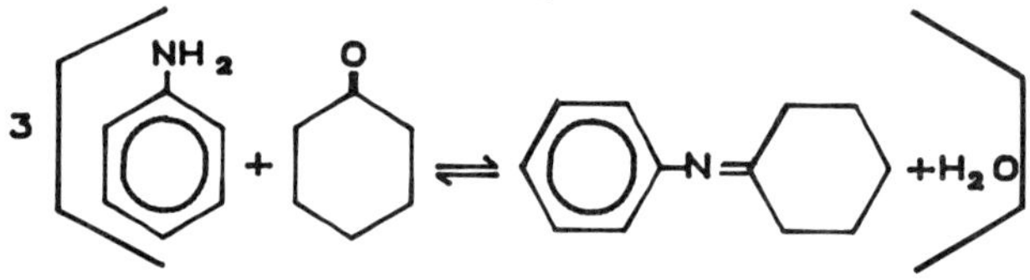

(19)

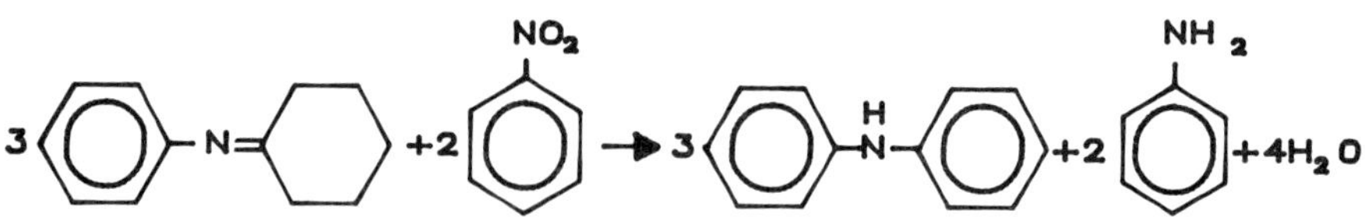

(20)

Our main interest in this type of chemistry was in the production of phenyl-alpha-napthylamine (PANA). The stoichiometry of this process is shown in Eqs. 21 and 22 and summed in equation 23. Alpha-tetralone is an

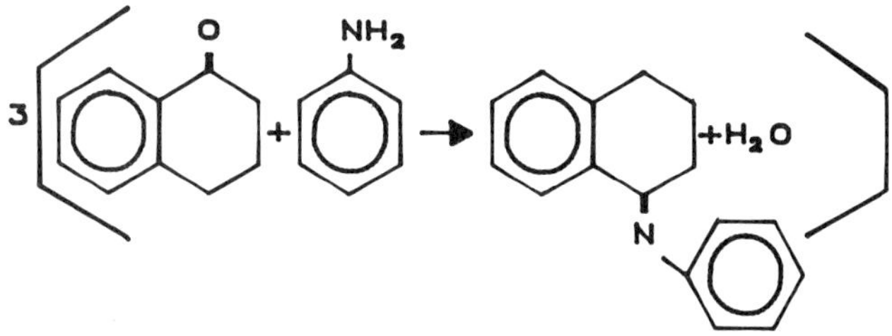

(21)

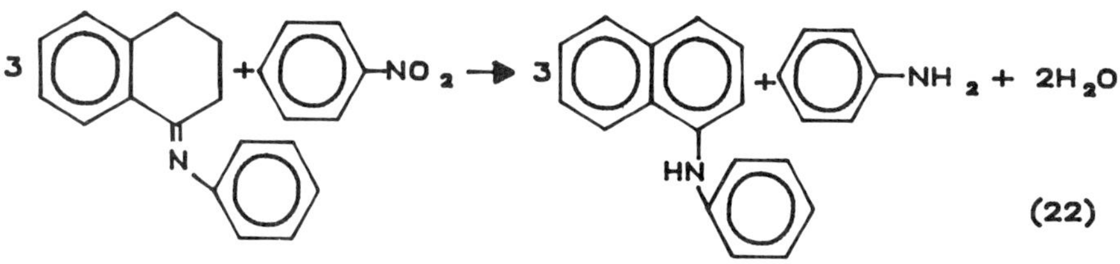

(22)

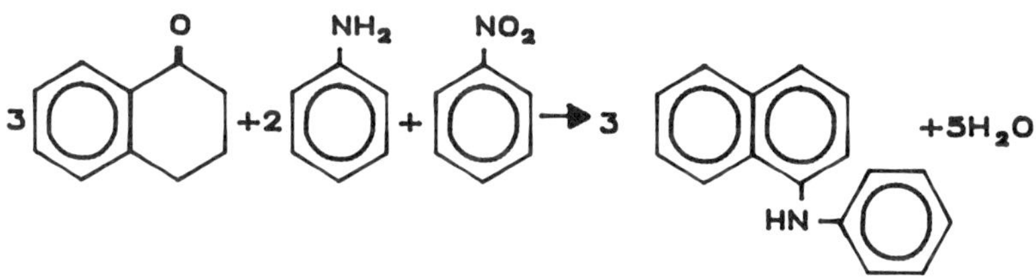

(23)

benzylic ketone. This could be expected to give different chemistry. The literature gives only one reference[2] to the reaction of a benzylic ketone and the yield is not given. In that reference 12.3 g nitrobenzene is hydrogenated using 2.0 g 5% Pd/C in p-cymene. On completion of 15% of the reaction, the reaction mixture is transferred and 16.1 g of 1-tetralone and additional solvent is added. This is heated under N_2 at reflux (183-185°C) for 6 hrs. After 1/2 the water had evolved, 2.0 g of an acid clay was added.

The known literature[2] used uneconomical amounts of catalyst. In an attempt to reduce this catalyst level we found[3] we could promote the reaction with trace amounts of acid. Table IV shows the effect of various acids on the formation of phenyl-naphthylamine (PANA).

Using 1.2 mmoles of acid under the conditions of Table IV, less than 5% yield of PANA was detected with acetic acid, hydrochloric acid and a sulfonic acid exchange resin.

TABLE IV

The Effect of Various Acids on the Production of PANA[a]

Acid Type	Millimole	Yield PANA[b] Mole %
None	0	0
Sulfuric	6	41
Phosphoric	6	18
p-Toluenesulfonic	6	20
Trifluoroacetic	6	20

a Each experiment was run for 5 hours using 0.20 g 5% Pd/C on a dry weight basis in a 200 ml, 3-necked flask equipped with a thermometer and Dean-Stark apparatus filled with p-cymene and having a condenser. To this was added 7.45 g (80 mmole) of aniline, 4.82 g (39 mmole) of nitrobenzene, 14.6 g (100 mmole) of 1-tetralone and 65 ml of p-cymene.

b The reaction mixtures were analyzed by gc using a weight percent internal standard method.

TABLE V

The Effect of Catalyst Concentration on PANA Yield[a]

5% Pd/C[b] wt. (g)	Yield PANA[c] mole %
0.20	31
0.40	68
0.60	88
0.80	87

a Each experiment was run for 5 hours using a 200 ml, 3-necked flask equipped with a thermometer and Dean-Stark apparatus filled with p-cymene and having a condenser. To this was added 7.45 g (80 mmole) of aniline 4.82 g (39 mmole) of nitrobenzene, 14.6 g (100 mmole) of 1-tetralone, 65 ml of p-cymene, 0.6 mmole of sulfuric acid and the indicated amount of catalyst.

b The catalyst was used on a dry weight basis.

c The reaction mixtures were analyzed by gc using wt % internal standard method. Further experiments on the interaction of catalyst level and sulfuric acid promotor level are shown in Table VI.

The acid of choice is sulfuric acid. If we reduce the amount of sulfuric acid to 10% of the value in Table IV, we will lower the yield of PANA to 31%. This still exceeds the yield with other acids.

The interaction of the amount of catalyst and this lower amount of sulfuric acid is shown in Table V. Further experiments on the interaction of catalyst level and sulfuric acid promotor level are shown in Table VI.

The relationship of promoter to catalyst level is different at the two catalyst levels we explored. For the lower catalyst level a ten-fold increase

TABLE VI

Effect of Promotor and Catalyst Levels on PANA Yield[a]

		PANA	
5% Pd/C[b] wt. (g)	mmoles H_2SO_4	Yield[c] mole %	%Yield Based on Tetralone Conv.[d]
0.20	0	0	0
0.20	0.6	31	32
0.20	6.0	41	57
0.60	0	7	28
0.60	0.6	88	88
0.60	6.0	81	87

a Each experiment was run for five hours using a 200 ml, 3-necked flask equipped with a thermometer and Dean-Stark apparatus filled with p-cymene and having a condenser. To this was added 7.45 g (80 mmole) of aniline, 4.82 g (39 mmole) of nitrobenzene, 14.6 g (100 mmole) of 1-tetralone, 65 ml of p-cymene, and the indicated amount of acid and catalyst.

b The catalyst was used on a dry weight basis.

c The reaction mixtures were analyzed by quantitative gc using wt % internal standard method.

d Yield of PANA calculated on the basis that 1-tetralone is the limiting reagent. Recovered 1-tetralone was also determined by quantitative gc.

TABLE VII

Effect of Acid Concentration of the Yield of DPA[a]

5% Pd/C[b] wt. (g)	H_2SO_4 mmoles	Time, hr.	Yield DPA[c] mole %
0.20	0	5.2	2
0.20	0.6	5.0	5
0.20	0.6	6.0	6
1.0	0	6.5	7
1.0	3.0	6.0	15

a Each experiment was run using a 200 ml, 3-necked flask equipped with a thermometer and Dean-Stark apparatus filled with p-cymene and having a condenser. To this was added 7.45 g (80 mmole) of aniline, 19.7 g (160 mmole) of nitrobenzene, 19.6 g (200 mmole) of cyclohexanone, 65 ml of p-cymene and the indicated amount of catalyst.

b The catalyst was used on a dry weight basis.

c The reaction mixtures were analyzed by gc using wt % internal standard method.

in promoter level raises the yield ten percent with the yield based on conversion rising to 25%. At the higher catalyst level we find a maximum promoter level as observed by Marschik and Rylander.[14]

We studied the effect of concentration using the simplest system for this reaction, i.e. the generation of DPA from aniline, nitrobenzene and cyclohexanone. The data are shown in Table VII.

The use of 0.20 g of 5% Pt/C catalyst with 3 mmoles of H_2SO_4 under identical conditions gave no detectable DPA ($<$0.5%).

The yields of DPA are much lower than those with PANA. Cyclohexanone, the limiting reagent under these conditions, undergoes a side reaction[15,16,17] as illustrated in Eq. 24. However we conclude that the effect of catalyst level is as important. The formation of DPA requires more forcing conditions than the formation of PANA. We confirmed that this reaction was occurring in our system by identifying, with gc-ms, the reaction products shown in Eq. 25. The effect of this condensation prohibits following of the reaction by water evolution whenever cyclohexanone is a reactant.

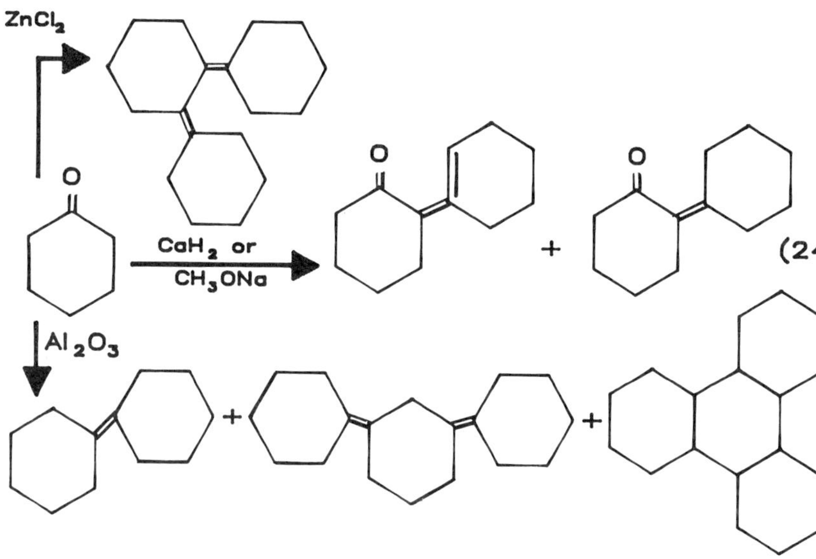

(24)

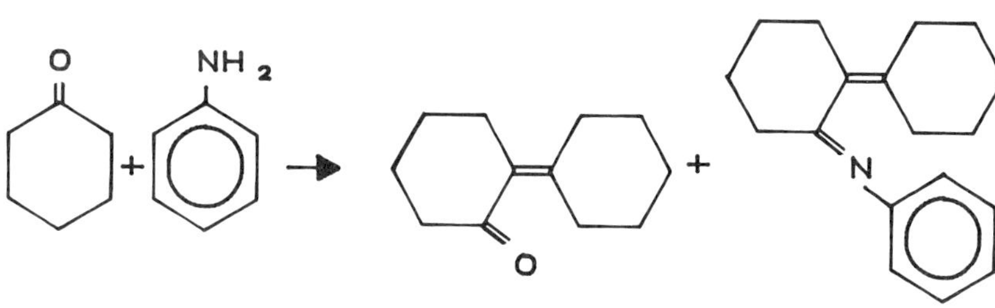

(25)

VIII. SUMMARY

The dehyrogenative condensation reaction to give aryl amines is promoted by traces of acids. It is not clear whether this promotion is due to:

1. the effect of acid on the catalyst.
2. the effect of acid on ketimine formation.
3. a combination of the above.

REFERENCES

1. P.N. Rylander, "Hydrogenation Methods", Academic Press, Orlando, Florida, (1984), p. 16.
2. J.E. Van Verth and Gene R. Wilder, U.S. 3,219,702, (to Monsanto), November 23, 1965.
3. R.E. Malz, Jr. and Harold Greenfield, U.S. 4,4331, 841, (to Uniroyal), February 14, 1984.
4. R.P. Linsted and K.O.A. Michaelis, J. Chem. Soc., 1134 (1940).
5. E.A. Braude, R.P. Linstead, and K.R.H. Wooldrige, J. Chem. Soc., 3586 (1954).
6. E.A. Braude and R.P. Linstead, J. Chem. Soc., 3544 (1954).
7. E.A. Braude, R.P. Linstead and P.W.D. Mitchell, J. Chem. Soc., 3578 (1954).
8. E.A. Braude, R.P. Linstead, P.W.D. Mitchell and K.R.H. Wooldridge., J. Chem. Soc., 3595 (1954)..
9. G.S. Pavlov and N.D. Zelinski, Ber., **66**, 1420, (1933).
10. H. Wieland, Ber., **45**, 484, (1912).
11. G. Brieger and T.J. Nestrick, Chem. Revs., **74**, 567, (1974).
12. S. Siegel and G.V. Smith, J. Am. Chem. Soc., **82**, 6087, (1960).
13. I. Horiuti and M. Polyani, Trans. Faraday Soc., **30**, 1164, (1934).
14. J.F. Marschik and P.N. Rylander, U.S. 3,517,021 (to Engelhard), May 20, 1968.
15. N. Kunze, Ber., **59B**, 2086, (1926).
16. F. Taboury and M. Godchot, Compt. Rend, **169**, 62-64 (1919).
17. A.D. Petrov, Bull. Soc. Chem. Fr., **43**, 1272, (1928).

14

Isomerization and Hydrogenation of Isomeric Pentenes on Copper Catalysis

Arpad Molnar, Geza Sirokman, Mihalay Bartok

Department of Organic Chemistry
Jozsef Attila University
Szeged, Hungary

ABSTRACT

Isomerization and hydrogenation of six isomeric C_5 alkenes, 1-pentene (1P), cis-2-pentene (c2P), trans-2-pentene (t2P), 2-methyl-1-butene (2Me1B), 2-methyl-2-butene (2Me2B) and 3-methyl-1-butene (3Me1B) were studied on copper-on-silica and copper-on-Cab-O-Sil catalysts, prepared by ion-exchange and precipitation, as well as on Cu powder. On the basis of the kinetic data obtained by a conventional closed circulation reactor, an associative mechanism is proposed for the hydrogenation of the isomeric pentenes. Lebedev's rule is found to hold for both isomerization and hydrogenation of the alkenes in pulse experiments on a 3.91% Cu-on-Cab-O-Sil and on a 2.01% Cu-on-silica catalyst prepared by ion-exchange. 2Me1B exhibited a very high reactivity for isomerization on the latter catalyst. A carbenium ion mechanism is proposed to interpret this transformation.

I. INTRODUCTION

Although the hydrogenation activity of copper was reported long ago,[1,2] numerous uncertainties concerning the mechanism have prevailed throughout the years. In many studies because of its much lower activity, copper has been used as a diluent in the investigations of alloy catalysts. But recently, several papers have dealt with the hydrogenation of ethene and propene on copper itself.[4-10]

The long-standing debate about whether hydrogen is chemisorbed on copper surfaces now seems to be resolved,[11,12] and a thorough examination of the hydrogenation of a variety of alkenes, as previously proposed,[13] is of topical interest. The literature data concerning the hydrogenation of alkenes over copper reveal that these reactions are basically similar to those observed on other metal catalysts, but no emphasis is given to the specific nature of copper catalysts. Not only alkenes, but also, perhaps preferably, alkynes can be hydrogenated over copper.[14-16] Since the scope of these reactions is reported to be wide,[17] an extension of the investigations of the mechanism and catalytic properties is indicated. Additionally, copper is a good industrial catalyst for methanol synthesis[18-20] and, accordingly great efforts are being made to develop materials for this and other purposes. And finally, a new method of preparation of copper catalysts by ion-exchange[21,22] has made it possible to prepare active catalysts with carefully controlled dispersion.

Therefore, the purpose of this present study is to clarify certain aspects of the surface reaction, and to compare new and conventional types of catalysts with regard to their surface characteristics, kinetics of hydrogenation, and reactant structure.

II. EXPERIMENTAL

A. Materials

Alkenes. The six compounds studied, 1-pentene (1P), _cis_-2-pentene (c2P), _trans_-2-pentene (t2P), 2-methyl-2-butene (2Me1B), 2-methyl-2-butene (2Me2B) and 3-methyl-2-butene (3Me1B), were all Fluka products (puriss.); they were used without further purification.

Gases. Oxygen-free hydrogen was prepared in a Matheson 8326 generator, operating with a palladium membrane. Helium (99.99%) was further purified by passage through an Oxy-Trap + Indicating Oxy-Trap (Alltech).

Catalysts. Ion-free water and analytical grade reagents were used throughout the preparations. The copper contents of the catalysts were determined by atomic adsorption.

Copper-on-silica catalysts were prepared by ion-exchange (Cu/S-X): silica gel (Strem, large pore, 120-230 mesh) was immersed in a tetrammine copper (II) solution at pH 11.[21,22] After 24 hr the blue product was washed thoroughly with water, dried at 393°K (24 hr) and decomposed at 773°K in air (3 hr).

Copper-on-Cab-O-Sil catalysts were prepared by ion-exchange (Cu/CS-X): the above procedure was applied with a Cab-O-Sil (BDH) support.

Catalysts were also prepared by precipitation. Following a literature method,[23] copper oxide-hydroxide was precipitated on silica (Cu/S-P) or Cab-O-Sil (Cu/CS-P).

Cu powder was prepared by the reduction of CuO (Merck) in hydrogen (573°K, 2 hr).

Reduction of Catalysts. All the supported catalysts were reduced by a standard reduction procedure. During this procedure, the precursors were heated to 673°K in flowing hydrogen (20mL min^{-1}) at a heating rate of 8°K min^{-1}, kept at this temperature for 2 hr, and then cooled in hydrogen. In some cases, reductions other than this standard one were employed (see Table 1). The reduced catalysts were activated at 673°K for 0.5 hr before use. This activation was performed either in flowing hydrogen (20 mL min^{-1}) in the pulse microreactor or under 13.3 kPa hydrogen in the reactor of the closed circulation apparatus.

Characterization of the Catalysts. The catalysts were characterized through reaction with nitrous oxide at 363°K, using a single-pulse method.[24] An appropriate amount of catalyst (50-150 mg) was activated in flowing

TABLE 1

Characterization Of The Copper Catalysts

catalyst	copper loading (%)	reduction temperature[a] (K)	S[b] (m^2/g_{Cu})	mean particle size (nm)
Cu/S-X	2.01	523	20.9	32.2
		573	44.8	15.0
		623	67.2	10.0
		st[c]	93.0	7.2
	3.67	523	44.7	15.0
		st	78.9	8.5
	6.85	523	61.8	10.9
		st	83.8	8.0
Cu/CS-X	0.77	st	76.6	8.8
	3.91	st	41.2	16.3
Cu/S-P	7.05	st	18.4	36.5
Cu/CS-P	7.23	st	23.0	29.2
Cu powder		573	2.75	244.5

[a]2 hr in flowing hydrogen

[b]specific surface area, calculated with the surface density of 1.46×10^{19} Cu atom/m^2

[c]standard reduction, see Experimental

hydrogen (25 mL min^{-1}), then cooled to 363°K, and kept in flowing helium for 0.5 hr; next 400 μl nitrous oxide was passed through the sample and the quantity of nitrogen evolved was determined after separation by GC (column: 1.2m Chromosorb 101, room temperature), using a calibration curve. Table 1 shows the specific surface areas of the copper in the catalysts produced by the standard reduction and by reduction at lower temperatures.

B. Apparatus

A pulse microreactor equipped with a Carlo Erba Fractovap Mod GV chromatograph (hot-wire detector) was used in the pulse experiments (catalyst sample = 10 mg, flow rate of hydrogen carrier gas = 20 mL min^{-1}, inlet pressure of hydrogen = 90 kPa). GC analysis: 10% ethyl N,N-dimethyloxamate (Supelco) on Chromosorb P (4 m), room temperature.

Kinetic data were determined in a conventional closed circulation apparatus with a volume of 250 mL (catalyst quantity = 10 mg of 2.01% Cu/S-X, hydrogen partial pressure = 1.33-13.3 kPa, alkene partial pressure = 1.33-13.3 kPa). GC analysis: Carlo Erba Fractovap 2150 chromatograph (flame ionization detector), 10% tri-o-tolyl phosphate on Chromosorb P (1.6 m) + 10% bis (2-methoxyethyl) adipate on Chromosorb P (4 m), room temperature.

III. RESULTS

The kinetic data determined by the closed circulation method for 1P, c2P and t2P are to be found in Tables 2 and 3.

Experimental data on the transformations of 1P and t2P in the pulse experiments, the calculated reactivities (R) and the turnover numbers (TN) are listed in Table 4, and the corresponding results for 2Me1B, 2Me2B and 3Me1B in Table 5.

The reactivities in pulse experiments of all six model compounds in hydrogen addition and isomerization on 2.01% Cu/S-X and 3.91% Cu/CS-X catalysts are plotted as functions of temperature in Figs. 1-4.

TABLE 2

Initial Rates As A Function Of Partial Pressure Of Hydrogen (P_{alkene} = 1.3 kPa, T = 348°K)

P_H x133 Pa	Products	Reactants initial rates ($mol\ s^{-1}\ g_{Cu}^{-1}$) 1P	t2P	c2P
10	nP	10.2		
	1P			
	t2P	85		
	c2P	35		
20	nP	15.7	2.9	0.83
	1P		11.3	5.0
	t2P	177		48.2
	c2P	51		
30	nP	29		
	1P			
	t2P	284		
	c2P	88		
50	nP	42	8.2	1.7
	1P		14.9	5.4
	t2P	513		57
	c2P	146	56	
75	nP	70	8.2	4.9
	1P		14.9	10.5
	t2P	622		147
	c2P	161	93	
100	nP	131	8.6	5.0
	1P		16.0	8.3
	t2P	676		141
	c2P	178	97	

TABLE 3

Initial Rates Of Formation Of Products As A Function Of Partial Pressure Of 1P (p_H = 6.7 kPa, T = 348°K)

P_{1P} x133 Pa	Initial rates ($mol\ s^{-1}\ g_{Cu}^{-1}$) nP	t2P	c2P
10	42	513	146
30	19.5	150	40.6
50	13.9	101	27.8
100	2.35	12.1	4.2

TABLE 4

Transformations of Isomeric Pentenes In Pulse Experiments (Pulse Size: 3 μl for 1P, 1 μl for t2P T = 423°K; p_H = 90 kPa)

catalyst	red. temp.[a] (K)	nP	product composition 1P	t2P	c2P	(mol %) add	iso	R_{add}[b]	R_{iso}[b]	TN_{add}[c]	TN_{iso}[c]
						1-pentene					
2.01% Cu/S-X	523	2	92.5	4.5	1	2	5.5	2.87	7.9	5.38	14.8
	573	4	85	8.5	2.5	4	11	5.74	15.9	5.02	13.8
	623	6	73	17	4	6	18	8.62	28.7	5.02	15.1
	st[d]	9	68	18	5	9	23	12.3	31.4	5.44	13.9
3.67% Cu/S-X	523	8	69	18	5	8	23	6.34	18.3	5.53	15.9
	st	19	46	27	8	19	35	14.9	26.2	7.82	13.7
6.85% Cu/S-X	523	15	48	30	7	15	37	6.48	16.0	4.01	9.89
	st	32	25	34	9	32	43	12.8	17.2	6.79	9.12
0.77% Cu/CS-X	st	4	92	3	1	4	4	14.2		7.66	
3.91% Cu/CS-X	st	12	61	22	7	12	29	8.42		8.69	
7.05% Cu/S-P	st	10	60	23	7	10	30	3.89		8.69	
7.23% Cu/CS-P	st	16	65	16	3	16	19	6.07		10.91	
Cu powder	573	31	51	13	5	31	18	10.6		15.99	
						<u>trans</u>-2-pentene					
2.01% Cu/S-X	st	4.5	3	81.5	11	4.5	14			0.92	
3.67% Cu/S-X	st	9.5	4	74.5	12	9.5	16			1.25	
6.85% Cu/S-X	st	16	5	66	13	16	18			1.14	
3.91% Cu/CS-X	st	5	3	82	10	5	13				

[a] 2 hr in flowing hydrogen

[b] mol/g_{Cu} per pulse x 10^3

[c] molecule/surface Cu atom per pulse

[d] standard reduction, see Experimental

TABLE 5

Transformations Of Isomeric Methylbutenes In Pulse Experiments (Pulse Size = 1 µl; T = 423 K; p_H = 90 kPa)

catalyst	red. temp.[a] (K)	product distribution (mol %) iP	3Me1B	2Me1B	2Me2B	add	iso	R_{add}[b]	TN^c_{add}
				2-methyl-1-butene					
2.01% Cu/S-X	523	0.8	0.2	47	52	0.8	52.2	3.89	0.31
	st[d]	3	0.3	17	79.7	3	80	13.9	0.61
3.67% Cu/S-X	523	3.1	0.25	23.65	73	3.1	73.25	8.34	0.73
	st	6	0.35	15.75	77.9	6	78.25	15.2	0.80
6.85% Cu/S-X	523	8	0.3	12.1	70.7	8	71	11.7	0.72
	st	9.2	0.4	14.6	76.3	9.2	76.7	12.5	0.66
0.77% Cu/CS-X	st	1.2	0.02	97.58	1.2	1.2	1.22		
3.91% Cu/CS-X	st	4.5	0.05	88.45	7	4.5	7.05		
7.05% Cu/S-P	st	6.25	0.05	91.7	2	6.25	2.05		
7.23% Cu/CS-P	st	3.3	0	96.2	0.5	3.3	0.5		
Cu powder	573	49.4	0	38.1	12.5	49.4	12.5		
				2-methyl-2-butene					
2.01% Cu/S-X	st	1.9	0.3	13	85	1.9	13.3	8.94	0.40
3.67% Cu/S-X	st	3.8	0.2	13.7	82.3	3.8	13.9	9.79	0.51
6.85% Cu/S-X	st	6.1	0.35	13.3	80.25	6.1	13.65	8.42	0.45
0.77% Cu/CS-X	st	0.18	0.42	0	99.4	0.18	0.42		
3.91% Cu/CS-X	st	0.65	0.35	1	98	0.65	1.35		
7.05% Cu/S-P	st	0.6	0.3	0.65	98.45	0.6	0.95		
7.23% Cu/CS-P	st	0.55	0.4	0.3	98.75	0.55	0.7		
Cu powder	573	8.3	0.2	1.9	89.6	8.3	2.1		
				3-methyl-1-butene					
2.01% Cu/S-X	st	20	44	6	30	20	36		
3.67% Cu/S-X	st	28	28	8	36	28	44		
6.85% Cu/S-X	st	37	14	9	40	37	49		
0.77% Cu/CS-X	st	13	72	1	14	13	15		
3.91% Cu/CS-X	st	28	40	2	29	28	31		

[a] 2 hr in flowing hydrogen

[b] mol/g_{Cu} per pulse x 10^4

[c] molecule/surface Cu atom per pulse

[d] standard reduction, see Experimental

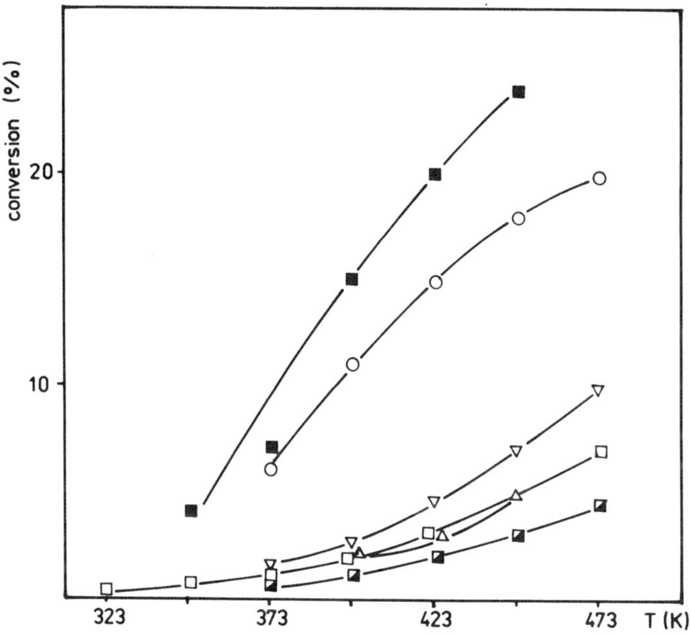

Fig. 1. Dependence of hydrogen addition to C_5 alkenes on temperature (2.01% Cu/S-X, standard reduction, 1 μl pulses. ○, 1P; Δ, c2P, ∇, t2P; ■, 3Me1B; □, 2Me1B; ◪, 2Me2B).

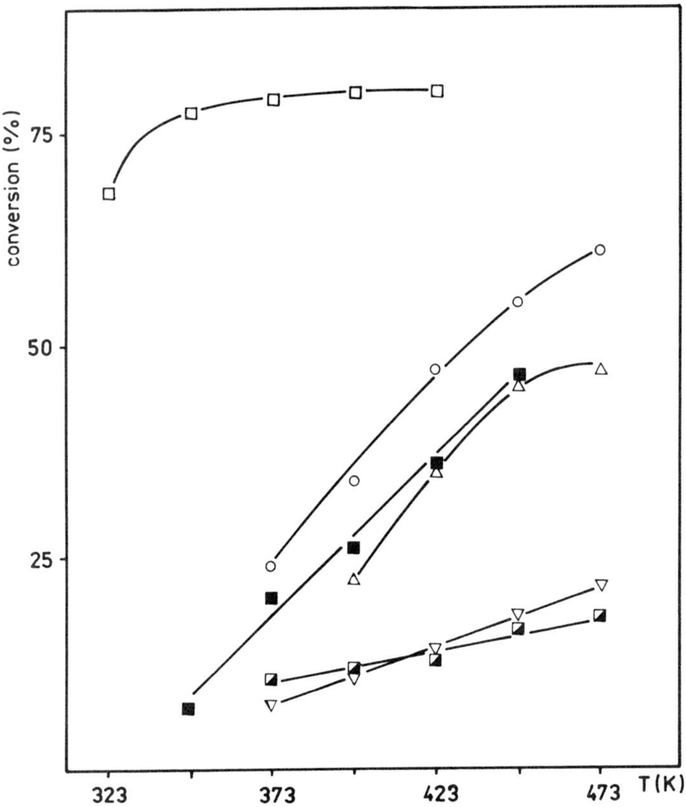

Fig. 2. Dependence of isomerization of C_5 alkenes on temperature (2.01% Cu/S-X, standard reduction, 1 μl pulses. For symbols, see Fig. 1).

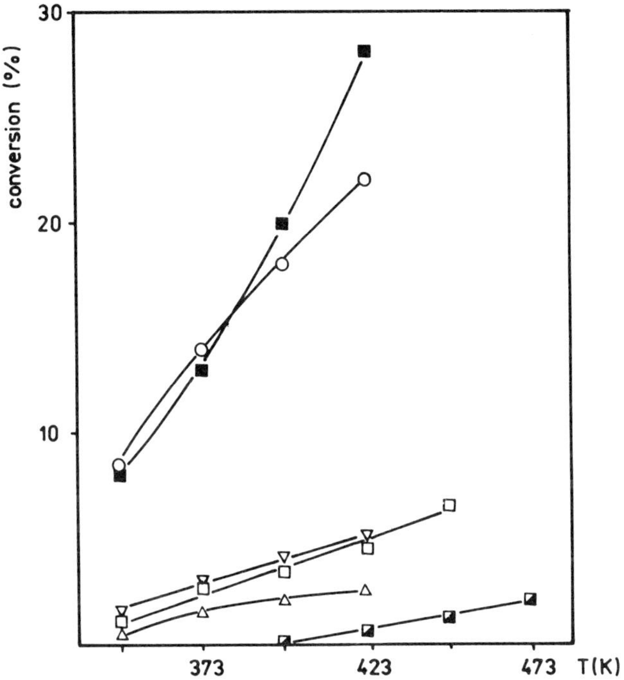

Fig. 3. Dependence of hydrogen addition to C_5 alkenes on temperature (3.91% Cu/CS-X, standard reduction, 1 μl pulses. For symbols, see Fig. 1).

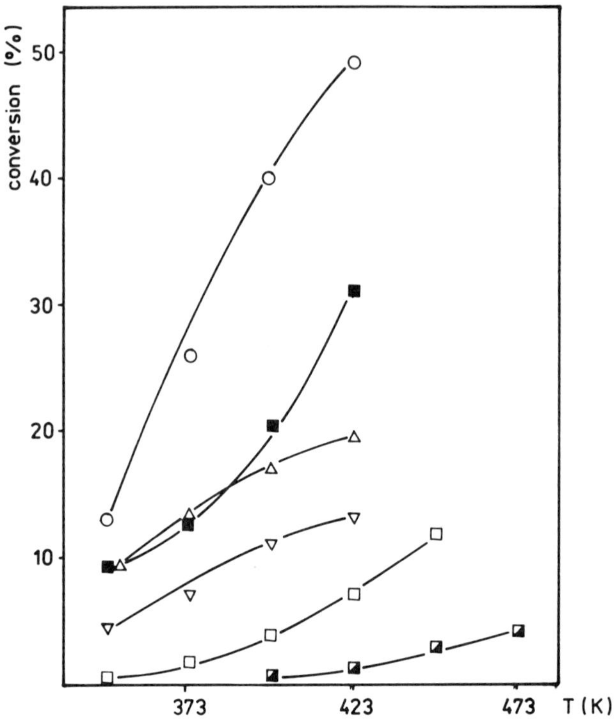

Fig. 4. Dependence of isomerization of C_5 alkenes on temperature (3.91% Cu/CS-X, standard reduction, 1 μl pulses. For symbols, see Fig. 1).

IV. DISCUSSION

A. Kinetics of Transformations of Pentenes

The initial rates of formation of products in the reactions of pentenes were determined at different hydrogen and reactant partial pressures.

Scheme 1 illustrates the complexity of the transformations, as compared to the hydrogenations of ethene and propene; this complexity arises from the rapid isomerization of the pentenes. A plot of the initial rate against the 1P pressure (Fig. 5) reveals that at the pressure range of 4-14 kPa, all three products formed have negative orders, varying between -1.8 and -2.1, indicating a strong inhibitory effect of the alkenes. The

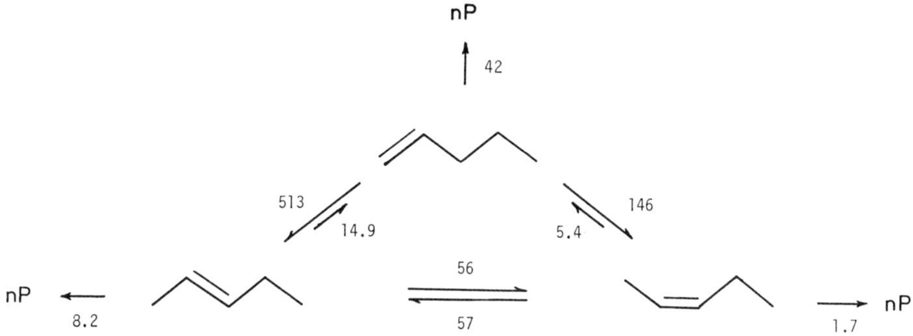

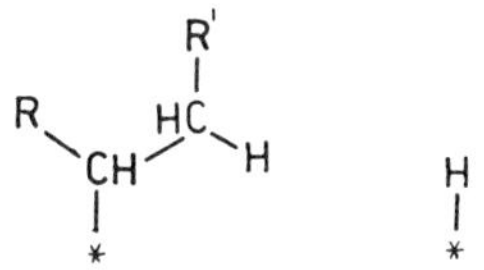

Scheme 1 Initial reaction rates of transformations of pentenes (Rates expressed as 10^3 mol s^{-1} g^{-1}_{Cu}, partial pressures; 1.3 kPa alkene, 6.7 kPa hydrogen; T = 348°K

theoretical value for very strongly adsorbing inhibitors is -2. All the products formed in the reaction of 1P (Fig. 6) as well as the formation of pentane in the transformations of 2-pentenes (Fig. 7), gave first-order kinetics (experimental values between 0.9 and 1.2) with respect to hydrogen, similar to other observations.[2,4,9]

At high hydrogen partial pressures, the rates of isomerization of 1P and 2-pentenes are practically independent of the hydrogen pressure (Figs. 6 and 8) (the experimentally determined kinetic order is less than 0.08).

The decreasing sequence of reactivity, 1P, t2P, c2P, is in accordance with the experimentally observed increasing sequence of apparent activation energy: 32.5 (1P), 43 (t2P) and 54 kJ/mol (c2P).

The observed kinetic results can be explained by assuming the extended associative mechanism proposed for the hydrogenation of propene over copper.[4] The slowest step is the reaction of the half-hydrogenated

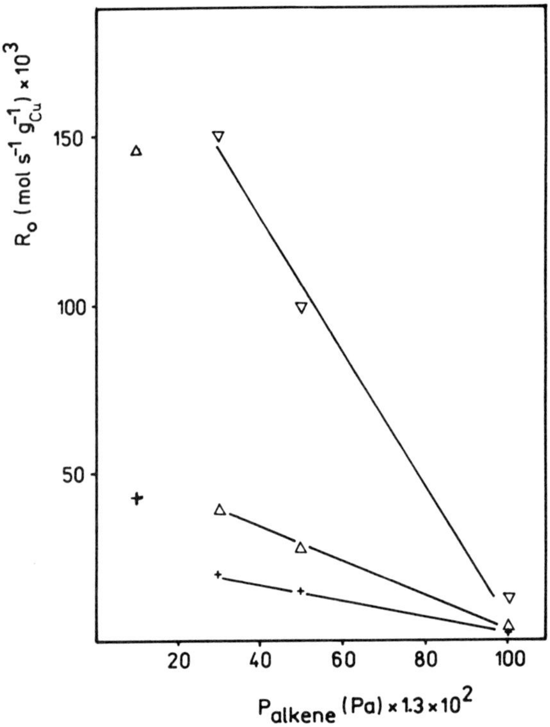

Fig. 5. Initial reaction rates of formation of products as a function of partial pressure of 1P (p_H = 6.7 kPa, T = 348°K. +, nP; ∇, t2P; Δ, c2P).

species and adsorbed hydrogen.[4,6] Interconversion of the half-hydrogenated states via diadsorbed alkenes may be very fast, resulting in rapid isomerization. Actually, the determined rates for cis-trans isomerization are unreasonably low and almost equal. This, together with the fact that the initial selectivities for the formation of c2P and t2P in the reaction of 1P are practically independent of the hydrogen pressure, indicates that a rapid interconversion of these products proceeds before desorption. The zero order dependence of alkene isomerization on hydrogen is a consequence of the transformations of the half-hydrogenated states into each other requiring only a small amount of adsorbed hydrogen. For alkane formation,

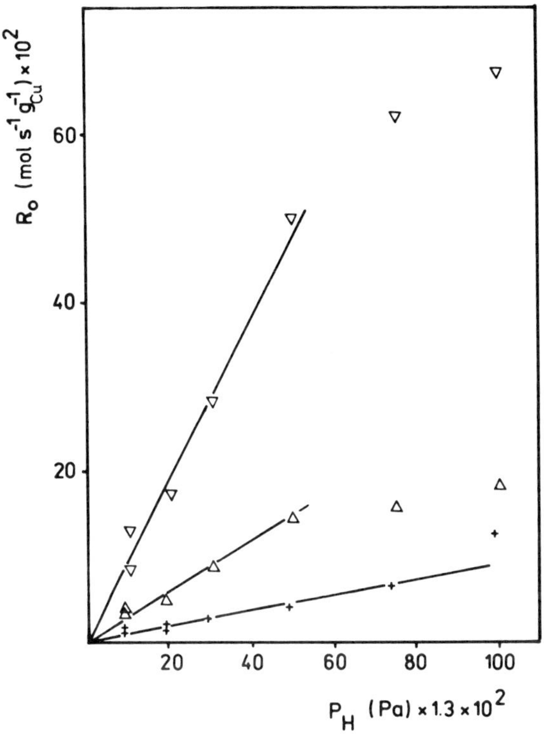

Fig. 6. Initial reaction rates of formation of products as a function of partial pressure of hydrogen (Reactant: 1.3 kPa 1P, T = 348°K. +, nP; ∇, t2P; Δ, c2P).

a high hydrogen coverage is necessary, but hydrogen adsorption is inhibited by the more strongly adsorbing alkene, as shown by the negative order with respect to 1P.

B. Reactivity of Alkenes in Isomerization and Hydrogenation

A comparison of the reactivities of the six isomeric alkenes in hydrogen addition on two different catalysts (2.01% Cu/S-X and 3.91% Cu/Cs-X) reveals that Lebedev's rule holds in the copper-catalyzed hydrogenation of alkenes (Figs. 1 and 3). The rule states[25] that in catalytic hydrogen

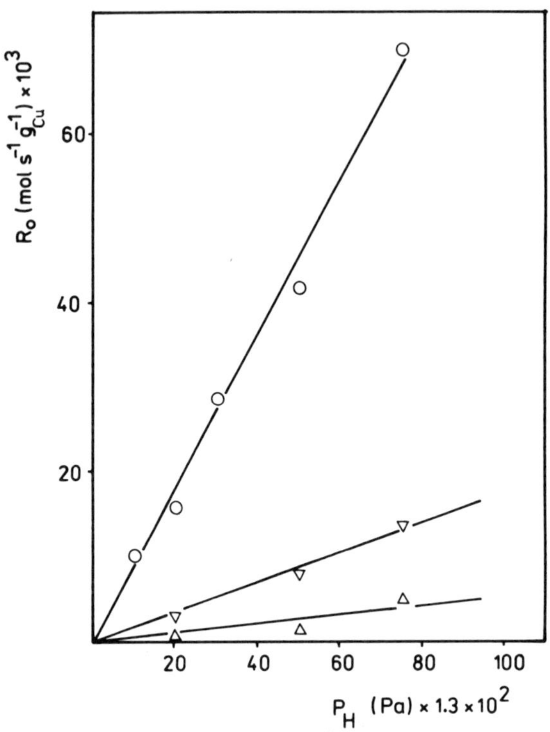

Fig. 7. Initial reaction rates of formation of pentane as a function of partial pressure of hydrogen (p_{alkene} = 1.3 kPa, T = 348°K, O, 1P; ▽, t2P; △, c2P).

addition, the more substituted the olefinic bond, the slower the process of addition. As the extremes, the monosubstituted compounds 1P and 3Me1B show the highest, and the trisubstituted 2Me2B shows the lowest reactivity.

The isomerization taking place in parallel with addition (Figs. 2 and 4) does not seem to affect the validity of Lebedev's rule. The sequence of reactivities of the alkenes for isomerization is the same as that of their reactivities for hydrogen addition, with one exception. The transformation of 2Me1B, leads to a surprisingly large amount of isomeric alkene (2Me2B) on the silica-supported catalyst. A saturation-type curve is obtained, which indicates the establishment of thermodynamic equilibrium. 2Me1B has such

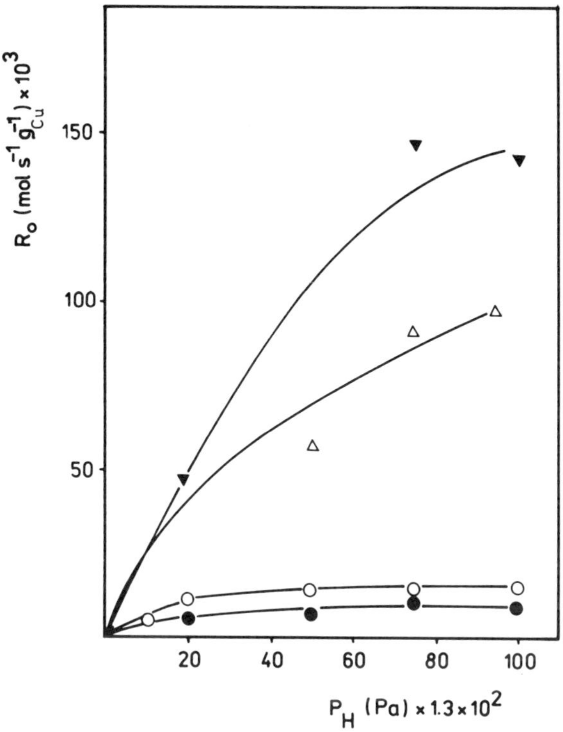

Fig. 8. Initial rates of formation of products as a function of partial pressure of hydrogen (Reactant partial pressure: 1.3 kPa c2P or t2P. Products: ●, 1P; ∇, t2P from c2P; ○, 1P; Δ, c2P from t2P).

a high tendency for isomerization on this catalyst that it gives a substantial amount of isomeric products even under very unfavorable experimental conditions (low temperature, high pulse size, i.e. low contact time) (Fig. 9). It should be noted that under similar conditions, 1P, which otherwise exhibits high reactivity in this transformation (see Figs. 2 and 4), undergoes only negligible isomerization. Detailed studies in connection with this observation, and also on the relation of addition and isomerization, revealed that changes of the different experimental variables give rise to similar changes in the two processes. Increases of the reaction temperature, the copper loading, or the activation temperature of the catalysts result in

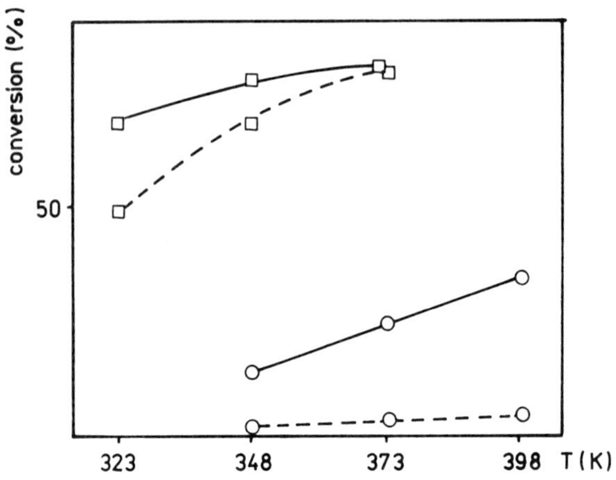

Fig. 9. Dependence of isomerization of 1P (○) and 2Me1B (□) on reaction temperature and pulse size (2.01% Cu/S-X, standard reduction. Solid lines, 1 μl pulses; broken lines, 5 μl pulses.

increases in the extent of both addition and isomerization for every compound studied.

The effects of the pulse size on 1P and 2Me1B reactions differ (Fig. 10). The change in hydrogen addition (not shown in the Figure) is similar for the two compounds, and there are similar decreases in the isomerization of 1P on both catalysts and in that of 2Me1B on Cu/CS-X. However, the decrease in the isomerization with increasing pulse size is much smaller in the case of 2Me1B on Cu/S-X (Fig. 10, upper line). This behavior means that the increasing hydrogen deficiency resulting from the increasing pulse size has a much smaller effect on the isomerization of 2Me1B than on isomerization of the other compounds. Additional careful and detailed investigations led to the following findings:

(a) The supports (silica and Cab-O-Sil) pretreated in the same way as the loaded catalysts displayed negligible isomerizing activities in hydrogen.

(b) The reduced catalysts, except Cu/S-X, were inactive in He.

(c) The unreduced Cu/S-X precursor and the reduced Cu/S-X catalysts exhibited no isomerizing activity in He in the transformation of five of the six compounds.

(d) 2Me1B, however, did undergo about 60-70% isomerization on both the unreduced and the reduced Cu/S-X catalysts in He.

The above observations support the earlier findings (see kinetic studies) that, in general, both hydrogen and metallic copper are needed for

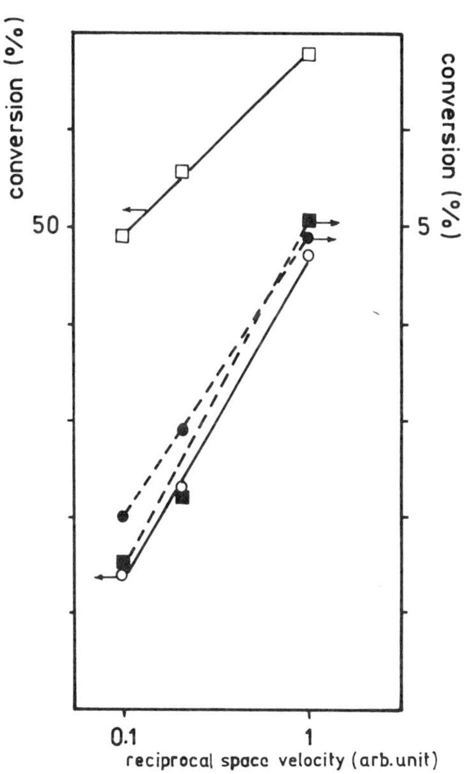

Fig. 10. Dependence of isomerization of 1P and 2Me1B on reciprocal space velocity (Pulse technique; circles, 1P; squares, 2Me1B; solid lines with open symbols, 2.01% Cu/S-X; broken lines with shaded symbols, 3.91% Cu/CS-X).

the two transformations, with the exception that isomerization of 2Me1B can also take place in the absence of Cu^{0} and hydrogen. Since the acidic centers of the supports do not take part in this process, it must be the copper ions which catalyze the isomerization of 2Me1B.

It is known that a proportion of the copper ions in Cu-on-silica catalysts prepared by ion-exchange are very difficult to reduce. Different spectroscopic techniques (XPS, AES) and TPR measurements have shown[26,27] that isolated Cu^{2+} ions need a very high temperature for reduction, and in some cases (at low copper loadings) only partial reduction ($Cu^{2+} \rightarrow Cu^{+}$) can be achieved. The unique behavior of 2Me1B and the possible catalytic effect of copper ions can be rationalized by presuming an ionic mechanism for isomerization, which takes place in parallel with the metal-catalyzed mechanism. Of the six C_5 isomers, 2Me1B is the only one which can transform to the most stable tertiary carbenium ion in a single step. This may explain why it exhibits this type of isomerization over unreduced Cu/S-X. That is, because proton transfers within the tertiary carbenium ion can lead to the formation of 2Me2B, the thermodynamically most stable alkene.

C. The Question of Structure-sensitivity

The results of the study on the relation between the rate of hydrogen addition to 1P and the surface area of the metal are presented in Table 1 and Fig. 11. The data fit two different lines. One line is for the silica-supported ion-exchanged catalysts with different copper loadings and different reduction conditions, while another line is for all the other catalysts, except Cu powder. Similar linear correlations exist for the isomerization of 1P (Fig. 12) and the hydrogenation of 2Me1B. However, no correlation was found for the isomerization of 2M2B. On the basis of the unique isomerization behavior of this compound over unreduced copper, it is understandable why there is no such correlation for the isomerization of 2Me1B.

The graphical plots in Figs. 11 and 12 do not reveal whether hydrogen addition to the alkenes studied is a structure-sensitive reaction. Therefore, we examined the plot of turnover number (TN) versus metal surface area or

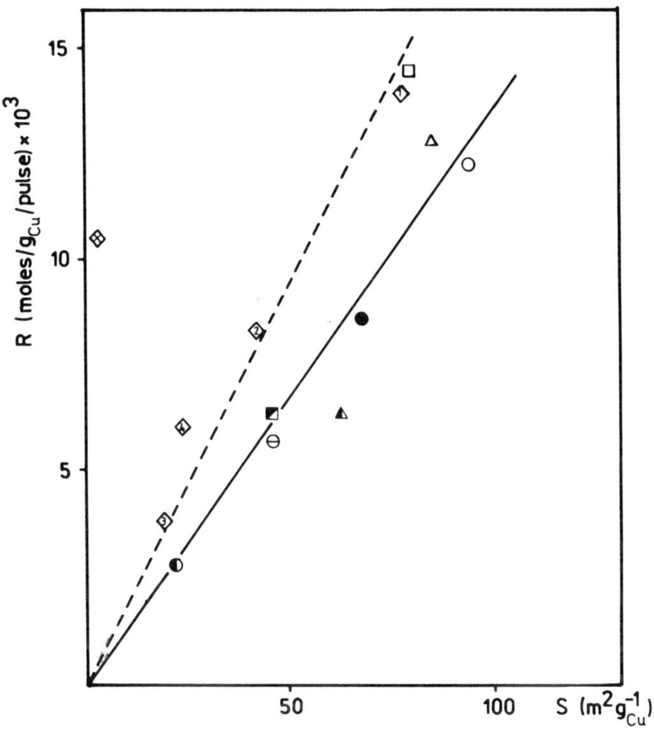

Fig. 11. Catalytic reactivity (R) of 1P in hydrogen addition versus copper surface area (S) (1 μl pulses, T = 423°K. Circles, 2.01% Cu/S-X; squares, 3.67% Cu/CS-X; triangles, 6.85% Cu/S-X; ◈ , 0.77% Cu/CS-X; ◈ , 3.91% Cu/CS-X; ◈ , 7.05% Cu/S-P; ◈ , 7.23% Cu/CS-P; ◈, Cu powder; Open symbols, standard reduction; half-shaded symbols, 523 K/2 hr; ⊕ , ◈; 573 K/2 hr; ● , 623 K/2 hr).

mean particle size. From this we found that the data on the silica-supported catalysts (Fig. 13) demonstrate that both addition and isomerization are structure-insensitive on a single catalyst (2.01% Cu/S-X) reduced under different conditions (lines I and II respectively). However, on catalysts with different copper loadings, activated under identical conditions, addition and isomerization exhibit structure-sensitivity, but in the wrong direction (III). That is, their turnover numbers should increase

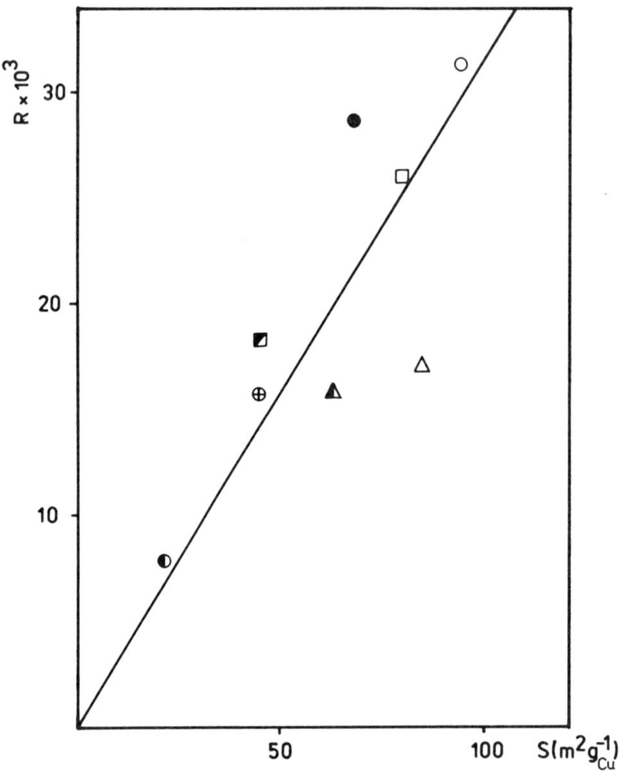

Fig. 12. Catalytic reactivity of 1P in isomerization on Cu/S-X catalysts versus copper surface area (1 μl pulses, T = 423°K. For symbols, see Fig. 11).

with increasing surface area for demanding reactions.[28] The plots of turnover numbers of addition for other compounds show similar changes (Figs. 14). Plots of turnover numbers as a function of mean particle size (Fig. 15) give lines corresponding to increasing turnover numbers with increasing particle size, a result which is also opposite to expectations.

Teichner and coworkers[10] studied the hydrogenation of cyclopentadiene on copper catalysts deposited on silica and alumina, and observed similar inverse correlations. They explained this behavior by assuming that the actual active surface area of the catalysts (S_o) is only a

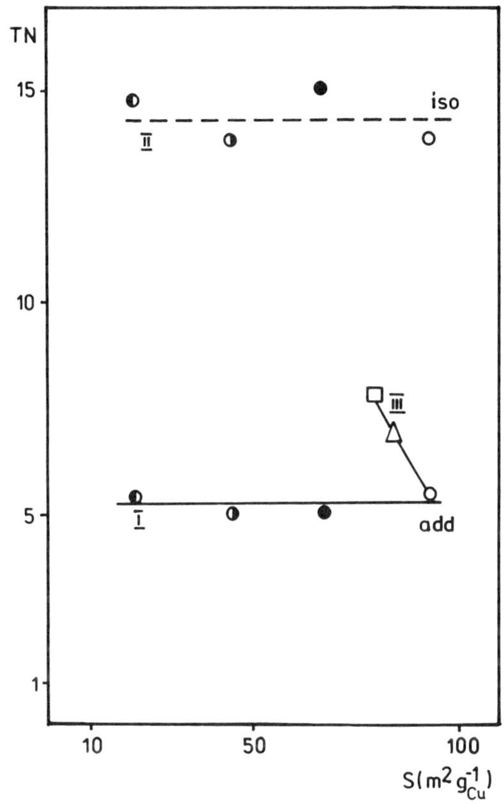

Fig. 13. Turnover numbers of addition and isomerization of 1P on Cu/S-X catalysts against copper surface area (1 μl pulses, T = 423 K. Circles, 2.01%; □, 3.67%; Δ, 6.85% copper loading; Open symbols; standard reduction; ◐, 523 K/ 2 hr; ⊕, 573 K/2 hr, ●, 623°K).

fraction of the total surface area (S) measured by the decomposition of N_2O. The ratio of the active and total surface areas changes with the support and with the experimental variables of the catalyst preparation. As a result, the turnover numbers calculated from these values must be apparent and not true values.

If the turnover numbers of hydrogen addition to 1P determined on all the catalysts made by standard reduction are plotted against S, the data

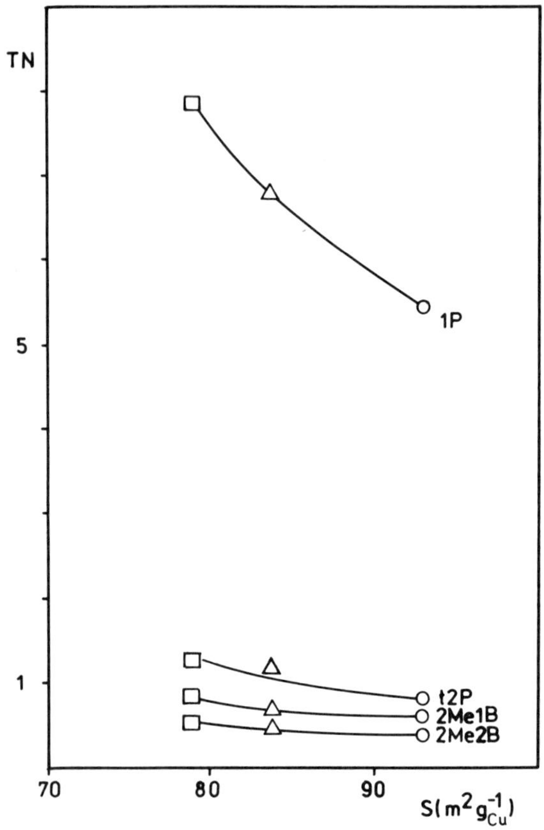

Fig. 14. Turnover numbers of hydrogenation of alkenes on Cu/S-X catalysts against copper surface area (1 µl pulses, standard reduction, T = 423°K; ○ , 2.01%; □ , 3.67%; Δ, 6.85% copper loading).

give two different curves (Fig. 16). Curve A fits the family of ion-exchanged, silica-supported catalysts, while curve B fits all other data, even those for the Cu/CS-X catalysts. It follows from Teichner's observations that each family of catalysts has something common in nature, thereby resulting in a similar S_0/S ratio. It has been shown[26,27] that copper atoms exist on silica in the form of highly-dispersed clusters. The common feature in the second group of catalysts (curve B) might be the presence of bulk

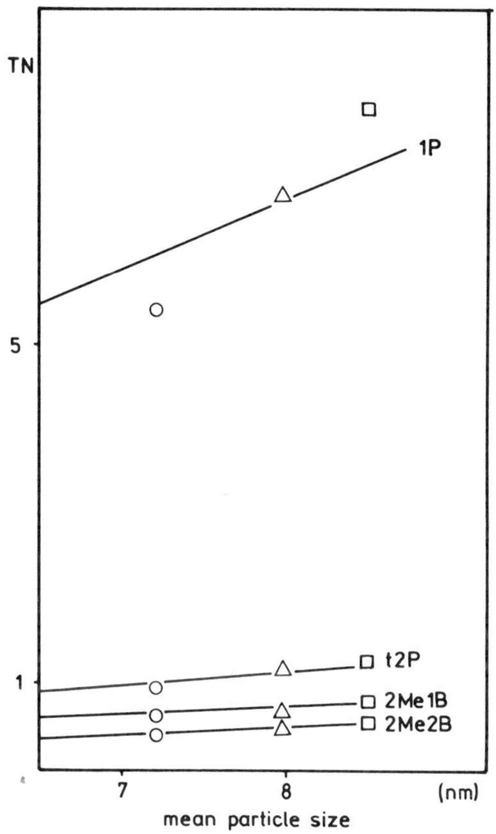

Fig. 15. Turnover numbers of hydrogenation of alkenes on Cu/S-X catalysts versus copper surface area (1 μl pulses, standard reduction, T = 423 K. For symbols, see Fig. 14).

copper as the active phase. If this is true, the "ion-exchange" on the Cab-O-Sil support is not a real ion-exchange process. Although Cab-O-Sil is able to retain a substantial amount of Cu^{2+} ions, because of the very low number of surface OH groups, the process is a strong surface adsorption rather than ion-exchange. As a result, the decomposition of the copper (II) complex leads to the formation of bulk copper oxide, which, in turn, gives bulk copper on reduction. This conclusion is supported by the above observation

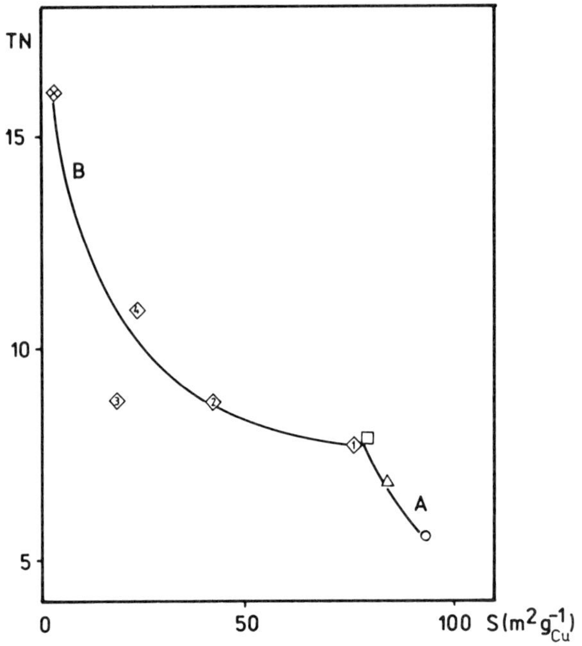

Fig. 16. Turnover numbers of hydrogen addition to 1P versus copper surface area (1 μl pulses, T = 423 K. Standard reduction, except for Cu powder (◈), 573 K/2 hr: ○ , 2.01% Cu/S-X; □ , 3.67% Cu/S-X; Δ; 6.85% Cu/S-X; ① , 0.77% Cu/CS-X; ② , 3.91% Cu/CS-X; ③ , 7.05% Cu/S-P; ④ , 7.23% Cu/CS-P).

that 2Me1B does not exhibit copper ion-induced isomerization on the two Cu/CS-X catalysts but does on the silica-supported catalysts. This indicates the ready reducibility of Cu^{2+} ions, because no strong interaction occurs between the support and the copper ions.

V. Conclusions

Results on the isomerization and hydrogenation of six isomeric C_5 alkenes revealed that copper is an active metal in both transformations. The most important difference between copper and other metal hydrogenation catalysts is that the reaction of the half-hydrogenated alkene and adsorbed hydrogen is much slower on copper, very probably due to the weak adsorption of hydrogen on copper surfaces. As a consequence, isomerization may become the dominant reaction, depending on the reaction conditions and alkene structure. The transformation of 2Me1B shows that copper ions can also be active in isomerization of alkenes which are able to form the most stable tertiary carbenium ion.

REFERENCES

1. G.C. Bond, Catalysis by Metals, Academic Press, London, p. 244 (1962).
2. R.N. Pease and C.A. Harris, J. Am. Chem. Soc., **49**, 2503 (1927).
3. J.H. Sinfelt and J.A. Cusumano, in Advanced Materials in Catalysis (J.J. Burton, R.L. Garten, eds.), Academic Press, New York, p. 1 (1977).
4. K. Hirota, N. Yoshida, S. Teratani and T. Kitayama, J. Catal., **13**, 306 (1969).
5. K. Hirota, Y. Yasuda, S. Teratai and N. Yoshida, Z. Phys. Chem. N.F., **71**, 195 (1970).
6. Sh. Sato and K. Miyakara, J. Res. Inst. Catal., Hokkaido Univ., **22**, 51 (1974).
7. M.C. Schachter, P.S. Gajardo and S.C. Droquett, J. Phys. Chem., **79**, 1698 (1975).
8. P. Gajdaro, M.C. Lartiga and S.C. Droquett, J. Phys. Chem., **82**, 2323 (1978).
9. M.B. Taghavi, G. Pajonk and S.J. Teichner, Bull. Soc. Chim. Fr., 180 (1978).
10. M.B. Taghavi, G.M. Pajonk and S.J. Teichner, J. Coll. Interface Sci., **71**, 451 (1979).
11. J. Pritchard, T. Catteric and R.K. Gupta, Surf. Sci., **53**, 1 (1975).
12. P. Mandhavan and J.L. Whitten, J. Chem. Phys., **77**, 2673 (1982).
13. J.J.F. Scholten, A.P. Pijpers and A.M.L. Hustings, Catal. Rev.-Sci. Eng., **27**, 151 (1985).

14. K.K. Showa Denko, Jpn. KokaiTokkyo Koho JP, **57**, 185, 228; Chem. Abstr., **98**, 142935f (1983).

15. M.C. Couvillion, U.S. Pat. 4,440,956; Chem. Abstr., **100**, 216459n (1984).

16. M.C. Couvillion, U.S. Pat. 4,493,906; Chem. Abstr., **102**, 101286c (1985).

17. G. Sirokman, A. Molnar and M. Bartok, Third Belgian Hungarian Colloquium on Catalysis, Csopak, Hungary. Abstracts, p. 8 (1985).

18. H.H. Kung, Catal. Rev.-Sci. Eng., **22**, 235 (1980).

19. K. Klier, Adv. Catal., **31**, 243 (1982).

20. (a) G.C. Chinchen and K.C. Waugh, J. Catal., **97**, 280 (1986). (b) T.H. Fleisch and R.L. Mieville, J. Catal., **97**, 284 (1986), and references therein.

21. H. Kobayashi, N. Takezawa, C. Minochi and K. Takashi, Chem. Lett., 1197 (1980).

22. M. Shimokawabe, N. Takezawa and H. Kobayashi, Appl. Catal., **2**, 379 (1982).

23. M. Hajek and K. Kochloefl, Coll. Czech. Chem. Commun., **34**, 2739 (1969).

24. J.W. Evans, M.S. Wainwright, A.J. Bridgewater and D.J. Young, Appl. Catal., **7**, 75 (1983).

25. S.V. Lebedev, G.G. Kobliansky and A.O. Yakubchik, J. Chem. Soc., 417 (1925).

26. N. Takezwa, H. Kobayashi, Y. Kamegai and M. Shimokawabe, Appl. Catal., **3**, 381 (1982).

27. M. Shimokawabe, N. Takezawa and H. Kobayashi, Bull. Chem. Soc. Jpn., **56**, 1337.

28. M. Boudart, Adv. Catal. **20**, 153 (1969).

Selective Reactions

15

The Use of Hydrogenolysis in the Synthesis of Surfactants

Roger A. Grey and Lawrence J. Karas

ARCO Chemical Company
Division of Atlantic Richfield Company
Newtown Square, Pennsylvania

ABSTRACT

The hydrogenolysis of the carbon-oxygen bond with molecular hydrogen using Group VIII metal catalysts is a key step in the organic synthesis of a series of model surfactants which have utility in enhanced oil recovery applications. The hydrogenolysis of tertiary benzyl alcohols, formed by the reaction of bromomagnesium 4-ethylbenzene with ketones, is accomplished in high yields and selectivities using Pd/C as the catalyst in glacial acetic acid (50°C, 200 psi of H_2). Under these conditions hydrogenation of the aromatic ring is a minor byproduct which competes with hydrogenolysis. Selectivity to ring hydrogenated products, however, increases with higher reaction temperatures and less acidic solvents.

The hydrogenolysis of the tertiary benzyl alcohol formed by the reaction of bromomagnesium 4-methoxybenzene and 6-dodecanone, in contrast to the above ethylbenzene analogues, is not selective using Pd/C under the same conditions. Instead of a complex mixture of predominantly ring hydrogenated and ether cleavage products are obtained. The hydrogenolysis of this benzyl alcohol is selective using the cationic metal complex COD Ir(pyridine)$P(C_6H_{11})_3^+PF_6^-$ as the catalyst in heptane solvent.

The hydrogenolysis of the tertiary alcohol moiety of haloalcohols are also discussed. 1-Chloro-5-hydroxy-5-propyloctane is converted to 1-chloro-5-propyloctane in high selectivity using Ru/C as the catalyst. The conversion of 1-chloro-4-(2-propyl-2-ol)benzene to 1-chloro-4-(2-propyl)benzene is best accomplished using a Pt/charcoal catalyst. For both of these compounds, the hydrogenolyses of the carbon-chlorine and carbon-oxygen bonds are competing reactions. The selectivities obtained using other metals and supports are discussed.

I. INTRODUCTION

Catalytic hydrogenation and hydrogenolysis using molecular hydrogen as the reductant are frequently employed in organic synthesis.[1,2,3] Hydrogenation involves the addition of molecular hydrogen to π-bonds to produce a more saturated molecule. Eq. 1 illustrates the transformation with the addition

$$\rangle C{=}C\langle \quad + \quad H{-}H \longrightarrow -\underset{|}{\overset{H}{\overset{|}{C}}}-\underset{|}{\overset{H}{\overset{|}{C}}}- \qquad (1)$$

of hydrogen to a carbon-carbon double bond. This transformation is general in nature and can be used to saturate π-bonds between carbon-oxygen, carbon-nitrogen, conjugated carbon-carbon and aromatic systems. Hydrogenations are catalyzed by many different metals, on solid supports, as oxides and as metal complexes, but the Group VIII metals are usually the most efficient.[2,4]

In contrast to hydrogenation, hydrogenolysis is the reductive cleavage of σ-bonds with molecular hydrogen as depicted in Eq. 2. This report is

$$C{-}X \quad + \quad H{-}H \longrightarrow C{-}H \quad + \quad HX \qquad (2)$$

primarily concerned with the hydrogenolysis of tertiary benzyl-oxygen and tertiary alkyl-oxygen bonds.

There are many reports in the literature of benzyl-oxygen hydrogenolysis.[5,6,7,8] For the hydrogenolysis of 1° and 2° benzyl alcohols, it has been suggested that a hydride attacks the benzyl carbon in a manner which resembles an S_N2 displacement.[7] The supporting evidence is the observed order of reactivity for the leaving group, $CF_3CO_2 > OAc > ArO \gg RO$, OH and the similarity between the substituent effects (i.e. alkyl substitution on the benzyl carbon) of benzyl alcohol hydrogenolysis and solvolysis.[7,9]

The hydrogenolysis of tertiary benzyl alcohols, however, is reported to resemble more of an S_N1 type mechanism since hydrogenolysis with substituents on the aromatic ring exhibit negative Hammett ρ-values and substituent effects on the benzyl carbon are smaller than for S_N2 displacements.[7] However, the hydrogenolysis of chiral tertiary benzyl alcohols and ethers have been found to occur with a high degree of stereospecificity. Both inversion and retention of configuration at the benzyl carbon have been observed, depending on the catalyst. Pd/C has been found to effect hydrogenolysis with 84 to 100% inversion while Raney Ni catalysts have been found to effect hydrogenolysis with 70 to 100% retention of configuration.[8,10] The difference in stereospecificity between these two catalysts has been attributed to the greater affinity of nickel for the oxygen of the hydroxyl group than palladium and therefore a different conformation of the chiral alcohol is adsorbed onto the catalysts surface.

The solvent has been found to have a large effect on the rate of hydrogenolysis with the rate found to decrease in the following order for benzyl alcohol: acetic acid > water > methanol > hexane. For most hydrogenolyses, Pd/C is the most active catalyst. In a comparison of Group VIII catalysts, the following order was observed for the hydrogenolysis of benzyl alcohol in acetic acid: Pd/C > Rh/C > Pt/C > Ru/C.[6] Pd/C has an additional advantage for benzyl alcohol hydrogenolysis since competing ring hydrogenation is relatively slow compared to the other catalysts.

This report discusses the hydrogenolysis of carbon-oxygen bonds as a key step in the organic synthesis of a series of model surfactants which have utility in enhanced oil recovery applications.

II. RESULTS AND DISCUSSION

A. Hydrogenolysis In the Synthesis of Alkyl Aryl Sulfonates

A series of surfactants of the general formula shown below have been prepared in which the alkyl aryl portion of the molecule is isomerically pure.

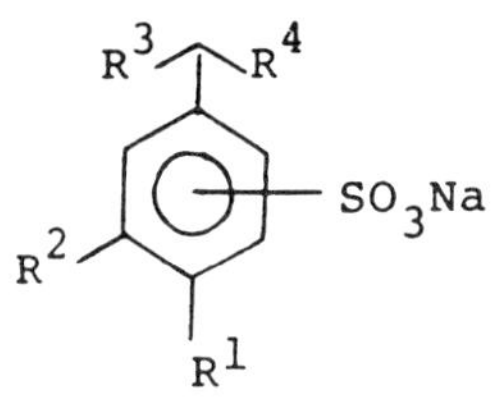

R^1 = H, CH_3, C_2H_5, i-C_3H_7

R^2 = H, CH_3, C_2H_5

R^3 = H, CH_3 to C_6H_{13}

R^4 = CH_3 to $C_{13}H_{27}$

The physical properties of the surfactants of well defined structure could then be investigated in order to establish structure-property relationships for enhanced oil recovery.[11] The variations in structure of the hydrocarbon investigated were; the length of the lipophilic chain as determined by the identity of R^3 and R^4, the point of attachment of the aryl along the lipophilic chain, the position of attachment of the lipophilic chain on the ring and the identity of the short groups R^1 and R^2.

An example of a para-ethylbenzene sulfonate with the aryl attached at an internal carbon atom of the lipophile is shown in Eqs. 3-5 for the

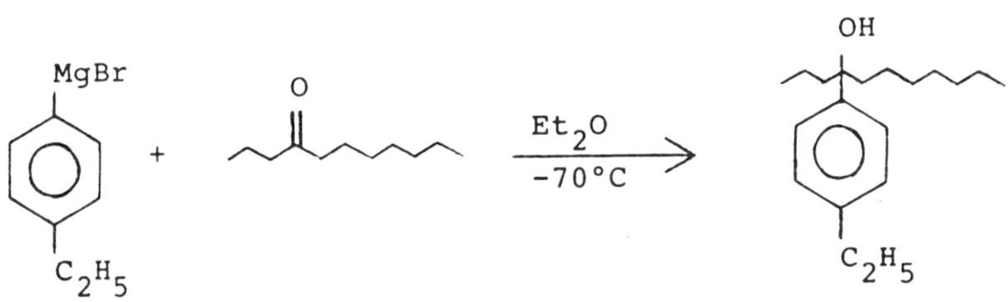

(3)

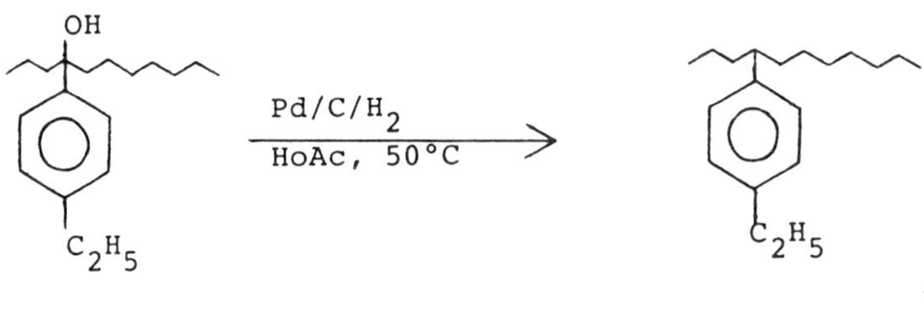

(4)

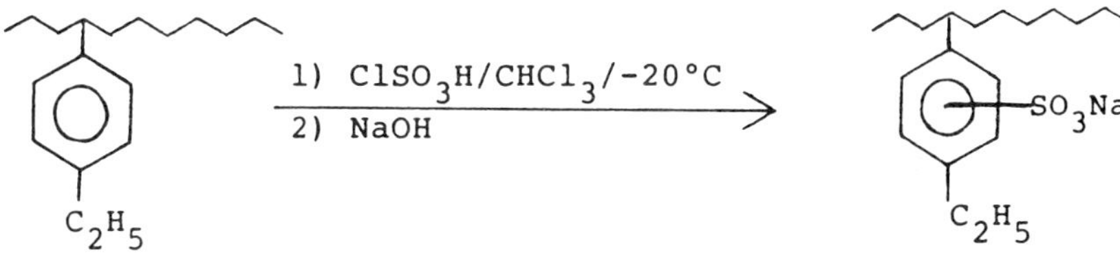

(5)

synthesis of sodium 4-(4-ethylphenyl)undecane sulfonate. The exclusive <u>para</u> orientation of the lipophilic and ethyl groups was established by preparing the Grignard of bromo-4-ethylbenzene. The reaction of this Grignard with 4-undecanone yielded the lipophilic precursor with exclusive C-4 attachment of the ring on the long chain. Loss of yield of the tertiary alcohol due to enolization by the reaction of the Grignard with the hydrogen on the carbon atom alpha to the carbonyl was minimized by conducting the reaction at low temperature (-70°C) and in diethyl ether solvent instead of tetrahydrofuran (THF). The yield in this case was approximately 80% but yields varied depending on the structures of the Grignard and ketone. The hydrogenolysis of the alcohol as shown in Eq. 4 was conducted using an acetic acid:alcohol:5% Pd/C weight ratio of 30:30:1. At 50°C and 200 psi of hydrogen, conversion of the alcohol to the corresponding hydrocarbon was complete in 16 hrs. in high isolated yield (~95%).

The hydrocarbon, after purification by distillation, was sulfonated by the reaction with excess chlorosulfonic acid at -20°C in chloroform as shown in Eq. 5. The reaction of the sulfonyl chloride thus formed with aqueous sodium hydroxide gave the desired sodium alkyl sulfonate which was purified by silica gel chromatography. Since the sulfonation of the hydrocarbon is a standard procedure applied to all of the aromatic hydrocarbons synthesized, the sulfonation step will not be discussed further.

B. The Effect of Solvent on Reactivity and Selectivity

In the hydrogenolysis step of all of the tertiary benzyl alcohols prepared during the syntheses of alkyl aryl sulfonates, ring hydrogenation was found to be a competing reaction. To minimize ring hydrogenation, glacial acetic acid is the preferred solvent and low temperatures and low hydrogen pressures are used. A comparison of glacial acetic acid and ethanol as the

TABLE 1

The Effect Of Solvent On Hydrogenolysis Reactivity And Selectivity For 3-(4-ethylphenyl)octan-3-ol

(a)Alc:Cat	(b)Solvent	Temp (°C)	(c)Selectivity (%) Arene	Cycloalkane	Time (hrs)
30	HOAc	50	96	4	20
30	EtOH	100	67	33	48

(a)Cat = 5% Pd/C, H_2 = 200 psig, wt ratio
(b)Solvent:Alc = 1:1
(c)Arene = 3-(4-ethylphenyl)octane; Cycloalkane = cis and trans 3-(4-ethylcyclohexyl)octane. GC analysis; 6' 10% sp 2100, 150 to 300°C at 10°C/min, std = hexadecane

solvent on the hydrogenolysis rate and selectivity to 3-(4-ethylphenyl)octane using 5% Pd/C as the catalyst is shown in Table I. The hydrogenolysis conducted in glacial acetic acid not only was faster but gave considerably less ring hydrogenated product than the reaction conducted in ethanol. The much higher temperature for the ethanol reaction had to be used in order to get complete reaction in 48 hrs. The ring hydrogenated cyclohexane derivative was found to consist of both cis and trans isomers in approximately equal amounts.

C. The Preparation of Alkyl Aryl Hydrocarbons with Terminally Attached Lipophiles

One method utilized for the preparation of alkyl aryl hydrocarbons with terminally attached lipophiles is shown in the two step acylation/hydrogenolysis synthesis (Eqs. 6 and 7) for a para-ethylbenzene derivative. The acylation of ethylbenzene with tetradecanoyl chloride used a stoichiometric of amount of $AlCl_3$ and gave exclusive para substitution to form the alkyl aryl ketones in high yield. Under hydrogenolysis conditions

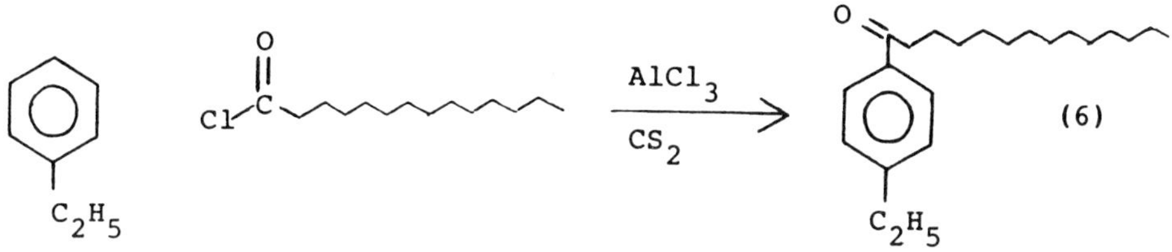

(6)

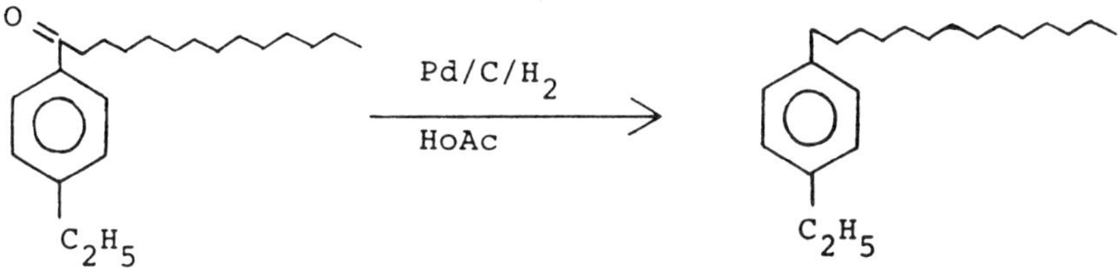

(7)

similar to those used for the tertiary alcohols, 1-(4-ethylphenyl)tetradecane was formed in high yield with 3% ring hydrogenated byproducts.

This method, however, was later replaced by a one step route using a palladium catalyzed cross coupling between bromo-4-ethylbenzene and the Grignard of the corresponding terminal bromide as shown in Eq. 8 for the preparation of 1-(4-ethylphenyl)tridecane. The coupling reaction, a

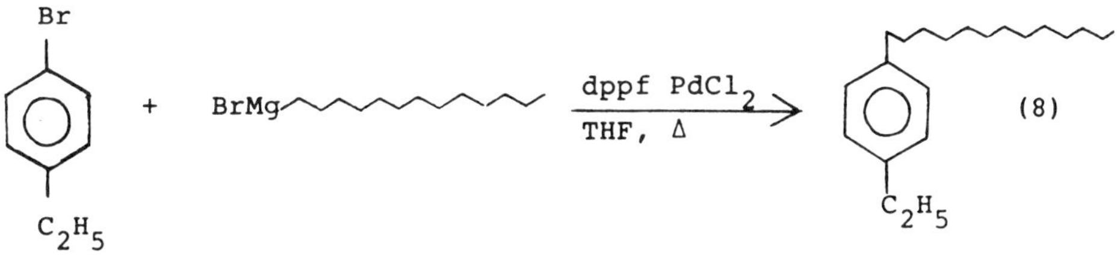

(8)

modification of Kumada's procedure, worked best using dppf $PdCl_2$[12] as the catalyst in THF giving the desired product in 70% yield.[13]

D. The Preparation of an Alkyl Aryl Hydrocarbon with a Terminally Attached Lipophile Containing Remote Branching

A combination of the palladium catalyzed Grignard coupling reaction and hydrogenolysis was used in the synthesis of 1-(4-ethylphenyl)-5-

propyloctane. This structure was of interest to determine the effect of the point of branching being four carbons removed from the arene on the physical properties of the corresponding sulfonate surfactant. The strategy used in the synthesis was to prepare the lipophilic structure, with a chlorine on the terminal carbon, as a unit and then to couple it with bromo-4-ethylbenzene. As shown in Eqs. 9-11, the first step involved in reaction of

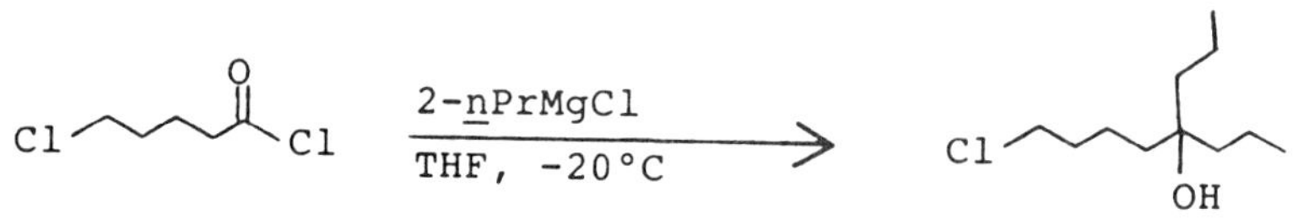

(9)

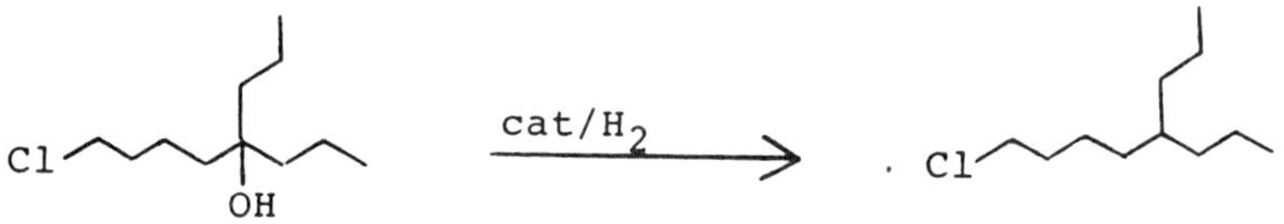

(10)

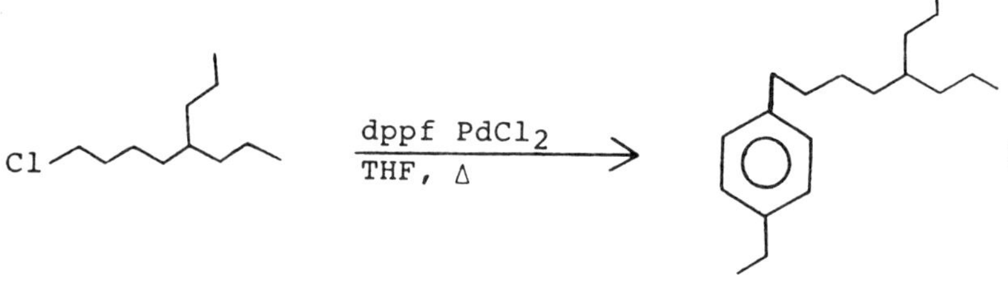

(11)

excess propylmagnesium chloride with 4-chlorobutyryl chloride in THF at -20°C to yield 1-chloro-5-hydroxy-5-propyloctane in 60% yield. Selective hydrogenolysis of the tertiary alcohol leaving the carbon-chlorine bond intact gives 1-chloro-5-propyloctane which after conversion to the Grignard could be coupled with bromo-4-ethylbenzene as was previously done to prepare the aromatic hydrocarbon with the linear terminally attached lipophile 1-(4-ethylphenyl)tridecane. The hydrogenolysis of 1-chloro-5-hydroxy-5-propyloctane using 5% Pd/C as the catalyst under the usual conditions resulted in considerable hydrogenolysis of both the chloride and hydroxyl groups to give 4-propyloctane along with the desired 1-chloro-5-propyloctane (Table II). Several other catalysts were screened in order to determine which catalyst would give the highest yield of the chloroalkane. After investigating Rh, Ru, Pd and Pt catalysts supported both on carbon and alumina, 5% Ru/C, although considerably less active than Pd/C, was found to give the highest selectivity to 1-chloro-5-propyloctane (97%) with only 3% of 4-propyloctane being formed. Ru/Al_2O_3 also gave very high selectivity to the chloroalkane but the rate was too slow to be preparatively useful. Pd/Al_2O_3, also less active than Pd/C, gave better selectivity to the chloroalkane than Pd/C. Rh/C and Rh/Al_2O_3 were very similar in their selectivities and rates to the chloroalkane. Platinum on coconut charcoal[14]

TABLE II

Tert Alkanol vs. Halo Alkyl Hydrogenolysis

[a]Cat.	[b]Sel. (%) Chloroalkane	[c]Sel. (%) Alkane	Conv. (%)	Time (hrs)
5% Ru/C	97	3	99	20
5% Ru/C	98	2	35	3
5% Ru/Al_2O_3	96	0	26	64
5% Pd/C	65	35	99	3
5% Pd/C	84	16	85	3
5% Pd/Al_2O_3	80	20	99	20
5% Rh/C	87	13	97	3
5% Rh/Al_2O_3	87	13	92	3
5% Pt/Char	90	10	97	3
5% Pt/Al_2O_3	82	18	97	3

[a]Alc:Cat = 20:1 by wt, 100°C, 100 psi H_2, HOAc:alcohol = 1:1 (vol:wt)
[b]Chloroalkane = 1-chloro-5-propyloctane, GC analysis = 6' 10% sp 2100, 100 to 300°C at 10°C/min, std = hexadecane
[c]Alkane = 4-propyloctane

gave high selectivity to the chloroalkane (90%) while Pt/Al_2O_3 in contrast to the other alumina supported catalysts gave less selectivity to the chloroalkane than the corresponding carbon supported catalyst.

The product distributions of reaction mixtures for several of the less active catalysts at incomplete conversions showed the presence of unsaturated products (Table III) which upon further hydrogenation are converted to saturated products. The mechanism for at least part of the hydrogenolysis of the tertiary alcohol involves the elimination of water to form the chloroalkene which is subsequently hydrogenated. The elimination of water is probably catalyzed by the acetic acid solvent. Reactions using the two least active catalysts, Ru/Al_2O_3 and Raney Ni, showed 5-propyl-1,4-octadiene in addition to the chloroalkene. Raney Ni was completely inactive under these conditions for the hydrogenolysis of 1-chloro-5-hydroxy-5-propyloctane even though the alcohol was totally converted to the trisubstituted alkene, 1-chloro-5-propyl-4-octene.

TABLE III

Haloalkanol Hydrogenolysis At Incomplete Conversion

	(b)Product Distribution (%)					
(a)Cat.	Chloroalkane	Chloroalkene	Alkane	Diene	Conv.	(hrs)
5% Ru/C	70	26	4	0	40	20
5% Ru/Al_2O_3	72	24	trace	4	34	64
5% Pd/Al_2O_3	79	7	14	0	96	20
(c)RaNi	0	97	0	3	100	20

(a)Alc:cat = 20:1, HOAc:Alc = 1:1, Temp = 100°C, H_2 = 100 psi.

(b)GC analysis = 6' 10% sp 2100, 100 to 300°C at 10°C/min., std = hexadecane, chloroalkene = 1-chloro-5-propyloctane, chloroalkene = 1-chloro-5-propyl-4-octene, alkane = 4-propyloctane, diene = 5-propyl-1,4-octadiene.

(c)Temp = 150°C

E. The Hydrogenolysis of a Tertiary Benzyl Alcohol Containing an Aryl Halide

The hydrogenolysis of 1-chloro-4-(2-propyl-2-ol)benzene was investigated to compare the relative selectivities of the hydrogenolyses of tertiary benzyl alcohols and aryl chlorides versus the above 1-chloro-5-hydroxy-5-propyloctane which contains a tertiary alkyl alcohol and an alkyl chloride.

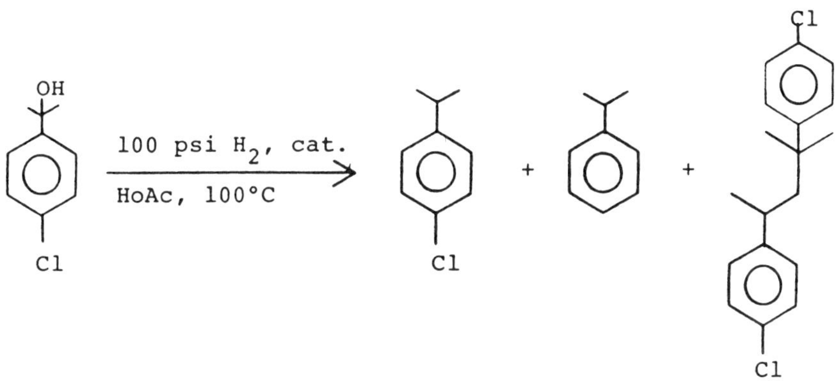

(12)

The hydrogenolysis of the aromatic molecule was found to be more complex than the aliphatic molecule since in addition to carbon-oxygen and carbon-chlorine bond hydrogenolyses products, dimers of 1-chloro-4-(2-propyl)benzenes were formed using certain catalysts (Table IV).

Equation 12 shows the structure of one of the dimeric species identified from the reaction. Chlorine hydrogenolysis products and unsaturated derivatives of this dimer are also formed but are listed in Table IV as dimer. The dimer shown above is most likely formed by the elimination of water from the starting alcohol to form 1-chloro-4-(2-propenyl)benzene which is protonated by glacial acetic acid, the carbonium ion thus formed adds to another chloroalkene. This adduct is then hydrogenated to the observed dimer.

Table IV shows the product distribution of the hydrogenolysis of 1-chloro-4-(2-propyl-2-ol)benzene using several different catalysts. Pt/charcoal in glacial acetic acid was found to give a very high ratio of aryl

TABLE IV

Hydrogenolysis Of Chloro-4-(2-Propyl-2-ol) Benzene

[a]Cat.	[b]Chloro-4-(2-propyl)benzene (%)	Cumene (%)	Dimer (%)	Solvent
5% Pt/Char	86	1	13	HOAc
5% Pt/Char	89	10	0	EtOH
5% Pt/Char	72	24	1	Heptane
5% Rh/C	85	12	0	HOAc
5% Ru/C	64	2	33	HOAc
5% Pd/C	54	46	0	HOAc
[c]COD Ir(Pyr)(L)+	88	0	12	Heptane

[a]Alc:cat = 20:1 by wt, Solv:Alc = 1:1 (vol:wt), Temp = 100°C, H_2 = psi.

[b]GC analysis = 6' 10% sp 2100, 150 to 300°C at 10°C/min, std = hexadecane.

[c]L = $(C_6H_{11})_3P$, Temp = 130°C, Alc:cat = 20:1 by wt, solv:alc = 2:1 (vol:wt).

halide to arene, however, the selectivity to the aryl halide was limited to 86% due to considerable dimer formation. Since dimer formation is believed to be due to acid catalysis, the hydrogenolysis was investigated using ethanol and heptane as the solvent. These solvents virtually eliminated dimer formation as a byproduct but the selectivity to cumene increased to 10 and 24% for the ethanol and heptane reactions, respectively. With Rh/C as the catalyst an 85% selectivity to the aryl halide was observed. The double hydrogenolysis product was found to be mixture of cumene and isopropylcyclohexane while the ring hydrogenated product from the aryl halide was not observed. Ru/C showed a very high ratio of aryl halide to arene but the very high dimer formation resulted in a lower overall yield to the aryl halide. In contrast Pd/C showed no dimer formation but the ratio of aryl halide to arene is almost 1:1.

The homogeneous catalyst COD $Ir(pyr)P(C_6H_{11})_3^+PF_6^-$[15] did not produce any hydrogenolysis of the carbon-chlorine bond. The selectivity loss to the desired aryl halide is exclusively due to dimer formation. The activity of this catalyst for the hydrogenolysis of the alcohol moiety is believed to be due to the cationic nature of the complex which enables it to act as a Lewis acid and thus weaken the carbon oxygen bond. In order for the Lewis acid nature of the catalyst to be effective the best solvent was a non-coordinating and water immiscible solvent such as heptane. The iridium cation can then promote the elimination of water to form the corresponding alkene which is either hydrogenated to the aryl halide by the iridium complex, known to be a powerful hydrogenation catalyst,[15] or converted to the dimer by the Lewis acid properties of the iridium cation. Cationic rhodium complexes, such as, $[(CH_3)_2PhP]_3RhNBD^+PF_6^-$ also catalyzed the selective hydrogenolysis of the hydroxyl functionality but was much slower than the iridium catalyzed reaction.[16] In contrast to the above cationic complexes, the neutral rhodium complex, $(Ph_3P)_3RhCl$ did not effectively catalyze this reaction.

F. The Use of Hydrogenolysis in the Synthesis of a Nonionic Surfactant

The synthesis of the branched alkylphenolic nonionic surfactant, 3,6,9,12,18,21-hexaoxa-21-(4-(6-dodecyl)phenyl)heneicosan-1-ol, is shown in Eqs. 13-17. The lipophilic chain is attached to the <u>para</u> position of the protected phenol at the 6-carbon by the reaction of bromomagnesium 4-methoxybenzene with 6-dodecanone to form 6-(4-methoxyphenyl)dodecan-6-ol. The hydroxyl group was selectively removed by hydrogenolysis using the cationic iridium complex, COD $Ir(pyr)P(C_6H_{11})_3^+PF_6^-$ as the catalyst, at

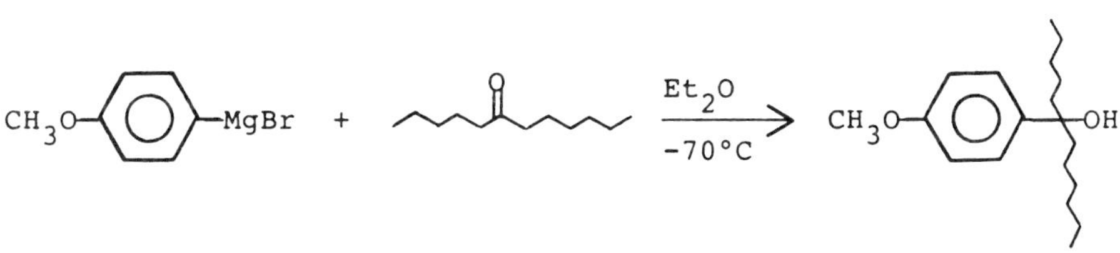

(13)

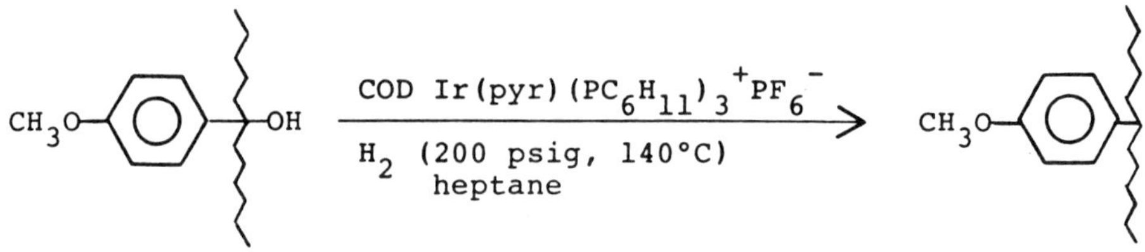

(14)

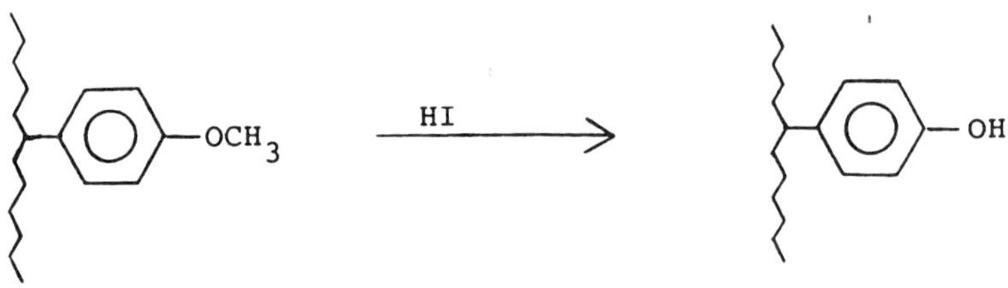

(15)

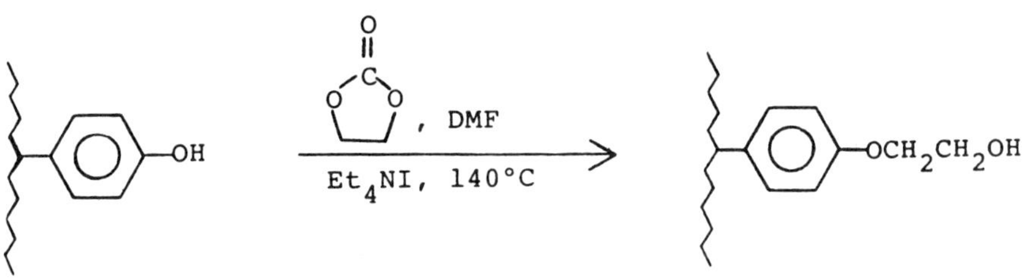

(16)

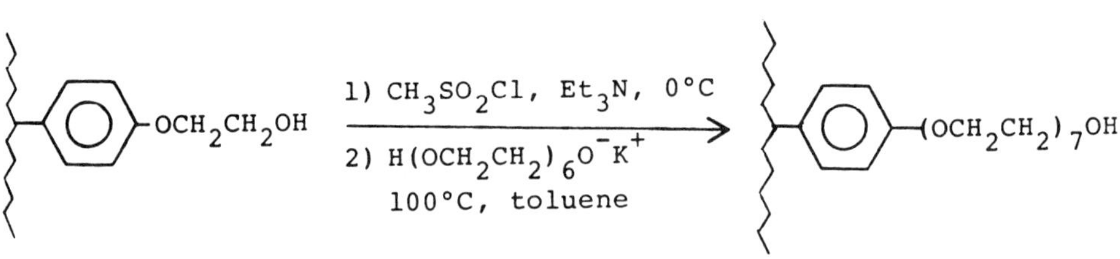

(17)

140°C in heptane under 200 psi of hydrogen. This catalyst is very selective giving only the desired product with no ring hydrogenation or ether cleavage products being formed which predominated when supported Group VIII catalysts such as Pd/C, Raney Ni, Rh/C or Pt/C were used for this transformation.

After isolating the 6-(4-methoxyphenyl)dodecane from the hydrogenolysis, the protecting methyl group was removed by heating with 57% hydroiodic acid in acetic anhydride. In order to add the hydrophilic heptaethyleneoxyol portion of the surfactant, the phenol thus produced was converted to the monoethylene glycol adduct very selectively by the Et_4NI catalyzed reaction with ethylene carbonate.[17] Hexaethylene glycol was then added to form the final product in 75% yield by the conversion of 2-(4-(6-dodecyl)phenyl)ethanol to the mesylate which was reacted with the mono potassium hexaethylene glycoxide at 95°C in toluene.

ACKNOWLEDGMENT

The authors gratefully acknowledge the skillful technical assistance of Mr. Joseph Travetti, Miss Mary Carol Marateo, Mr. Diandre Armstead and Mr. Andrew Gavula.

REFERENCES

1. A.P.G. Kieboom and F. van Rantwijk, "Hydrogenation and Hydrogenolysis in Synthetic Organic Chemistry", Delft University Press, Rotterdam, 1974.
2. P.N. Rylander, "Catalytic Hydrogenation Over Platinum Metals", Academic Press, New York, 1967.
3. P.N. Rylander, "Catalytic Hydrogenation in Organic Synthesis", Academic Press, New York, 1979.
4. B.R. James, "Homogeneous Hydrogenation", Wiley-Interscience, 1973.
5. W.H. Hartung and R. Siminoff, Org. React., **1**, 263, (1953).
6. P.N. Rylander and D.R. Steele, Engelhard Ind. Tech. Bull., **6**, 41, (1965).
7. A.P.G. Kieboom, J.F. deKreuk, and H. van Bekkum, J. Catal., **20**, 58-66, (1971).
8. S. Mitsui, Y. Kudo and m. Kobayashi, Tetrahedron, **25**, 1921-1927, (1969).
9. A. Streitwieser, Jr., "Solvolytic Displacement Reactions", McGraw-Hill, New York, pp. 95, 140, (1962).

10. S. Mitsui, S. Imaizumi, and Y. Esahi, *Bull. Chem. Soc. Japan*, **43**, 2143, (1970) and references therein.

11. S.J. Salter, ARCO Exploration and Technology Company, Plano, TX, private communication.

12. The ligand dppf = 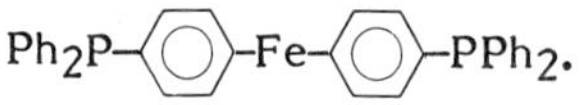.

13. K. Tamao, K. Sumitani, Y. Kiso, M. Zembayashi, A. Fujioka, S. Kodama, I. Nakajima, A. Minato, and M. Kamada, *Bull. Chem. Soc. Japan*, **49**, 1958-1969 (1976).

14. A "special" catalyst as obtained from Strem Chemical. All other supported catalysts were obtained from Engelhard Corp.

15. R. Crabtree, *Accts. Chem. Res.*, **12**, 331-338, (1979), and references therein.

16. R.R. Schrock and J.A. Osborn, *J. Am. Chem. Soc.*, **98**, 2134-2143, (1976) and references therein.

17. T. Yoshino, S. Inaba, and Y. Ishido, *Bull. Chem. Soc. Japan*, **46**, 553-556, (1973).

16
The Hydrogenation of Methylenedianiline

G.F. Allen

Mobay Corporation
New Martinsville, WV

ABSTRACT

Methylenedianiline (MDA) is a key intermediate in the production of a "specialty" isocyanate that is used in the coatings area. To make this diisocyanate, the MDA is first hydrogenated to form the completely saturated cycloaliphatic diamine, and this amine is then phosgenated to give the corresponding cycloaliphatic diisocyanate. Although MDA is hydrogenated on a commercial scale and is the subject of many patents, much of the chemistry has not yet been described. It is the purpose of this paper to present a summary of our studies on the hydrogenation of MDA.

The hydrogenation of MDA was shown to be a series, first order process involving the intermediate, 4-(4-aminocyclohexylmethyl) benzenamine (1/2 r-MDA), which is then converted to the final product, 4,4'-methylenebiscyclohexylamine (r-MDA). Although the hydrogenation of MDA yields predominantly the cis isomer, some trans isomer is always present.

Throughout the hydrogenation reaction, the aliphatic rings of the half r-MDA and the r-MDA are undergoing continual isomerization. The cis-trans isomerization rate was shown to be much more sensitive to temperature than the hydrogenation rate. The effects of catalyst and reaction temperature on the relative rates of hydrogenation and isomerization are described.

INTRODUCTION

The hydrogenation of 4,4'-methylenebisbenzenamine (MDA) and the characterization of the three isomers formed (see Fig. 1) was described in 1950 and 1953[1,2,3]. Since then over 35 patents have been issued describing catalytic processes for the hydrogenation of MDA. Specific isomeric compositions have been produced commercially and used as intermediates in the fibers and coatings areas. However, there is still very little information in the literature on the reaction pathway of the hydrogenation, or on the isomerization processes. We were interested in determining how the relative rates of hydrogenation and isomerization varied with catalyst and with reaction temperature.

The hydrogenation of MDA is known to proceed through the half-hydrogenated intermediate 4-(4-aminocyclohexylmethyl)benzenamine (1/2

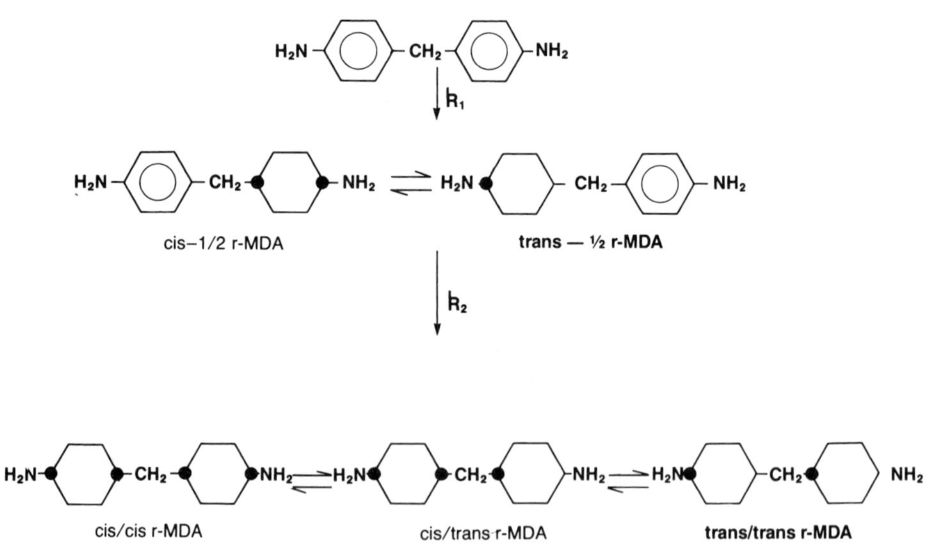

Fig. 1. Reaction Pathway

r-MDA) as shown in Fig. 1.[3] Watanabe, et al.[4] have used an iridium catlyst to produce reaction mixtures which have a 1/2 r-MDA content as high as 80%. The preferential, complete hydrogenation of one ring before the second ring is attacked, is typical behavior for such compounds.[5,6]

The cis-trans stereoselectivity of heterogeneous hydrogenations of disubstituted aromatics has been described by several groups: xylenes,[7,8] di-t-butylbenzenes,[9] and toluidines[10] have all produced the cis isomer as the major component by means of a cis addition of hydrogen. The formation of the trans isomer was explained by the readsorption and the subsequent hydrogenation of olefin intermediates that desorbed from the catalyst before the ring was completely saturated.[11] The relative amount of trans isomer increased with increasing reaction temperature. These studies were carried out under very mild conditions such that the source of the trans isomer was established to be from hydrogenation mechanisms only.[7]

However, at much higher temperatures, aliphatic products can undergo a cis-trans isomerization. Brake[12] demonstrated that the aliphatic diamine, r-MDA, containing low levels of the trans isomers could be isomerized in the presence of a ruthenium catalyst and hydrogen at temperatures of 150° to 300°C to an r-MDA containing up to 51% of the trans-trans isomer. We have shown, as have others, that the isomer composition of the r-MDA is dependent upon the reaction temperature of the hydrogenation, but it was not known what role isomerization played in determining the isomer ratio of the final product. We wanted to determine whether the isomer content of the r-MDA was determined by the hydrogenation mechanism, or whether it was also a function of the isomerization rate of r-MDA. In addition, we wanted to characterize the changes in selectivity when using a ruthenium, rhodium, or iridium catalyst.

II. EXPERIMENTAL

All experiments were carried out in a one liter autoclave (Autoclave Engineers) fitted with internal cooling coils, thermowell, magnetic agitator (1750 rpm), and an external electrical heating mantel.

4,4'-Methylenebisbenzenamine (MDA) contained 3% of the 2,4'-methylenebisbenzenamine as an impurity. t-Butyl alcohol was reagent grade, and $LiNO_3$ was of high purity. The catalysts, all fine powders (325 mesh) were obtained from either Engelhard or Johnson Matthey as 5% metal on γ-Al_2O_3. Lithium nitrate was used in all experiments to reduce the amount of condensation products typically formed in aromatic amine reductions.[13,14]

Reactions were carried out in the following manner: (1) The autoclave was sealed with 100 g of t-BuOH, 300 g of MDA, 1 g of $LiNO_3$, and 0.25 mole percent of supported metal catalyst (based on MDA) inside. (2) The autoclave was pressurized to 4000 psi and heated to the desired reaction temperature. (3) After the temperature lined out the agitator was started. The temperature was controlled manually (+/- 3°C) and the pressure was maintained constant by means of a regulator. Samples were pulled regularly through a special dip tube and were analyzed on an HP 5890A gas chromatograph. The response factors for all components discussed were very close to one; therefore area percent values were used as weight percent in all calculations.

III. RESULTS

To establish the reaction pathway, MDA was hydrogenated in batch reactions at 4000 psi as described above. Three metals, ruthenium, rhodium, and iridium, were evaluated; all were supported on γ-alumina at 5% by weight loadings. The mole percent of metal present in the reactor (based on moles of MDA) was kept constant in all experiments at 0.25%. Samples were taken from the reactor during the reaction and analyzed by gas chromatography (Fig. 2). The amount of condensation products was less than three percent for all experiments. The amount of hydrogenolysis (deamination) occurring during the reaction varied dramatically with catalyst. In the case of the 5% Ir/Al_2O_3 runs, the amount of 4-cyclohexylmethylcyclohexanamine was very high (Table I).

The rate constant k_1 was calculated from the slope of $\log \frac{(MDA)}{(MDA)_o}$ versus time data. The constant k_2 was determined from a regression analysis of the simple series first order rate equations described by Frost and Pearson for homogeneous reactions (see Table I).[15]

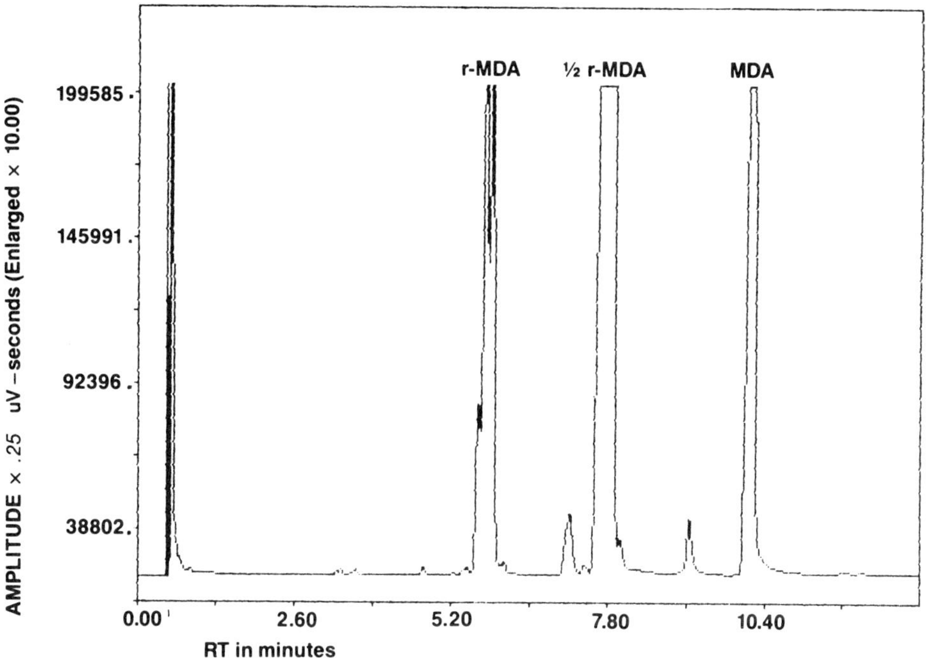

Fig. 2. Separation of MDA, 1/2 r-MDA, and r-MDA

TABLE I

Rate Constants for the Hydrogenation and the Hydrogenolysis of MDA

Catalyst	Temp(°C)	$k_1(Min^{-1})$	$k_2(Min^{-1})$	k_1/k_2	k_{MA}
5% Ru/Al_2O_3	140	0.013 (0.992)*	0.004	3.25	0.0054 (0.957)*
	160	0.050 (0.993)	0.016	3.13	0.0054 (0.984)
	180	0.088 (0.990)	0.027	3.26	0.0089 (0.994)
5% Rh/Al_2O_3	160	0.009 (0.955)	0.005	1.8	0.0014 (0.991)
	200	0.190 (0.999)	0.095	2.00	0.0271 (0.934)
5% Ir/Al_2O_3	200	0.017 (0.991)	0.008	2.13	0.045 (0.991)
	220	0.036 (0.952)	0.015	2.40	0.109 (0.965)

*correlation coefficient

$$(\text{MDA}) = (\text{MDA})_o\, e^{-k_1 t}$$

$$(1/2\ \text{r-MDA}) = \left[\frac{(\text{MDA})_o k_1}{k_2 - k_1}\right]\left[e^{-k_1 t} - e^{-k_2 t}\right]$$

$$(\text{r-MDA}) = (\text{MDA})_o - (\text{MDA}) - (1/2\ \text{r-MDA})$$

The error due to the effects of the relative adsorption coefficients of MDA and 1/2 r-MDA on the calculation of k_2 was small but could be seen in all of the hydrogenation experiments. The effect on the value of k_1 was within the error of the experiment. No attempt was made to incorporate the adsorption phenomena into the model.

Data on the isomerization of the 1/2 r-MDA and r-MDA were obtained by gc analysis of the above crude reaction samples after derivatization with

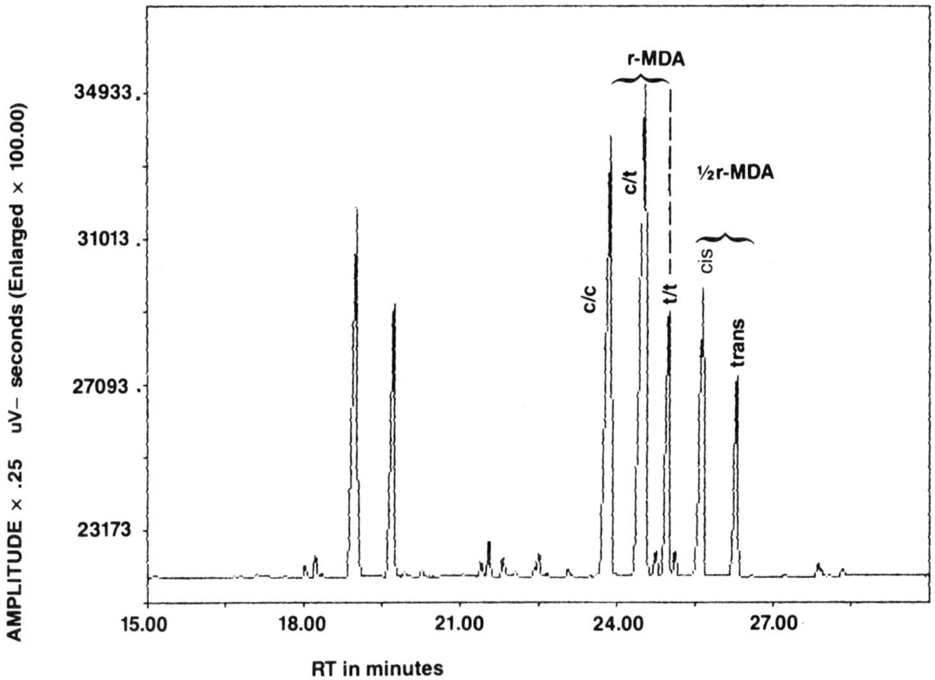

Fig. 3. Separation of Isomers

a sililating agent. Fig. 3 shows a typical analysis. To date the assignment of the cis and trans isomers of 1/2 r-MDA rests only on analogy to the isomers of r-MDA. The gas chromatographic elution order and isomerization behavior of the 1/2 r-MDA isomers correlate with results from the related r-MDA isomers. These results indicate that the cis isomer is the one in greater abundance. This agrees with the published results of similar compounds.[8,9,10] The assignment of the r-MDA isomers are based on previously reported work.[2,16] In addition, the pure trans/trans isomer has been isolated, characterized, and used as a reference material in these studies. The rate constants, k_{cis} for 1/2 r-MDA and k_{tt} for r-MDA were determined from the slope of concentration versus time curves and are listed for each catalyst in Table II. Catalyst poisoning was a problem in all

TABLE II

Zero Order Rate Constants For The Isomerization of 1/2 r-MDA and r-MDA Isomers

Catalyst	Temp (°C)	$(Cis)_o$ [a]	k_{cis}	$(t/t)_o$ [b]	k_{tt}
Ru/Al_2O_3	140	67.1	0.006 (0.32) [c]	11.6	0.006 (0.976) [c]
	160	64.7	0.007 (0.84)	12.1	0.029 (0.994)
	180	64.6	0.067 (0.96)	12.6	0.070 (0.995)
Rh/Al_2O_3	160	64.6	0.019 (0.976	13.3	0.012 (0.989)
	200	too fast		13.6	0.066 (0.999)
Ir/Al_2O_3	200	77.4	0.094 (0.999)	6.1	0.047 (0.999)
	220	74.9	0.143 (0.982)	6.6	0.102 (0.999)

a. $(Cis)_o$ is the cis isomer content of 1/2 r-MDA at time = 0.

b. $(t/t)_o$ is the trans/trans content at time = 0.

c. correlation coefficients.

of the work with iridium on alumina at the temperatures studied. Towards the end of each reaction, the hydrogen consumption prematurely stopped. This behavior may be related to the relatively high levels of ammonia formed by the hydrogenolysis reaction. However, the first half of the reaction models well, as is indicated by the correlation coefficients from the calculations of k_1.

IV. DISCUSSION

The reaction profile for the hydrogenation of MDA is shown in Figs. 4 and 5. The smooth curves in Fig. 4 were calculated from the rate expressions and superimposed on the actual data. The half-hydrogenated intermediate (1/2 r-MDA) first increases rapidly as the MDA is consumed, reaches a maximum, and then decreases. In the temperature range studied, the

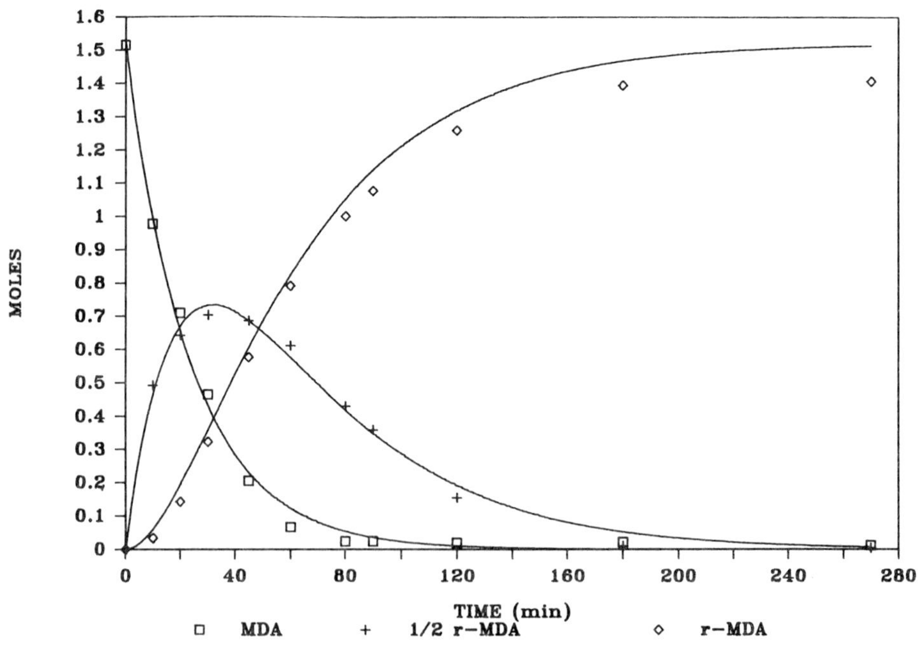

Fig. 4. Series First Order Kinetics

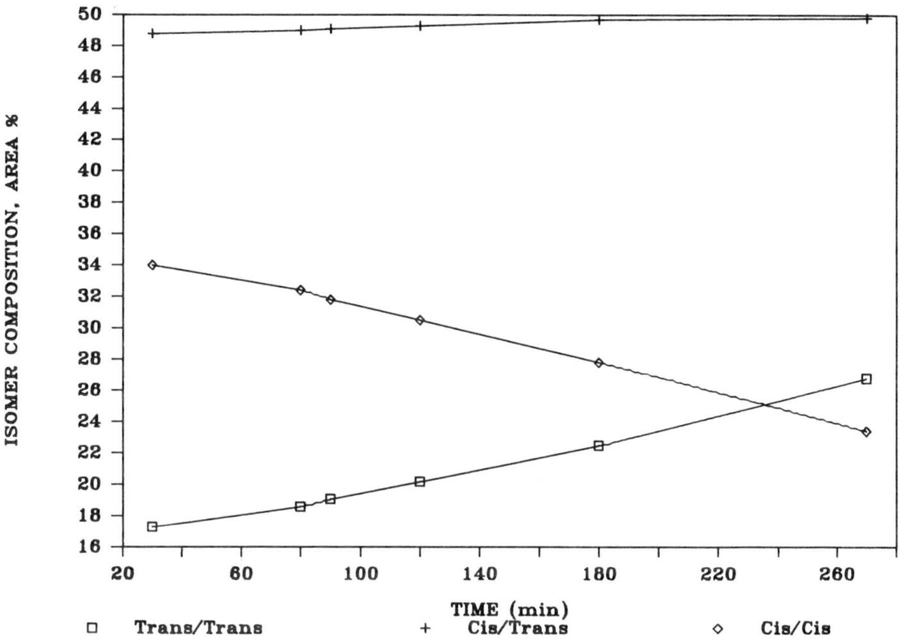

Fig. 5. Cis-Trans Isomerization

maximum level of 1/2 r-MDA remained in a very narrow range, 50-56 mole%. The reaction profile describes a series first order process in which k_1 is significantly greater than k_2. Similar profiles have been described for the reduction of phenol through the intermediate cyclohexanone[17] and for several substituted aromatic compounds.[18] Fig. 5 shows the change in the ratios of cis/cis, cis/trans, and trans/trans isomers of r-MDA during the hydrogenation. This will be looked at in greater detail in the following pages.

Figs. 6 and 7 show the first order disappearance of MDA. From these plots of ln (MDA/MDA_o) versus time, the value of k_1 was determined for each catalyst. Each catalyst is compared at various temperatures in Fig. 6. (The legend indicates by the first two letters the metal used, followed by the reaction temperature in degrees Centigrade). The ruthenium catalyst is by far the most reactive towards ring reduction. From prior work, we know

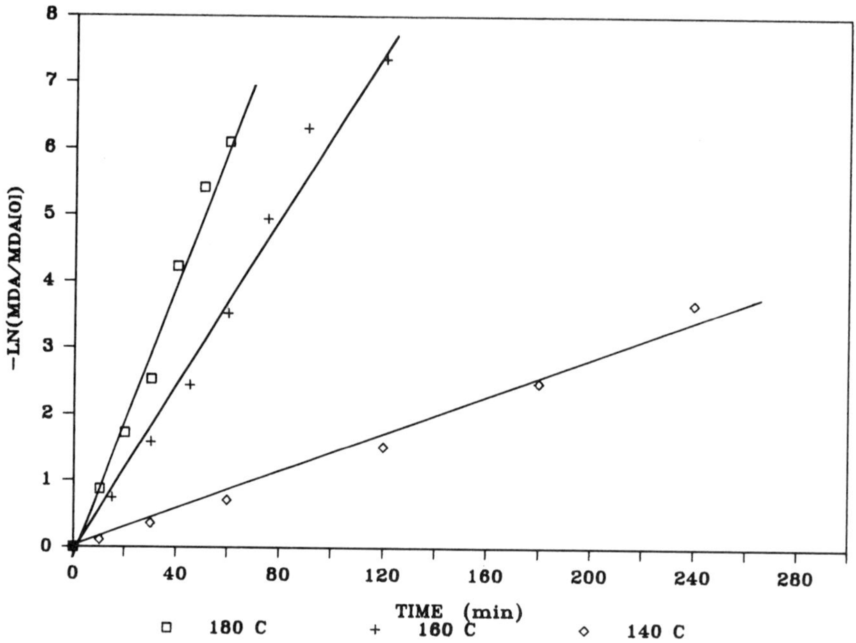

Fig. 6. 5% Ru/Al_2O_3 Catalyst

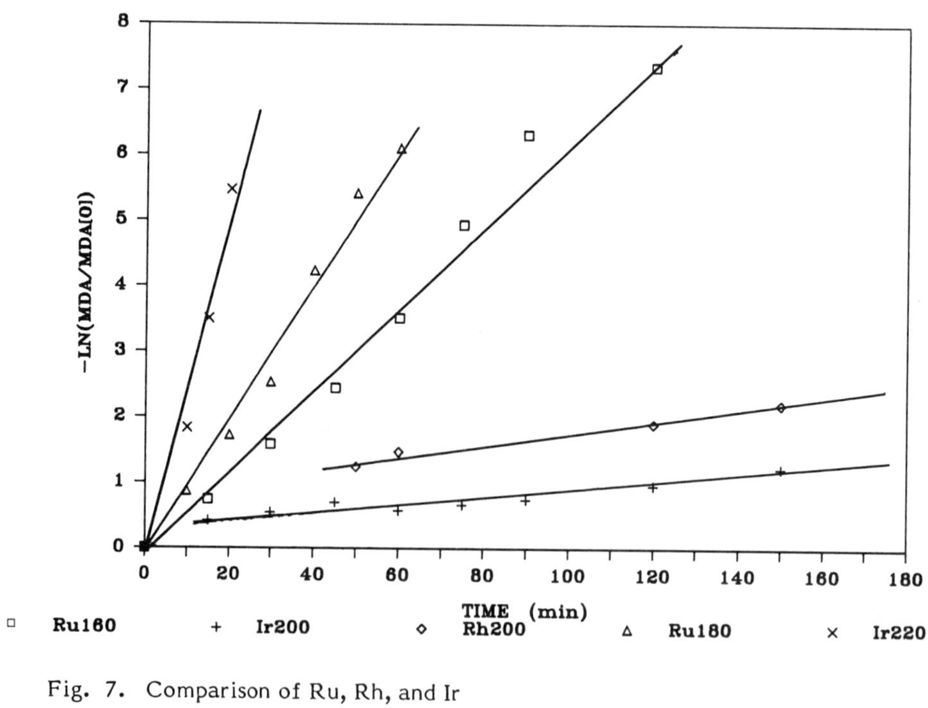

Fig. 7. Comparison of Ru, Rh, and Ir

that the rhodium catalyst requires much higher temperatures to get comparable reaction rates. This is clearly shown here for rhodium at 200°C where the k_1 value (Table I) is much lower than for ruthenium at 160°C or even at 140°C. The low temperature run (200°C) for rhodium and iridium, although clean first order processes, do not intercept the origin. This may be due to an initial inhibition of the hydrogenation as ammonia is produced. However, the iridium 220°C run does not indicate that inhibition is occurring. In any event, the relative reactivities of the metal-supported catalysts for the hydrogenation of MDA differ from one another by a factor of ten in the order of Ru > Rh > Ir.

The selectivity with regard to hydrogenolysis varied dramatically from catalyst to catalyst and was sensitive to temperature. The amounts of 4-cyclohexylmethylcyclohexanamine (MA) formed during the reaction for each catalyst is shown in Fig. 8 and zero order rate constants (k_{MA}) are listed in Table I. The order of hydrogenolysis activity is the reverse of hydrogenation

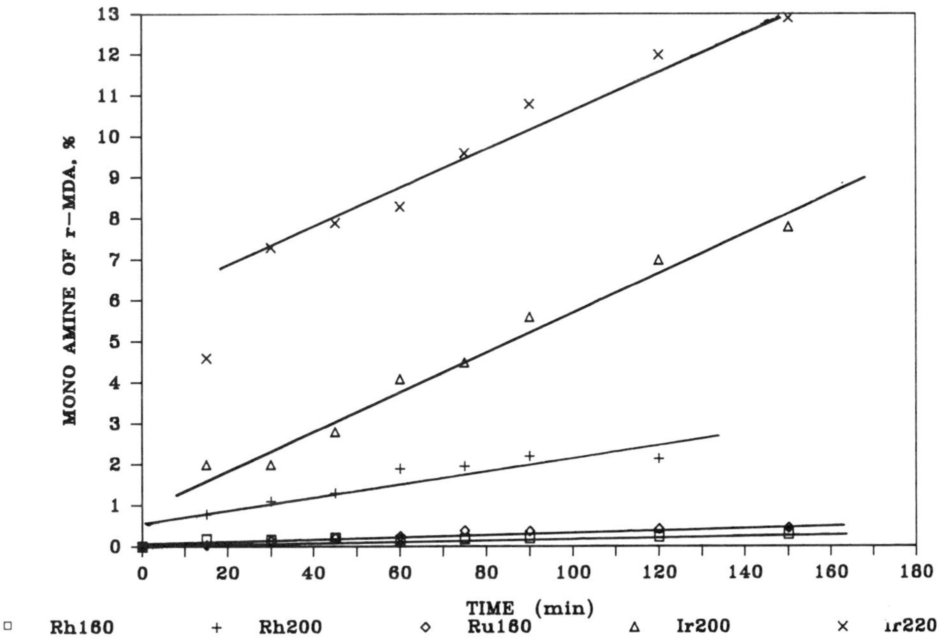

Fig. 8. Rate of Deamination

activity. The ruthenium catalyst at 160°C has a k_1/k_{MA} ratio of 9.1 while rhodium at 160°C has a ratio of 6.3. At 200°C the ratio for rhodium is 7.0, while for iridium the ratio at 200°C is 0.4. We used the ratio of k_1/k_{MA} as a relative measure of selectivity to demonstrate the tremendous swing in behavior between ruthenium and iridium under our hydrogenation conditions.

Having defined the reaction path and correlated the reactivity of the three catalysts with respect to k_1 and k_2 at various temperatures, we wanted to define the isomerization processes which determine the composition of isomers in the final product. Table II and Figs. 9 and 10 summarize the data collected from the hydrogenation experiments described above. The hydrogenation of MDA to 1/2 r-MDA produces predominantly the cis isomer (Fig. 9). The change in the relative concentration of the cis isomer during the reaction occurs by a zero order process which is not affected by the changing concentration of the 1/2 r-MDA during the

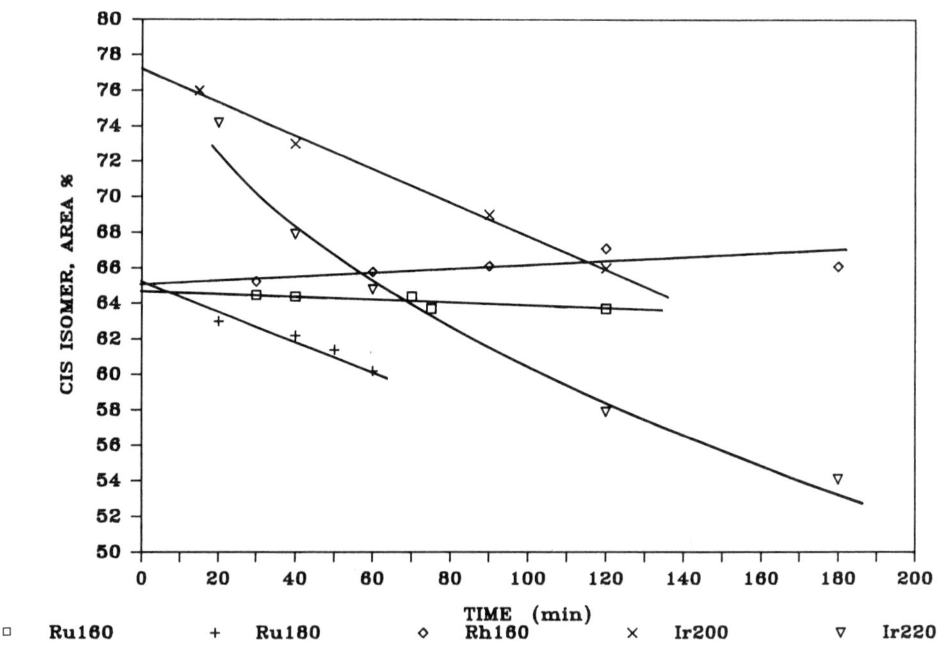

Fig. 9. Isomerization of 1/2 r-MDA

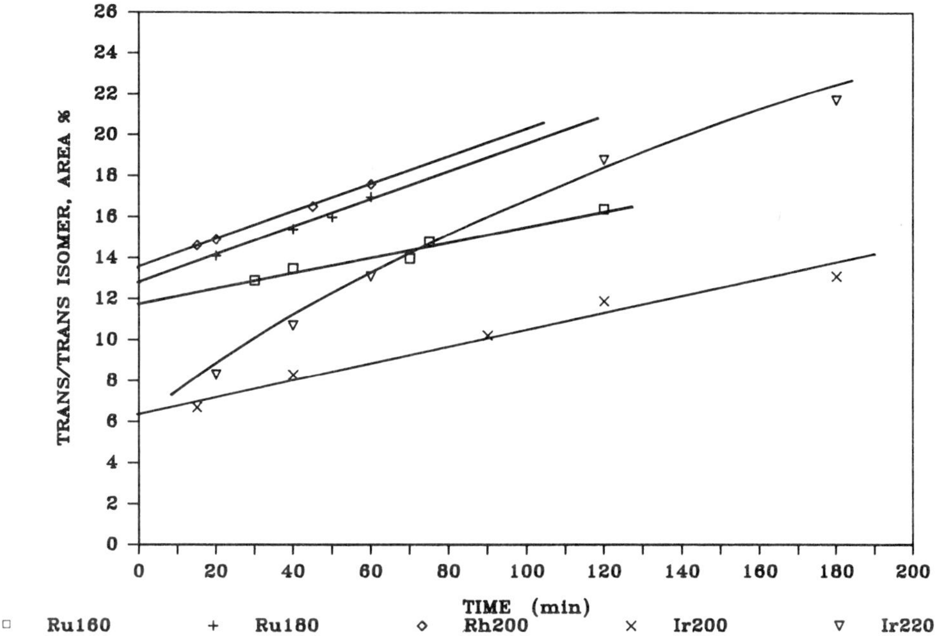

Fig. 10. Isomerization of r-MDA

reaction. The extrapolation of each curve to time equals zero (t_o) indicates the amount of cis isomer being formed by the hydrogenation process, and the slope of each curve is the zero order rate constant of the isomerization process. The cis isomer contents at t_o for the ruthenium and rhodium catalysts are in the range of 62-70% and appear to be relatively insensitive to reaction temperature. The cis isomer contents are in the range found for similar compounds. In reactions done under conditions that precluded isomerization, Siegel obtains 74% cis isomers from p-xylene using PtO_2 and acetic acid at ambient temperatures and 35 psi of hydrogen.[7] Rylander found similar selectivity in the hydrogenation of p-xylene using 5% rhodium on carbon or 5% ruthenium on carbon at 100°C and 1000 psi.[20] The hydrogenation of toluidine under similar conditions using ruthenium and rhodium catalysts yields 85-95% of the cis isomer.[10]

The 5% iridium catalyst displayed some very unusual characteristics. While yielding the highest level of cis isomer (ca. 80%) at 200°C, it had the greatest isomerization rate. The reason for this is unknown but is assumed to be specific to iridium. It is also interesting to note that the isomerization rates were affected only slightly by the poisoning of the hydrogenation reaction.

The three isomers of r-MDA, cis/cis, cis/trans, and trans/trans, are formed from the hydrogenation of the 1/2 r-MDA isomers. We found that the isomer composition at t_o can be calculated by using the isomer contents of the 1/2 r-MDA at t_o, and assuming the same isomer selectivity and the same reaction rate for both 1/2 r-MDA isomers. The calculated values are within a few tenths of a percent of the extrapolated values taken from curves in Fig. 10.

It can be seen from Fig. 3 that the cis/trans isomer remains relatively constant and that the cis/cis isomer content is inversely proportional to the trans/trans isomer. Therefore, we have reported only trans/trans values in this study. The rate constants for the isomerization (k_{tt}) are reported in Table II and key experiments are shown in Fig. 10. Again the most striking feature is the isomer selectivity of the iridium catalyst. At 200-220°C, the iridium catalyst produces an r-MDA with a trans/trans content of 6-7.5% initially. The isomerization rates, however, are similar to those found for ruthenium and rhodium.

Our studies have shown that the hydrogenation of this aromatic diamine displays an interesting combination of hydrogenation, isomerization, and hydrogenolysis chemistry, the fundamentals of which have analogies in the cited literature. It has been demonstrated that each catalyst produces an inherent mix of cis and trans isomers which is relatively insensitive to temperature in the range of 140 to 220°C. In contrast the rate of the isomerization process, which occurs independently of the hydrogenation process and continues after the hydrogenation is complete, increases with increased temperature. The composition of the final product at 100% conversion is dependent on the relative isomerization/hydrogenation rates. From a comparison of the change in rate constants with temperature, it can be seen that the isomerization rate increases more rapidly with increasing temperature than the hydrogenation rate, thus producing a product at a faster rate but with a higher trans-trans content.

ACKNOWLEDGMENTS

I would like to acknowledge the talents and expertise of two gentlemen whose efforts made this work possible: C.D. Ferrell, senior technician, who runs our high pressure lab, and J.W. Webster, from our analytical group.

REFERENCES

1. R.E. Barkdoll, D.C. England, H.W. Gray, W. Kirk Jr., and G.M. Whitman, J. Am. Chem. Soc., **75**, 1156 (1953).
2. R.E. Barkdoll, H.W. Gray, and W. Kirk Jr., J. Am. Chem. Soc., **73**, 741 (1951).
3. G.M. Whitman, U.S. 2,511,028 (June 13, 1950).
4. K. Watanabe, Y. Hirai, and K. Miyata, Japan Pat. 37,758 (1975).
5. H.A. Smith, D.M. Alderman Jr., C.D. Shacklett and C.M. Welch, J. Am. Chem. Soc., **71**, 3772 (1949).
6. P.N. Rylander, **Catalytic Hydrogenation in Organic Syntheses**, Academic Press, New York, p. 196 (1979).
7. S. Siegel and M. Dunkel, Advances in Catalysis, Academic Press, N.Y., **9**, 15 (1965).
8. S. Siegel and V. Ku, Proc. Int. Congr. Catal., 3rd, (Amsterdam, 1964), p. 1199 (1965).
9. B. Van De Graff, H. van Bekkum and B.M. Sepster, Rec. Trav. Chim. Pays-Bas., **87**, 777 (1968).
10. P.N. Rylander and D.R. Steele, Engelhard Ind. Tech. Bull., **4**, 20 (1963).
11. S. Siegel, Advances in Catalysis, Academic Press, New York, **16**, 23 (1966).
12. L.D. Blake, U.S. 3,766,272 (Oct. 16, 1973).
13. G.F. Allen, U.S. 4,448,995 (May 4, 1984).
14. H. Greenfield, J. Org. Chem., **29**, 3082 (1964).
15. A.A. Frost and R.G. Pearson, **Kinetics and Mechanism**, John Wiley & Sons, New York, p. 166 (1953).
16. M.W. Scoggins, L. Skurcenski, and D.S. Weinberg, J. Chromat. Sci., **10**, 678 (1972).

17. F. Coussemant and J.C. Jungers, *Bull. Soc. Chim. Belg.*, **59**, 295 (1950).

18. H. Van Bekkum, H.M. Buurmans, G. Van Minnen-Pathuis, and B.M. Wepster, *Rec. Trav. Chim. Pays-Bas.*, **88**, 779 (1969).

19. H.A. Smith, *Ann. N.Y. Acad. Sci.*, **145**, 72 (1967).

20. P.N. Rylander and D.R. Steele, *Engelhard Ind. Tech. Bull.*, **3**, 91 (1962).

17

Organometallics in Organic Synthesis: A New Diorganozinc Reaction

Alan P. Kozikowski, T. Konoike and A. Ritter

Department of Chemistry
University of Pittsburgh
Pittsburgh, Pennsylvania

ABSTRACT

The mechanistic and stereochemical features of a new organozinc-based substitution process first discovered during a total synthesis of the alkaloid mycotizin α-cyclopiazonic acid, are described. Phenylthio glycosides have served as valuable substrates in studying the nature of this reaction process. Since these sulfur compounds are converted to C-glycosides with some degree of stereoselectivity, the organozinc chemistry does provide a new entry to these biologically active substances. The possible use of this chemistry in the construction of the brevetoxins has also been probed.

I. INTRODUCTION

It is interesting to take note of the fact that the beginnings of organometallic chemistry are often associated with the work of Frankland published in 1848 on the preparation of the organozinc compounds, Me_2Zn and Et_2Zn.[1] While the role of organozinc compounds in organic chemistry dwindled once Grignard discovered the higher reactivity of the organomagnesium halides (especially toward carbonyl compounds), diorganozinc reagents do still play a modest role in chemistry today.[2] The chemistry of the diorganozinc reagents is, of course, guided by the fact that zinc possesses a completely filled low-lying shell of 3d electrons ($3d^{10}4s^2$). The hybridization in dimethylzinc can thus be considered to be sp, which therefore leaves two vacant orbitals for bonding with donor ligands. The presence of these empty orbitals on zinc plays a fundamental stereochemical role in one organozinc reaction which still finds considerable use today, the Simmons-Smith reaction.[3] This reaction, which provides an efficient route to cyclopropanes from alkenes, has been shown to be directed and facilitated by the presence of oxygen substituents. Cyclopenten-4-ol, for example, yields exclusively the cis isomer of bicyclo[3.1.0]hexan-3-ol upon cyclopropanation, a stereochemical result which can be explained by complexation between the hydroxyl group and the zinc reagent.[4] Later on, we will return to the Simmons-Smith reaction as we take up the discussion of our own chemistry.

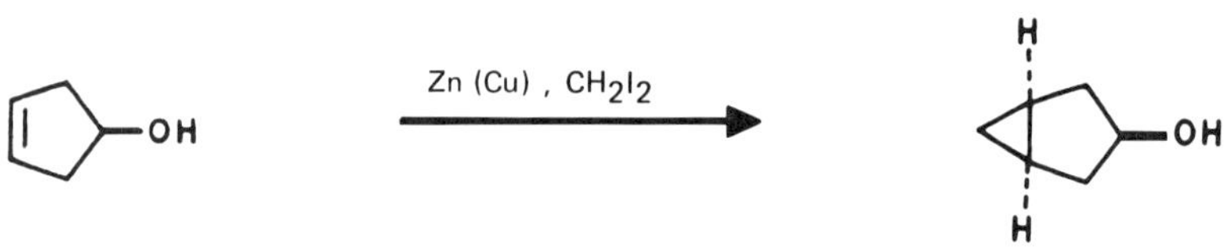

(1)

The second zinc reaction which still finds use in the laboratory today is the Reformatsky process. In this reaction an α-haloester is coupled with a carbonyl compound in the presence of zinc to deliver a β-hydroxy ester.[5] A host of other electrophilic groups can also serve as suitable substrates for this reaction (e.g., acid chlorides, imines, nitriles, ketenes, epoxides, and nitrones, <u>inter</u> <u>alia</u>). Related to this Reformatsky reaction is the Zaitsev process in which allylic, benzylic or propargylic halides are coupled with carbonyl compounds in the presence of zinc to provide alcohols.[6]

$$XCH_2CO_2R^1 + R^2C(O)R^3 \xrightarrow{Zn} R^2(R^3)C(OH)-CH_2CO_2R^1 \quad (2)$$

$$R^1CH=CHCH_2X + R^2C(O)R^3 \xrightarrow{Zn} R^1CH=CHCH_2-C(OH)(R^2)R^3 \quad (3)$$

More recently, Negishi has explored the allylzincation of alkynes in the development of a novel route to cyclopentenones.[7] The chemistry carried out in this work is summarized in the accompanying Eq. (4). The trimethysilyl group was shown to play an important role in the success of the carbometallation process.

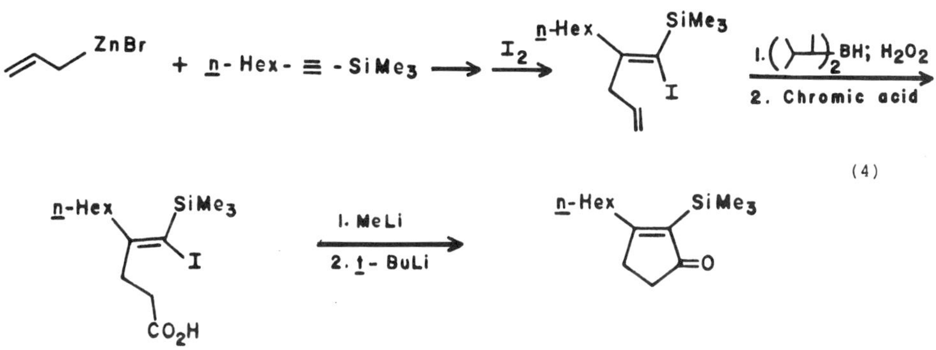

(4)

At this point we shall quit our discussion of the prior art in the area of organozinc chemistry and now turn to a synopsis of our own findings in this still fruitful area of chemistry.

II. THE INITIAL DISCOVERY COMING FROM AN ALKALOID SYNTHESIS

For several years we had engaged ourselves in an effort to synthesize the structurally and biochemically unique mycotoxin α-cyclopiazonic acid (structure work done by Holzapfel)[8] During this effort we had reached a stage at which the sulfur bearing tetracycle **1** could be easily procured through an intramolecular Michael reaction of a sulfenylated ketone to an α-acetamidoacrylate. To carry the synthesis to its end, we needed to convert the phenylthio group present in **1** to a methyl group, thus completing the installation of the geminal methyl groups at the C-10 quaternary carbon center. While a fairly diverse range of methylmetallics (e.g., derivatives of copper, aluminum, magnesium, and titanium) were examined to achieve this conversion, success was not found until we happened to expose **1** to a large

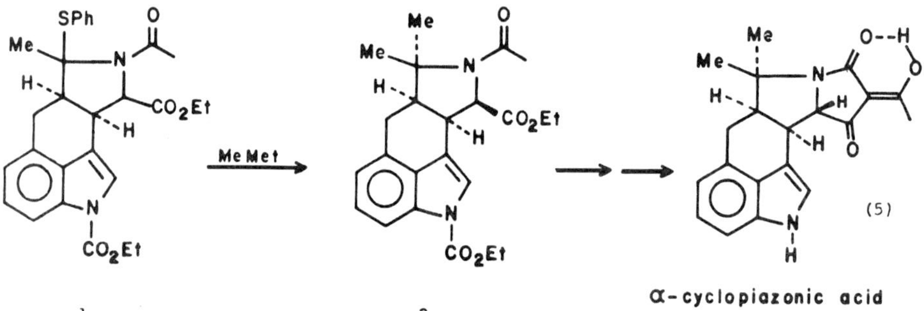

1 2

excess of dimethylzinc in chloroform. We happened to have some dimethylzinc on hand, for this was used in the preparation of Reetz's titanium reagent, Me_2TiCl_2, a compound reported to be useful for the conversion of carbonyl compounds to their geminal dimethyl derivatives.[9]

$$R^1(R^2)C{=}O \xrightarrow{Me_2TiCl_2} R^1(R^2)C(Me)_2 \qquad (6)$$

While the Me_2TiCl_2 failed to convert **1** to **2**, we found instead that by simply treating **1** with dimethylzinc in chloroform at 75°C, the desired geminal dimethyl derivative could be generated. While this result was very welcome, since it meant the end to a long struggle, we were somewhat mystified as to the mechanistic details of the overall process. Certainly, it did seem possible that the mildly electrophilic zinc center might sufficiently well tug on the phenylthio group so as to give rise to an acyliminium salt. This iminium salt could in turn be intercepted by one of the nucleophilic methyl groups of the zinc reagent to produce the gem-dimethyl compound. As we shall see, this notion was not entirely correct, for we subsequently discovered an important solvent effect on the reaction rate.

III. APPLICATION OF THE ZINC REACTION TO PHENYLTHIO GLYCOSIDES - A NEW C-GLYCOSIDE SYNTHESIS

In order to further probe both the synthetic and mechanistic aspects of this new diorganozinc reaction, we decided to try to extend our chemistry to phenylthio glycosides. Our purpose in choosing carbohydrates as substrates was twofold: (a) these substrates were expected to provide valuable stereochemical information on the course of the zinc reaction; and (b) if the yields and stereoselectivity observed in this reaction were good, then a valuable new entry to an important class of natural product substances, the C-glycosides, would be at hand. Much intensive effort has been expended in devising new entries to the C-glycosides over the past few years primarily because of the significant biological activity exhibited by these special carbohydrate structures (e.g., antibiotic, antiviral and antitumor).[10]

We also take note of the fact that by applying our zinc reaction to these carbohydrate derivatives, we are again employing a system, which like the cyclopiazonic acid intermediate, has the phenylthio group located on a carbon atom which bears an additional heteroatom. The importance of this feature will become apparent as we proceed.

IV. RESULTS AND DISCUSSION

To begin our study of the zinc reaction with carbohydrates as substrates, we examined first the reactions of the phenylthio glycoside **4**, a compound readily prepared by an oxidation-reduction condensation reaction between the protected lyxose derivative **3**,[11] diphenyldisulfide, and tri-n-butylphosphine.[12] The structure of this new phenylthio glycoside has been independently confirmed by a single crystal X-ray analysis. The fact that the anomeric hydroxyl group of **3** is replaced by the phenylthio group with apparent inversion of stereochemistry is suggestive of reaction through an S_N2 type mechanism. When a chloroform solution of **4** was treated subsequently with excess dimethylzinc at 75°C in a Kimax tube for 12 h, a single less polar product was obtained.[13] This product was assigned the structure **5α** (R = Me) based upon a detailed NMR analysis (vide infra).

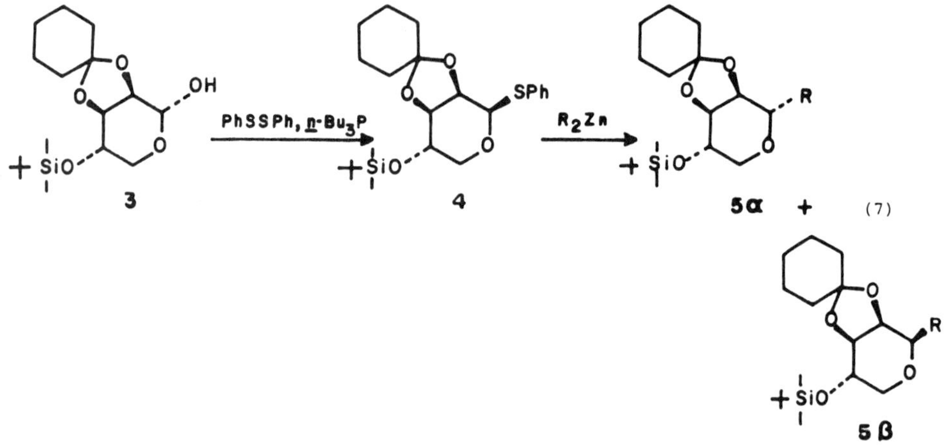

(7)

Interestingly, when the above reaction was carried out in toluene, benzene, or hexanes, in the absence of chloroform, no C-methyl glycoside was formed. Additionally, when a toluene solution of **4** was treated with five equivalents of methylene iodide and ten equivalents of diethylzinc at 65°C for 2 h, the C-ethyl glycoside was obtained in an isolated yield of 78%.[14] Likewise, by substituting carbon tetrachloride for the methylene iodide, the C-ethyl glycoside was again obtained, but under a milder set of reaction conditions (35°C for 2 h).

Early on in our studies we felt that the zinc reagent might be reacting with the phenylthio glycoside through a ligand exchange process in which the thiophilic zinc center might sufficiently tug on the sulfur atom to form an oxonium ion intermediate. However, our findings regarding the necessity of the halocarbon additives does logically suggest that the dialkylzinc is performing a dual role. First, it reacts with the methylene iodide or carbon tetrachloride (by a free radical pathway)[2b] to deliver a carbenoid, a reaction which represents a well known variant of the standard Simmons-Smith cyclopropanation procedure (methylene iodide plus zinc-copper couple). Subsequently, this zinc carbenoid interacts with the sulfur substituent to generate a sulfonium ion intermediate (this can also be viewed as a sulfur ylid if the zinc and its accompanying ligand are neglected). The formation of this "sulfonium ion" species may be additionally favored through complexation of the zinc center with the peripheral ether oxygens of the carbohydrate nucleus.

As an aside we note here that the direct formation of sulfur ylids from the reaction of sulfides with carbenes is well known. The copper catalyzed decomposition of dimethyl diazomalonate in the presence of the allyl sulfide **6** leads, for example, to the [2,3] sigmatropic rearrangement products **8** and **9** via the intermediacy of the sulfur ylid **7**.[15]

Sulfur Ylid Formation

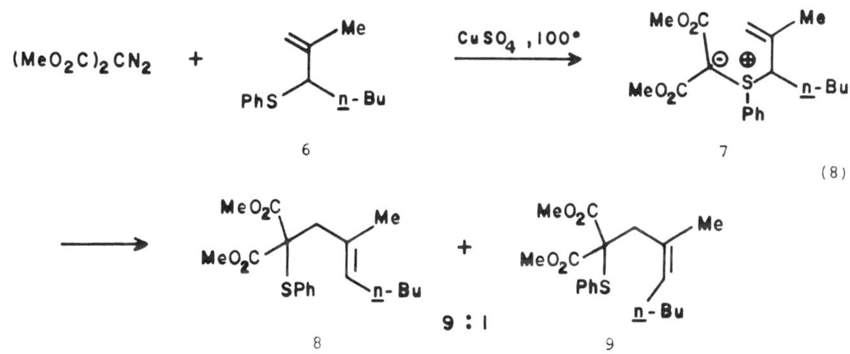

(8)

Allyl phenyl thioethers have also been converted in a single step to homoallyl phenyl thioethers by the use of diethylzinc in combination with methylene iodide or ethylidene iodide (Eq. 9). Apparently, this variation of

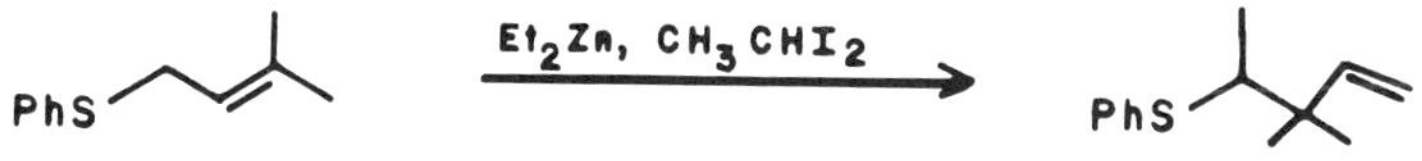

(9)

the Simmons-Smith reaction again provides for a sulfur ylid type intermediate which can undergo a subsequent sigmatropic rearrangement. The standard Simmons-Smith conditions (methylene iodide plus zinc-copper or zinc-silver couple) failed in this case to initiate the [2,3] sigmatropic rearrangement process, for only insoluble polymeric materials were produced.[16]

Returning to our carbohydrate derivatives, once the activated organozinc-sulfur complex has formed, it either reacts with the excess diorganozinc reagent present in solution by a direct displacement mechanism, or (better) it breaks down to a cyclic oxonium ion which then undergoes attack via the kinetically favored axial addition mode (Eq. 10).

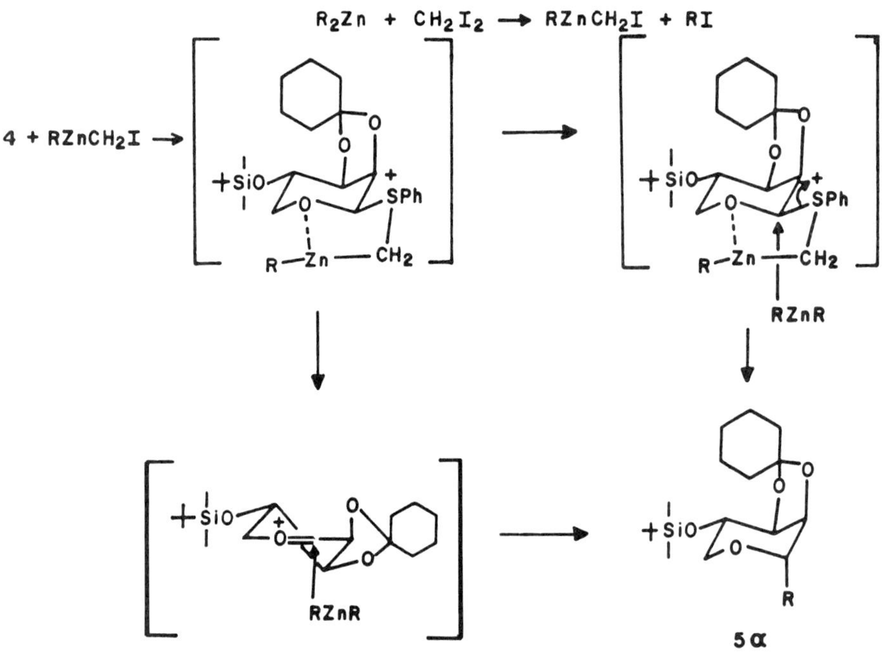

(10)

Conformational energetics and the kinetic anomeric effect thus act synergistically to control the addition stereochemistry in the latter case.[17]

As can be seen from the accompanying Table I, the dimethyl-, diethyl-, and di-n-propylzinc reagents provided within the limits of ^1H NMR and HPLC analysis only the α-C-glycosides. The assignment of structure to these materials is based upon detailed spin-spin decoupling experiments carried out on the desilylated products. The anomeric proton of these α-isomers (with R equatorial) exhibits a coupling constant of 8.2-9.2 Hz with the neighboring pyran ring proton. Further proof of structure for the C-n-

TABLE I

Reactions of Diorganozinc Reagents with the Phenylthio Glycoside 4

Entry	Diorganozinc Reagent R_2Zn (R)	R-C-Glycoside[a] % α	% β	Isolated Chemical Yield (%)
a	Me	>94	<6	72
b	Et	>94	<6	78
c	n-Pr	>94	<6	76
d	n-Bu	93	7	75
e	s-Bu	64	36	72
f	i-Pr	82	18	73
g	Allyl	60	40	35[b]

[a] A standard set of reaction conditions were adopted which consisted of treating **4** with 15 equivalents of the diorganozinc reagent and 8 equivalents of CH_2I_2 in toluene at 40 °C for 2 h.

[b] The low yield in this case probably stems from our difficulty in preparing the pure diallylzinc reagent.

propyl glycoside was obtained by comparing it with an authentic sample generated by hydrogenating the previously known C-allyl glycoside **5α** (R = $CH_2CH{=}CH_2$).[18] With di-n-butylzinc a 93:7 mixture of the α and β isomers was obtained.

The secondary alkylzinc reagents proved somewhat less selective in their reactions with **4**. While the α-isomer still predominated, more of the β-isomer was detected. Production of an increased percentage of the β-isomer might be accounted for by assuming reaction through a transition state which may look somewhat more like the starting oxonium ion (i.e., an earlier transition state) with therefore a somewhat lessened preference for α-face attack.

In order to test the applicability of this orgnozinc reaction to other carbohydrate substrates, as well as to further confirm our suspicions that this reaction was proceeding through an open oxonium ion intermediate, we have examined the chemistry of the additional phenylthio glycosides tabulated below. The methods by which these substrates were prepared are shown in the accompanying equations 11-17.

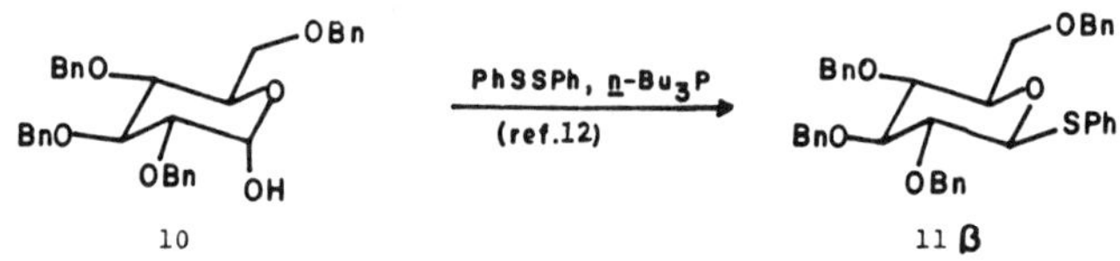

(11)

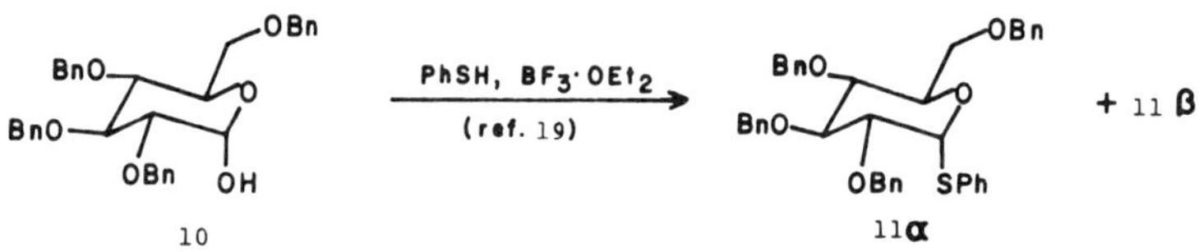

(12)

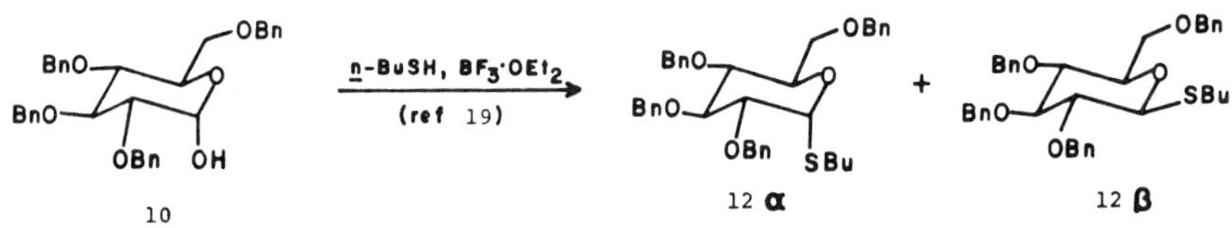

(13)

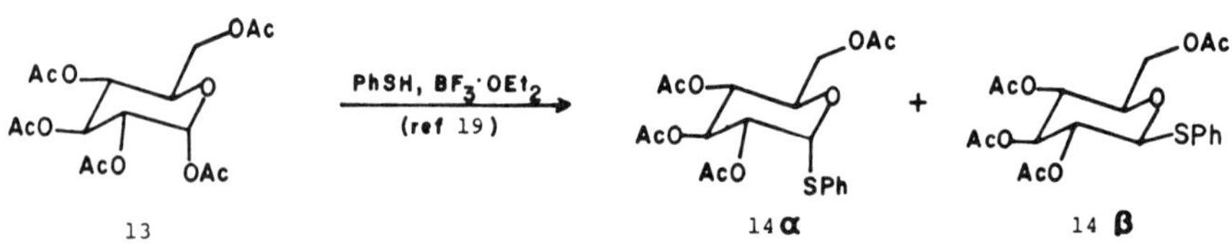

(14)

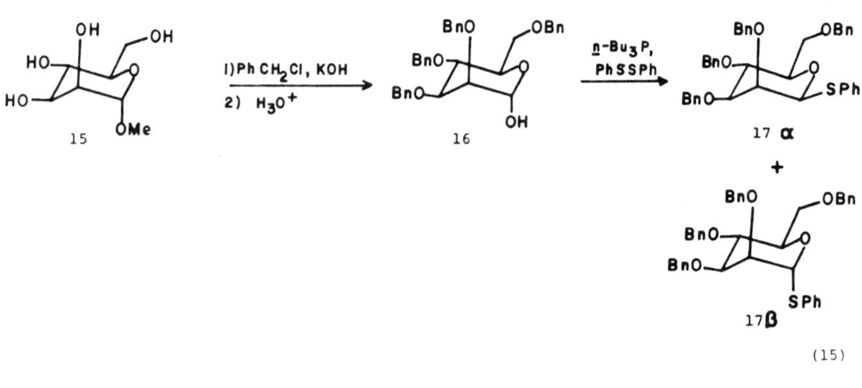

(15)

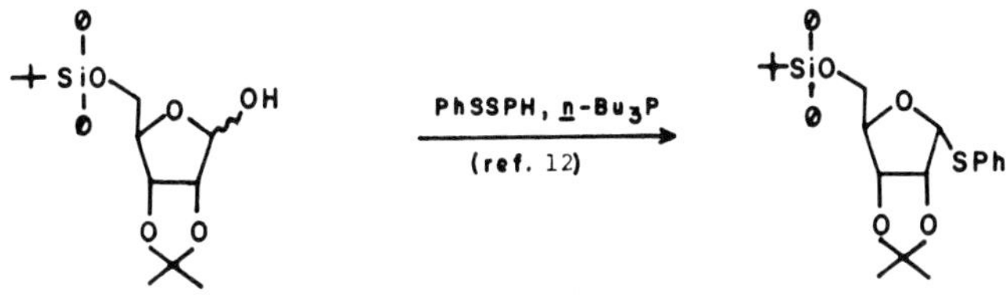

18 19α (16)

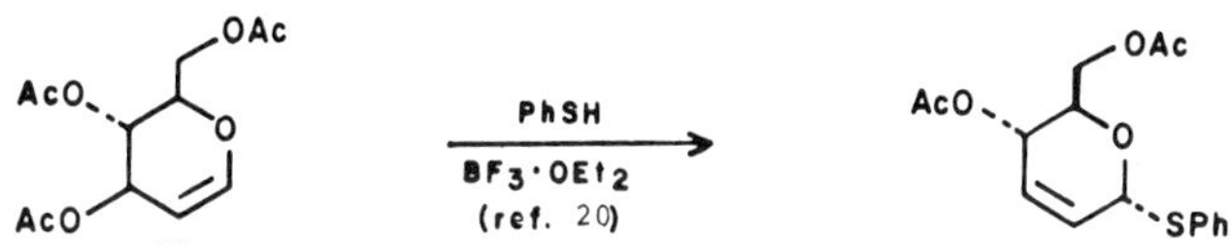

20 21α (17)

The α- and β-phenylthio substituted derivatives of glucose (**11α** and **11β**) provided us with an excellent opportunity to examine the effect of the starting stereochemistry on the product stereochemistry. If a cyclic oxonium ion intermediate was being formed in the course of the organozinc mediated reactions, then **11α** and **11β** should give rise to nearly identical product ratios. Indeed, as tabulated in Table II, the product ratios observed for the reactions of **11α** and **11β** with $Me_2/Zn/CH_2I_2$ were quite similar. As should be noted, the major isomer is that arising from an axial addition mode. On reacting **11β** with diethylzinc or diphenylzinc in the presence of methylene iodide, the product isomer ratios were again found to be approximately three parts α to one part β C-glycoside. The nature of the diorganozinc reagent thus does not dramatically influence the product distribution. The lower level of stereoselectivity observed with this glucose derivative compared to that observed for the lyxose case presumably reflects the ability of the 1,3-dioxolane ring in the latter to significantly shield the pyran ring from attack in the equatorial direction (i.e., concave face attack is hindered).

Interestingly, the butylthio derivatives of glucose differed considerably in their reactivity towards Me_2Zn/CH_2I_2. While the isomer **12α** reacted within 4 h at 50°C to provide a 73:27 mixture of α/β-C-methyl glycosides, the β isomer required heating at 90°C for 12 h in order to generate even small amounts of the desired products. On treating **12β** with Me_2Zn/CH_2I_2 at 50°C for 4 h followed by workup with HCl, only the

TABLE II

Reactions of Other Phenylthio Glycosides with Diorganozinc Reagents

Starting Phenylthio Glycoside	Organozinc Reagent	Reaction Conditions[a] Temp (°C)/time (hr)	Product Ratio (α/β)	Yield (%)
11α	Me_2Zn	50/3	69/31	60
11β	Me_2Zn	45/3.5	76/24	67
11β	Me_2Zn	5/72	77/23	76
11β	Et_2Zn	50/3	72/28	67
11β	Ph_2Zn	50/2	75/25	48
12α	Me_2Zn	50/4	73/27	55
12β	Me_2Zn	90/12	~67/33	24
12β	Me_2Zn	50/4	α-chloride (see text)	44
17α, 17β	Me_2Zn	40/3	84/16[b]	80
17α, 17β	Ph_2Zn	50/1	81/19	40
19α	Me_2Zn	48/20	37/63	91
21α	Et_2Zn	25/2	75/25	68

[a]These reactions were all carried out using 15 equivalents of the diorganozinc reagent and 8 equivalents of methylene iodide in toluene as solvent.

[b]Product ratio assigned after debenzylation and conversion to the tetraacetates.

corresponding glycosyl chloride could be isolated. Apparently, the sluggish character of **12β** in the zinc reaction reflects both the poorer leaving group ability of a butylsulfonium ion species relative to a phenylsulfonium ion group as well as the inability of the lone pairs on the pyran ring oxygen to participate in the cleavage process.[17]

On reacting a 1:1 mixture of the mannose derived phenylthio glycosides **17α** and 17β with either dimethylzinc or diphenylzinc, the products of axial attack again predominated. The presence of the β-axial benzyloxy group at C-2 may be responsible for the slight enhancement in axial selectivity relative to the glucose examples.

When the α-phenylthio derivative of ribose (**19α**) was subjected to Me_2Zn/CH_2I_2, a mixture of C-methyl furanosides was formed in excellent yield. The major product was assigned β stereochemistry based upon an examination of the chemical shift and coupling data of the α and β isomers vis-à-vis that reported for related compounds.[21]

On exposing the O-acetylated derivative **14β** of glucose to the standard reaction conditions, it was interesting to find that none of the expected C-methyl glycoside was formed. Instead, the 1,2-O-isopropylidene derivative **23** of glucose was isolated. This compound, whose structure was made apparent by ^1H NMR analysis, presumably results from attack of the C-2 acetate group on the anomeric center backside to the departing sulfur group. The resulting acyloxonium ion then suffers attack by the dimethylzinc leading to the isopropylidene derivative. Not unexpectedly, the α isomer of **14** gave rise to a complex mixture of products.

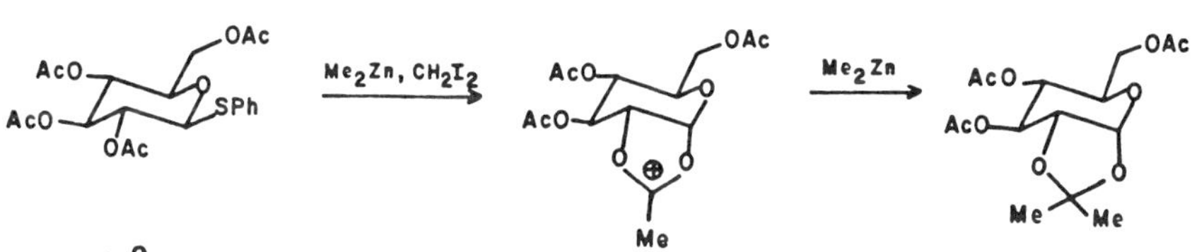

14β 22 23

(18)

The zinc reaction was also applied to the phenylthio derivative **21α** prepared from glucal triacetate **20** by the Ferrier rearrangement.[20] In this case, the new alkyl group entered into the position α to the pyran ring oxygen, and by literature comparisons the major isomer was found to have α stereochemistry.[10f]

At this point, we felt that we had studied a sufficient number of phenylthio glycosides to firmly establish the scope of the organozinc reaction for C-glycoside synthesis. We did, however, wish to probe the potential applicability of this chemistry to the installation of the angular methyl groups in the unusual marine natural products, the brevetoxins.[22]

We had imagined that an appropriately substituted phenylthio bearing pyranopyran, generated for a model study by the reaction of a 2-(3-hydroxypropyl)tetrahydropyran-3-one with thiophenol, might serve as a suitable synthetic precursor. A question remained, however, as to whether

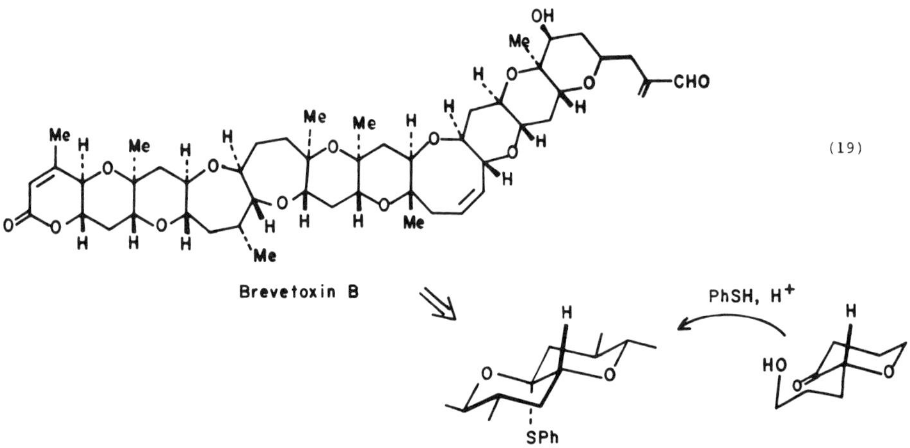

(19)

the required trans-fused pyranopyran would emerge from this substitution reaction. When the 2-(3-hydroxypropyl)tetrahydropyran-3-one prepared as shown in equation 20 was cyclized with thiophenol and acid, a 3:1 mixture of the cis- and trans-fused pyranopyrans **29** and **30** resulted. On exposing each of these compounds individually to dimethylzinc and methylene iodide, a good yield of the methylated pyranopyrans could be isolated. Unfortunately, in each case nearly a 10:1 mixture of the cis- and trans-fused products **31** and **32** was formed, respectively. Product stereochemistry was therefore again independent of the starting material stereochemistry. While we are at present uncertain as to what factors are responsible for this stereochemical result, i.e., product stability, steric approach control, or even some "kinetic gauche effect", the ultimate stereochemical result serves as a poor harbinger of the applicability of the zinc reaction to the synthesis of the brevetoxins.

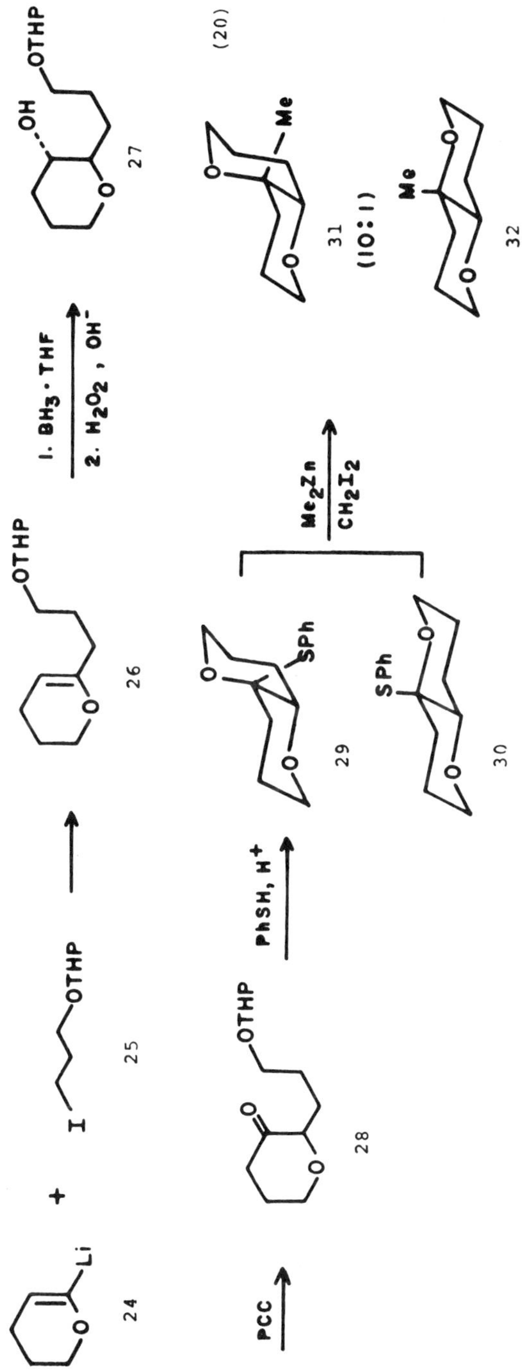

In one final effort, we have attempted to extend the foregoing chemistry to another type of α-hetero atom containing sulfur system, the thioacetals and thioketals. On subjecting these compounds to various diorganozinc reagents in the presence of different halocarbons, rather complicated mixtures of products were formed.

The bis(phenylthio)acetal **33** of phenylacetaldehyde reacted, for example, with diethylzinc in the presence of chloroform as solvent to give rise to the expected substitution product **34** in which one of the phenylthio groups was replaced by an ethyl group. Additionally, some of the product **35** of reduction was isolated as well. In other cases, elimination of one of the thio groups to afford a vinyl sulfide was also observed (e.g., **36** → **37**).[17] Due to the complicated mixtures which resulted from these reactions, we chose not to pursue this chemistry further.

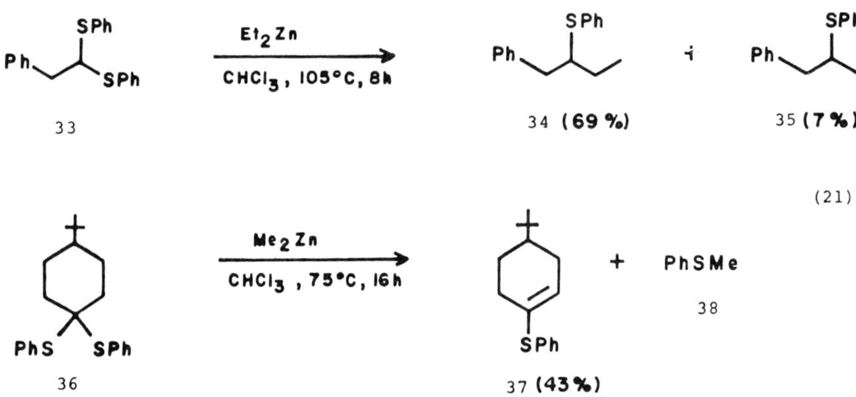

(21)

We do note, however, that Rodriguez and Nickon have recently reported that dithioketals can be converted to vinyl sulfides in good yield on exposure to diethyl zinc and methylene iodide (Eq. 22). A very reasonable

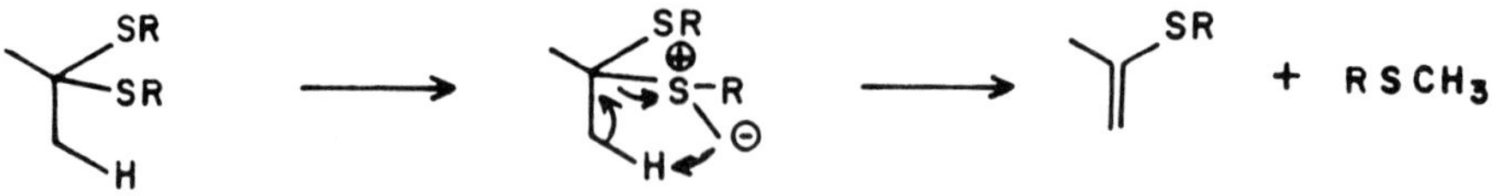

(22)

mechanism involving capture of the zinc carbenoid by one of the sulfur atoms followed by a cis-cycloelimination reaction of the intermediate sulfur ylid was proposed.[23] While their elimination reaction appears to proceed in better yield (67-92%) than our own, this difference may be due to the use of a dialkyl sulfide rather than an alkyl phenyl sulfide in the elimination process.

In summary, the new zinc reaction described herein would appear to provide a valuable method for the introduction of alkyl and aryl groups into organic molecules at carbon sites adjacent to a heteroatom substituent. In the case of the phenylthio glycosides, the stereochemistry of the substitution reaction appears to be consistent with axial attack on an oxonium ion intermediate. The zinc chemistry thus offers a further useful addition to the rapidly growing number of new methods available for C-glycoside construction. While the applicability of this chemistry in alkaloid construction has been subjected to less extensive tests, the very favorable observation made in the α-cyclopiazonic acid work would certainly seem to guarantee its successful application in related nitrogen containing systems.

REFERENCES

1. E. Frankland, Liebigs Ann. Chem., **71**, 171 (1848).
2. For reviews on the synthesis and applications of orgnozinc reagents, see: (a) J. Furukawa and N. Kawabata, Organozinc Compounds in Synthesis, **Advances in Organometallic Chemistry** (F.G.A. Stone and R. West, eds.) Academic Press, New York, Vol. 12, p. 83 (1974); (b) J. Boersma, Organozinc Compounds, **Comprehensive Organometallic Chemistry** (G. Wilkinson, ed.), Pergamon Press, Oxford, Vol. 2, p. 823 (1982); (c) K. Nutzel, Organo-zink-Verbindungen, **Methoden der Organischen Chemie**, (Houben-Weyl), (E. Müller, ed.), Georg Thieme Verlag, Stuttgart, XIII/2a, p. 553 (1973).
3. E.P. Blanchard and H.E. Simmons, J. Am. Chem. Soc., **86**, 1337 (1964); E.P. Blanchard, H.E. Simmons, and J.S. Taylor, J. Org. Chem., **30**, 4321 (1965); H.E. Simmons and R.D. Smith, J. Am. Chem. Soc., **80**, 5323 (1958); H.E. Simmons and R.D. Smith, J. Am. Chem. Soc., **81**, 4256 (1959).

4. E.J. Corey and R.L. Dawson, J. Am. Chem. Soc., **85**, 1782 (1963); W.G. Dauben and G.H. Berezin, J. Am. Chem. Soc., **85**, 468 (1963); W.G. Dauben and A.C. Ashcraft, J. Am. Chem. Soc., **85**, 3673 (1963).

5. M. Gaudemar, Organomet. Chem. Rev. (A), **8**, 183 (1972).

6. N.I. Sheverdina and K.A. Kocheshkov, **Methods of Elemento-Organic Chemistry**, North-Holland Publ., Amsterdam, Vol. 3 (1967).

7. E. Negishi and J.A. Miller, J. Am. Chem. Soc., **105**, 6761 (1983).

8. A.P. Kozikowski and M.N. Greco, J. Am. Chem. Soc., **106**, 6873 (1984).

9. M.T. Reetz, Organotitanium Reagents in Organic Synthesis, A Simple Means to Adjust Reactivity and Selectivity of Carbanions, Topics in Current Chemistry, **106**, 3 (1982).

10. For other methods of C-glycoside synthesis see: (a) G.D. Daves and C.C. Cheng, Prog. Med. Chem., **13**, 303 (1976); (b) S. Hanessian and A.G. Pernet, Adv. Carbohydr. Chem. Biochem., **33**, 111 (1976); (c) A.P. Kozikowski and K.L. Sorgi, Tetrahedron Lett., **23**, 2281 (1982); (d) T.L. Cupps, D.S. Wise, and L.B. Townsend, J. Org. Chem., **47**, 5115 (1982); (e) M.D. Lewis, J.K. Cha, and Y. Kishi, J. Am. Chem. Soc., **104**, 4976 (1982); (f) R.D. Dawe and B. Fraser-Reid, J. Chem. Soc., Chem. Commun., 1180 (1981); (g) S. Danishefsky and J.F. Kerwin, J. Org. Chem., **47**, 3803 (1982); (h) R.R. Schmidt and M. Hoffmann, Tetrahedron Lett., **23**, 409 (1982); (i) G.E. Keck, E.J. Enholm, and D.F. Kachensky, Tetrahedron Lett., **25**, 1867 (1984); (j) A. Hosomi, Y. Sakata, and H. Sakurai, Tetrahedron Lett., **25**, 2383 (1984); (k) A.O. Stewart and R.M. Williams, J. Am. Chem. Soc., **107**, 4289 (1985); (l) G.H. Posner and S.R. Haines, Tetrahedron Lett., **26**, 1823 (1985); (m) K.C. Nicolaou, R.E. Dolle, A. Chucholowski, and J.L. Randall, J. Chem. Soc., Chem. Commun., 1153 (1984); (n) J.-M. Beau and P. Sinaÿ, Tetrahedron Lett., **26**, 6185 (1985); (o) B. Giese and J. Dupuis, Ang. Chem. Int. Ed. Engl., **22** 622 (1983).

11. A.P. Kozikowski, K.L. Sorgi, B.C. Wang and Z.-b. Xu, Tetrahedron Lett. **24**, 1563 (1983).

12. I. Nakagawa and T. Hata, Tetrahedron Lett., 1409, (1975); R.M. Williams and A.O. Stewart, Tetrahedron Lett., **24**, 2715 (1983). The structure of **4** was confirmed by a single crystal X-ray analysis.

13. The dimethylzinc was prepared by reacting methyl iodide with a zinc-

copper couple. R.C. Krug and P.J.C. Tang, J. Am. Chem. Soc., **76**, 2262 (1954).

14. Diethylzinc is available commercially (Aldrich) as a 1.6 M solution in toluene.
15. W. Ando, S. Kondo, K. Nakayama, K. Ichibori, H. Kohoda, H. Yamato, I. Imai, S. Nakaido, and T. Migita, J. Am. Chem. Soc., **94**, 3870 (1972); W. Ando, M. Yamada, E. Matsuzaki, and T. Migita, J. Org. Chem., **37**, 3791 (1972).
16. T. Cohen and Z. Kosarych, Tetrahedron Lett., **23**, 3019 (1982).
17. P. Deslongchamps, **Stereoelectronic Effects in Organic Chemistry**, Pergamon Press, Oxford (1983); A.J. Kirby, **The Anomeric Effect and Related Stereoelectronic Effects at Oxygen**, Springer-Verlag, New York (1983); B. Capon, Chem. Rev., **69**, 407 (1969).
18. A.P. Kozikowski and K.L. Sorgi, Tetrahedron Lett., **25**, 2085 (1984).
19. R.J. Ferrier and R.H. Furneaux, Carbohydr. Res., **52**, 63 (1976).
20. W. Priebe and A. Zamojski, Tetrahedron, **36**, 287 (1980).
21. H. Ohrui, G.H. Jones, J.G. Moffatt, M.L. Maddox, A.T. Christensen, and S.K. Bryam, J. Am. Chem. Soc., **97**, 4602 (1975).
22. Y.-Y. Lin, M. Risk, S.M. Ray, D. VanEngen, J. Clardy, J. Golik, J.C. James, and K. Nakanishi, J. Am. Chem. Soc., **103**, 6773 (1981).
23. A.D. Rodriguez and A. Nickon, Tetrahedron, **41**, 4443 (1985).

Selected Topics

18

Catalysis from the Perspective of an Organic Chemist: Common Problems and Possible Solutions

Cynthia A. Maryanoff, John E. Mills,
Robin C. Stanzione and John T. Hortenstine, Jr.

McNeil Pharmaceutical
Spring House, Pennsylvania

ABSTRACT

Heterogeneous catalysis offers some distinct advantages over other chemical methods for the reduction of certain functional groups. In addition, different metal catalysts or the same metal on different supports offer selectivity in the reduction of different functional groups. However, heterogeneous catalysts are often susceptible to poisoning. Examples of problems encountered in the laboratory scale reduction of pyridine derivatives will be presented with methods used to solve them.

The large variety of reactions which can be achieved using homogeneous palladium catalysis under mild conditions is rapidly making this the method of choice for the preparation of certain carbon-carbon bonds. The demands for purity of bulk drug in the pharmaceutical industry introduce some constraints not necessarily found in other industrial applications. Problems encountered during the scale-up from laboratory to pilot plant of homogeneously catalyzed coupling reactions of aryl halides with phenylacetylene will be presented.

I. INTRODUCTION

Over the last thirty years, the field of organic catalysis has grown from predominantly heterogeneous hydrogenation reactions to a plethora of reactions mediated by organometallic species, heterogeneous or homogeneous hydrogenation reactions, and phase transfer catalyzed reactions. Despite the diversity of chemistry found in metal catalyzed reactions, many of these reactions are still plagued by the same problems of catalyst poisoning and unpredictability which were recognized early in the development of catalytic hydrogenations. This paper presents some uses of catalysis, some problems encountered in catalytic processes, and some methods to circumvent these problems at McNeil Pharmaceutical. Specifically, we will highlight work on the hydrogenation of 4-substituted pyridine derivatives to exemplify problems we have encountered in heterogeneous catalysis, and work on the preparation of diarylacetylene derivatives to illustrate difficulties in the use of homogeneous catalysis.

II. RESULTS AND DISCUSSION

A. Fenoctimine

Fenoctimine [4-(diphenylmethyl)-1-[(octylimino)methyl]-piperidine, McN-4097], Fig. 1, is an antisecretory compound discovered at McNeil Pharmaceutical which displayed a marked separation of anticholinergic and antisecretory effects in animal models.[1] Thus, this compound was expected to be useful in the treatment of ulcers without causing side effects such as dry mouth. Work on this project was divided into two synthetic objectives.

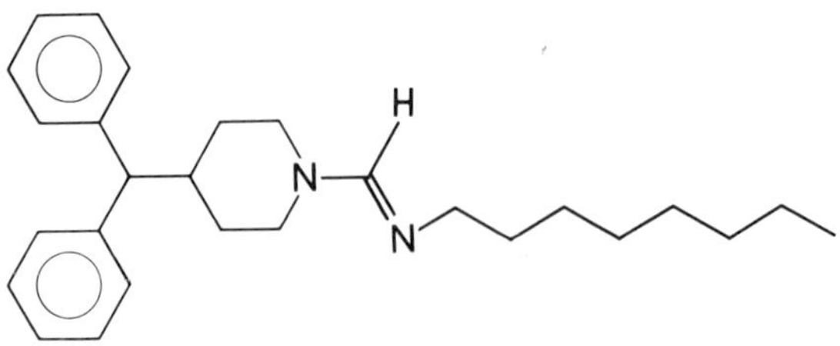

Fig. 1. Fentoctimine (McN-4097).

The first objective was to develop a viable large scale synthesis of fenoctimine. The second objective was to develop a synthesis of 4-diphenylmethylpiperidine using a low pressure hydrogenation.

1. Preparation of 4-Diphenylmethylpiperidine

In order to identify a route utilizing hydrogenation pressures of 50 psi or less, several synthesis of 4-diphenylmethylpiperidine were evaluated on a laboratory scale. One successful method investigated for its preparation was the reaction of methyl isonicotinate with phenylmagnesium bromide followed by multiple step reductions of the resulting α-4-pyridylbenzhydrol (Fig. 2). Although over 40 different sets of experimental conditions were used to try to convert the α-4-pyridylbenzhydrol directly to 4-diphenylmethylpiperidine, no one step low pressure reduction was found for this conversion. Table I shows some of the solvent/catalyst systems which were investigated for the hydrogenation of α-4-pyridylbenzhydrol.

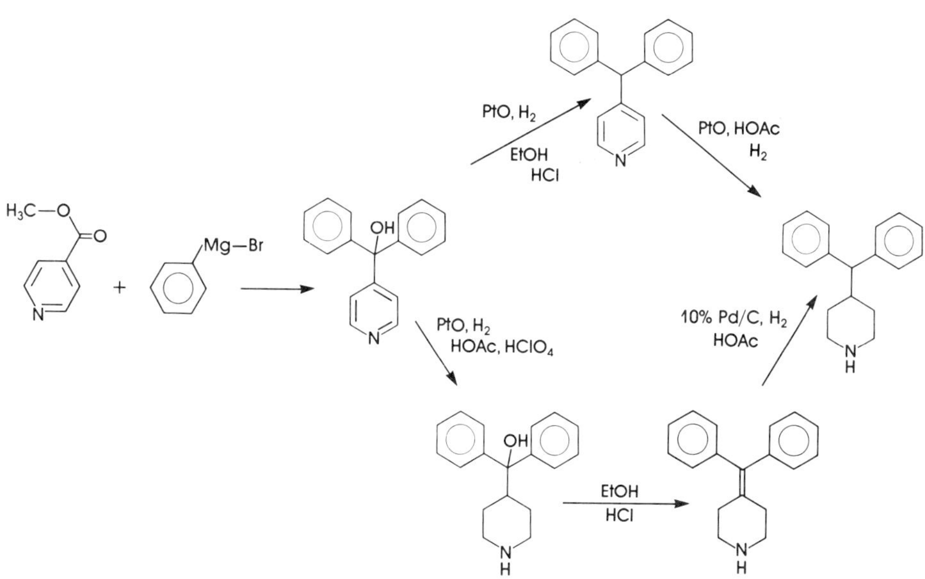

Fig. 2. Preparation of 4-diphenylmethyl piperidine.

TABLE I

Catalytic Hydrogenations of Alpha-4-Pyridylbenzhydrol

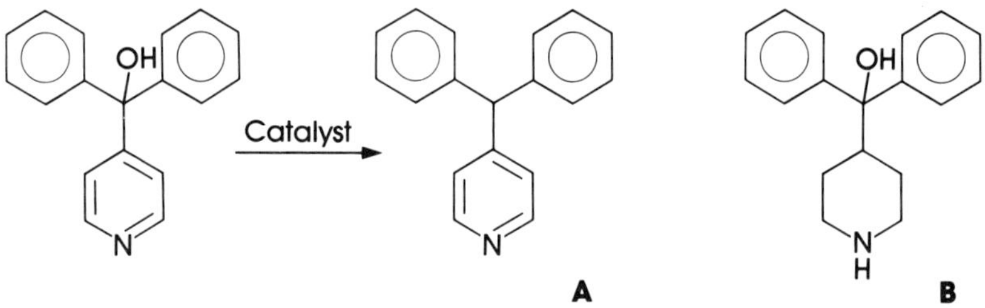

CATALYST	SOLVENT	PRODUCT	YIELD
PtO	Abs EtOH/HCl (g)	A	99%
PtO	HOAc/Perchloric Acid	B	100%
RaNi	Abs EtOH/HOAc (or NaOH)	NR	—
Pd/C	HOAc	B	95%
Pd/C	HOAc/HCl (g)	A	95%
Rh/Alumina	EtOH/conc. HCl	NR	—
Pd/$CaCO_3$	MeOH	NR	—

The results indicate how sensitive this system is to hydrogenation conditions. For example, addition of hydrogen chloride to an acetic acid solution of α-4-pyridylbenzhydrol resulted in hydrogenolysis of the benzylic hydroxyl while the same hydrogenation over palladium on carbon produced the piperidine carbinol in the absence of the hydrogen chloride.

A two-step procedure involved hydrogenolysis of the benzylic hydroxyl group followed by hydrogenation of the pyridine ring. Thus, reduction of α-4-pyridylbenzhydrol to 4-diphenylmethylpyridine was achieved at 50 psi of hydrogen using platinum oxide in absolute ethanol saturated with hydrogen chloride. This pyridine derivative could be hydrogenated to the desired piperidine using platinum oxide in acetic acid.

Alternatively, hydrogenation of α-4-pyridylbenzhydrol to the piperidine carbinol was achieved over platinum oxide at 50 psi in a mixture of acetic acid and perchloric acid under a hydrogen atmosphere. The piperidine carbinol could be converted to 4-diphenylmethylenepiperidine through treatment with ethanolic hydrogen chloride. The resulting tetrasubstituted double bond was hydrogenated using 10% Pd/Carbon in acetic acid under 50 psi of hydrogen.

It was also found that the reduction of α-4-pyridylbenzhydrol to 4-diphenylmethylpyridine could be easily performed on the kilogram scale using chemical methods (Fig. 3). Thus, addition of concentrated hydrochloric acid to a slurry of α-4-pyridylbenzhydrol and metallic tin in refluxing ethanol resulted in isolation by filtration of 4-diphenylmethylpyridine hydrochloride directly from the reaction mixture after cooling. The free base was isolated by dissolution of the filter cake in chloroform, filtration to remove excess tin, extraction with aqueous base, and concentration of the organic phase. However, subsequent lots of material prepared through this procedure would behave differently when hydrogenated over platinum oxide. It appeared that an unknown impurity in the 4-diphenylmethylpyridine prepared through the chemical reduction made catalytic hydrogenation of this compound irreproducible. It was presumed that the unknown impurity was a poison to the catalyst and that the impurity was present at very low concentrations. Elution of the chloroform solution of the 4-diphenylmethylpyridine free base through an alumina plug prior to concentration resulted in isolation after concentration of a solid which was hydrogenated reproducibly. Adsorption of trace impurities on the surface of alumina or silica gel prior to catalytic hydrogenation is a technique which is often effective in enhancing the reproducibility of heterogeneous hydrogenations.

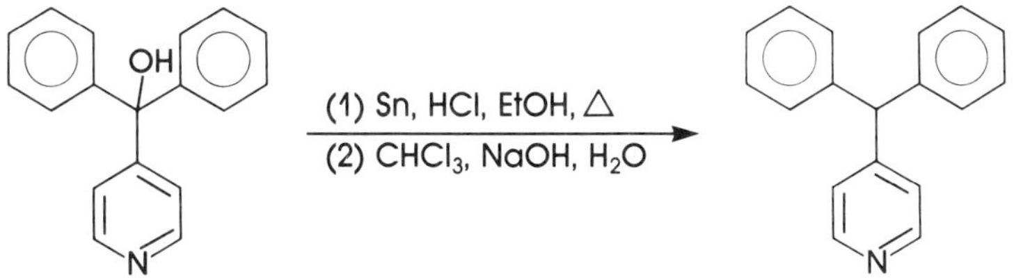

Fig. 3. Chemical reduction of the carbinol.

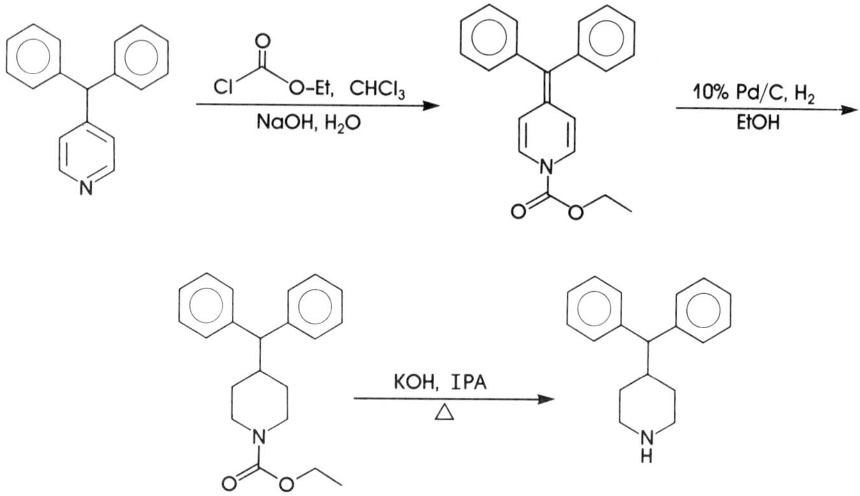

Fig. 4. Reduction of the pyridine ring via urethane derivative.

In a separate series of experiments, it was found that the 4-diphenylmethylpyridine could be easily hydrogenated over 10% Pd/Carbon after derivatization with ethyl chloroformate (Fig. 4). Thus, treatment at room temperature of a well-stirred mixture of the pyridine in chloroform and aqueous sodium hydroxide with ethyl chloroformate resulted in isolation of ethyl 4-(diphenylmethylene)-1,4-dihydro-1-pyridinecarboxylate as an orange solid. Hydrogenation of this compound was effected over 10% Pd/Carbon (2.5% by weight catalyst) in ethanol at 60°C under 50 psi hydrogen. Hydrolysis of the resulting urethane was effected using potassium hydroxide in refluxing isopropanol to give the desired 4-diphenylmethylpiperidine (98% yield from 4-diphenylmethylpyridine).

In the reduction of 4-diphenylmethylpyridine using platinum, reduction of the phenyl rings to saturated cyclohexane rings as well as reduction of the pyridine ring to a piperidine ring was observed when hydrogen-rich conditions were used. When platinum was used as catalyst, the desired reduction of only the pyridine ring was accomplished without hydrogenation of the phenyl rings, most readily at less than 100 psi, using a high molar ratio (10% by weight) of catalyst to substrate. These results may be indicative of a change of reaction mechanism between hydrogen rich and hydrogen poor catalytic systems.[2]

In the low pressure hydrogenations which were studied, the basic nitrogen atom in either the starting material or the product may have been interfering with the reduction of the pyridine ring system. Preparation of a urethane from the 4-diphenylmethylpyridine by treating a chloroform solution of the 4-diphenylmethylpyridine with ethylchloroformate in the presence of aqueous base resulted in making the nitrogen lone pair less accessible to the catalyst surface. In addition, aromaticity of the pyridine ring in the urethane derivative was lost. Consequently, this derivative was easily reduced in high yield using a more reasonable catalyst ratio.

2. Preparation of the Amidine

Retrosynthetic analysis of fenoctimine suggests five viable routes for its preparation (Fig. 5). Each of these routes are discussed briefly below. Only

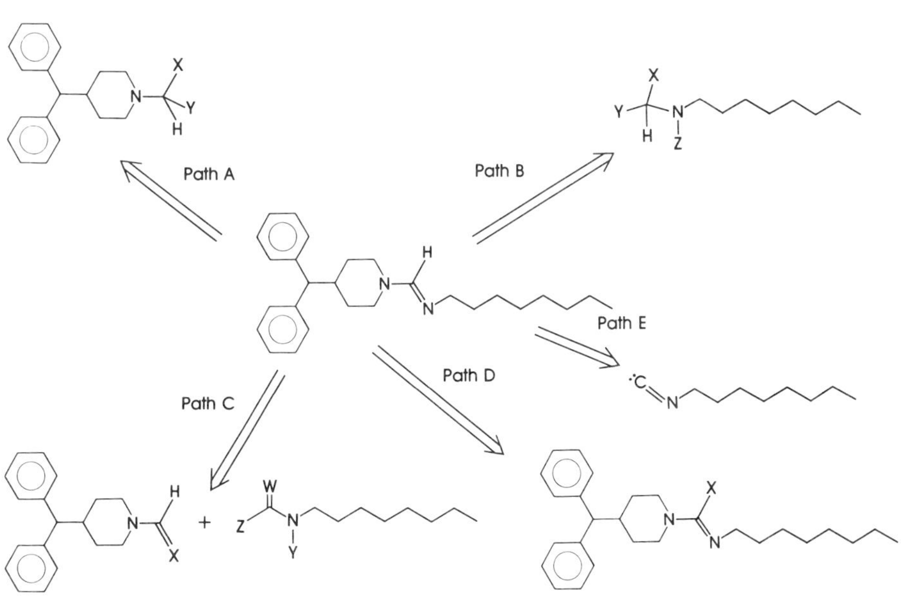

Fig. 5. Retrosynthetic analysis of Fentoctimine.

those methods which use a catalytic process in their preparation will be presented in detail.

Path A represents the condensation of n-octylamine with any of a number of 1-methyleniminopiperidine derivatives bearing a leaving group on the 1-methylene position. Such compounds are readily prepared and present a great diversity in their chemistry. Included in this class of compounds are Vilsmeier reagents,[3-5] isothioformamidinium salts,[1] O-alkylated formamidines,[6,7] and even amidines.[8]

Path B represents the same types of chemistry as depicted in Path A except that the leaving group is attached to the n-octylamine rather than the 4-diphenylmethylpiperidine.

Path C represents the reaction of a suitable piperidine derivative with an isocyanate, isothiocyanate, or other amine derivative capable of undergoing either an addition-elimination reaction or a cycloaddition-elimination reaction leading to the desired product. Reactions of this type are precedented: it has been shown that arylisocyanates react with amides at reflux to yield amidines.[9] Also included in this method is the reaction of amides with carbamoyl chlorides to yield amidines.[10]

Path D involves reduction of suitable formamidine derivatives and ureas to their corresponding formamidines. Literature precedent includes the reduction of Vilsmeier reagents,[11] thioureas,[12] and ureas.[13]

Path E can be envisioned to involve a catalytic insertion of n-octylisocyanide into the N-H bond of 4-diphenylmethylpiperidine. At the start of this work, it had been reported that such reactions were catalyzed either by copper of silver salts.[14,15] Preparation of the desired n-octylisocyanide was achieved through the reaction of chloroform with n-octylamine under phase transfer catalysis[16] (Fig. 6).

Coupling of the isocyanide to 4-diphenylmethylpiperidine to yield fenoctimine was achieved under the reported conditions[14,15] in good yield. Two obvious factors contributed to making this synthesis of fenoctimine impractical: (1) the stench of the isocyanide, and (2) chelation of the silver catalyst used in the reaction by the amidine product. Upon standing, colloidal silver precipitated from solutions of the amidine over a period of several days, even after the amidine was isolated as a salt. The fact that

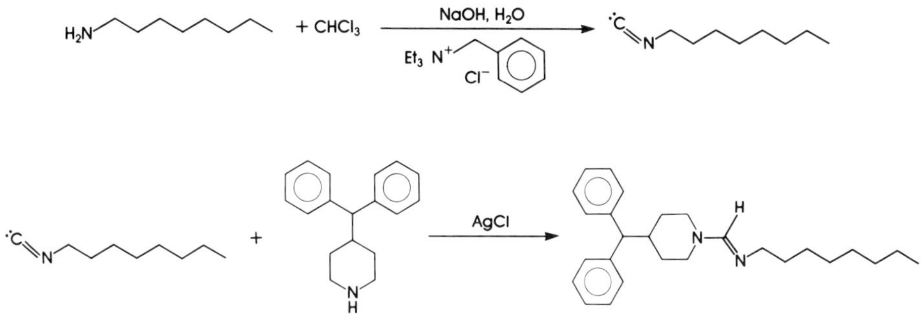

Fig. 6. Preparation via isonitrile.

homogeneous catalysts can be sequestered in reaction products is frequently overlooked. Since the metals commonly used in homogeneous catalysis are frequently present in the ppm range, it is not surprising that they are often not observed in elemental analyses when only C, H, N analyses are obtained. The selective removal of catalytic metals from a reaction mixture will be presented in a different context later.

A sequence based on Path B which was economically competitive with this synthesis was developed at nearly the same time.[17] Treatment of dimethylformamide with dimethyl sulfate yields a complex which is reacted with n-octylamine without isolation (Fig. 7). Addition of dimethylamine to this reaction drives an equilibrium toward the desired trisubstituted amidine, and produces the desired product in essentially quantitative yield. In the absence of added dimethylamine, the product was contaminated with high molecular weight by-products which could only be removed through distillation. The distillation however, caused decomposition of 10-20% of the desired product. Heating the trisubstituted amidine with 4-diphenylmethylpiperidine resulted in a transamination reaction to produce fenoctimine in high yield. The transamination reaction was run neat, resulting in very high volume efficiency for this reaction. Since this synthesis was not plagued by the problems outlined above for the silver catalyzed reaction, the non-catalytic synthesis was used to prepare clinical supplies of fenoctimine.

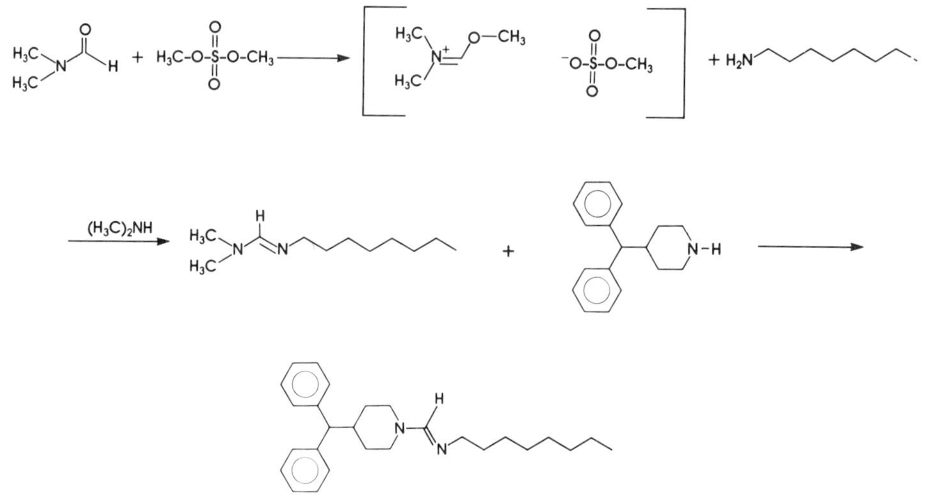

Fig. 7. Current synthesis of Fentoctimine.

B. McN-5691

N-[2-(3,4-dimethoxyphenyl)ethyl]-5-methoxy-N,-dimethyl-2-(phenylethynyl) benzenepropanamine, McN-5691 (Fig. 8) is a calcium entry blocker under development at McNeil Pharmaceutical as an antihypertensive agent. This compound shows no increase in heart rate, no significant effect on salt-water balance, no dysrhythmic events and a sustained decrease in blood pressure in spontaneously hypertensive rat (SHR) studies.[18] As was the case with fenoctimine, more than one step in the synthesis of this compound required the use of catalytic processes.

1. Hydrogenation of 4-(3-Methoxyphenyl)-3-butene-2-one

The first step in the synthesis of McN-5691 which required catalysis was the reduction of an α, β-unsaturated ketone substituted at the terminus by an aryl ring. Literature precedent indicated that such reductions are difficult to perform using chemical methods and that over-reduction of the ketone is frequently a competing reaction.[19] Literature reports indicated that the

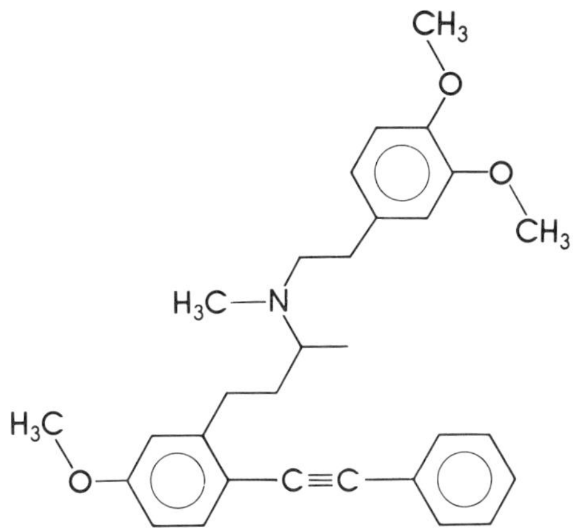

Fig. 8. McN-5691.

reduction of the ketone could be suppressed by addition of hydroxide.[20] Other investigators have recommended the addition of ferrous ion to the reaction media.[21]

Table II shows the results from a series of experiments performed at McNeil. All experiments were run on 11.4 mmole of substrate in ethanol (20mL). From this table, it can be seen that, without added base, 5% Pd/Carbon and 5% Pd/Alumina gave essentially the same amount of over-reduction to the undesired alcohol. Use of 5% Pd/Barium Sulfate or 5% Pd/Calcium Carbonate gave even more of the undesired alcohol than the previous catalysts. Addition of potassium hydroxide to the reduction media diminished the amount of over-reduction as a function of either the concentration or the number of equivalents of base. Addition of triethylamine or diethylamine to the reaction media also reduced the amount of over-reduction to the alcohol. On a mole equivalent basis, diethylamine appears to be most active in preventing the over-reduction, followed by triethylamine. Potassium hydroxide appears to be the least effective reagent to suppress the over-reduction.

TABLE II

Catalytic Reduction of 4-(3-Methoxyphenyl)-3-Butene-2-One

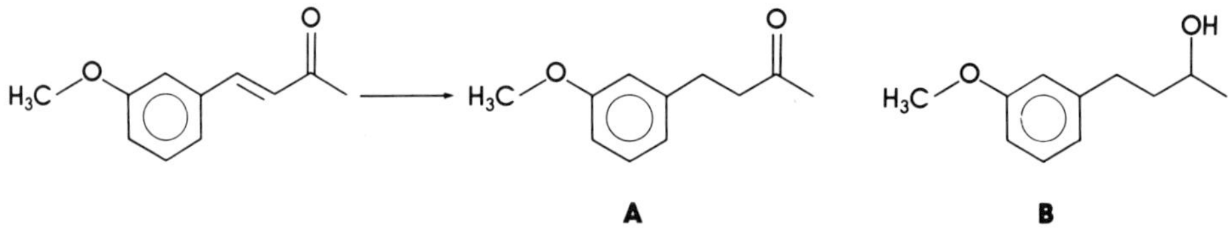

CATALYST	AGENT ADDED	YIELD A	YIELD B
5% Pd/C	None	85%	15%
5% Pd/Alumina	None	86%	14%
5% $Pd/BaSO_4$	None	78%	22%
5% $Pd/CaCO_3$	None	73%	27%
5% Pd/C	1 mM KOH, 1.75 eq	94%	6%
5% Pd/C	13 mM KOH, 23 eq	100%	—
5% Pd/C	Diethylamine, 0.72 eq	100%	—
5% Pd/C	Triethylamine, 1.3 eq	98%	2%

All hydrogenations were run in 20 mL of 95% ethanol at ca. 10 psi hydrogen.

With the exception of triethylamine, these species may be adding to the substrate through either a 1,4- or a 1,2-addition (Fig. 9). The triethylamine could conceivably be generating low concentrations of hydroxide or alkoxide which could undergo the same conjugate or carbonyl addition reactions. If 1,4-addition were occurring, the double bond would

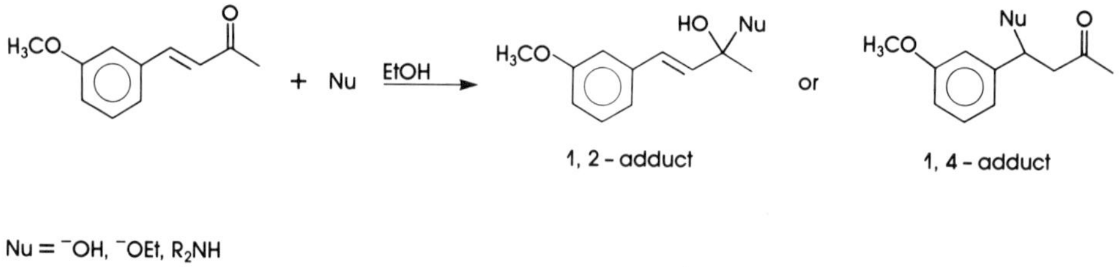

Fig. 9. Possible adducts.

effectively be reduced through hydrogenolysis of the new benzylic substituent. If 1,2-addition to the carbonyl group is occurring, the resulting hemiaminal or hydrated ketone should be less susceptible to reduction than the starting ketone. We currently have no evidence supporting either of these pathways; however, no 1,4-addition products have been isolated chromatographically after short reaction times.

Another explanation for the selectivity observed using any of the additives mentioned above is that the additives interact with the catalyst itself to modify the number or the location of active sites on the catalyst surface. A change in the number and relative location of the active catalyst sites may change the binding sites for the reaction substrate, resulting in enhanced selectivity for the desired hydrogenation.

2. Preparation of the Diarylacetylene

A survey of the literature indicates that the diphenylacetylene moiety in McN-5691 can be prepared through a variety of methods,[22-27] (Fig. 10). At McNeil, initial syntheses of this class of compounds utilized the coupling reaction of aryl iodide with copper acetylide.[28] Attempts to scale this reaction to 4 moles in a 22 L pot resulted in appreciably lower yields and tedious work-up procedures.

Homogeneous palladium catalysis is becoming a very important area in organic synthesis. Many of the reactions which have been developed have shown a high degree of selectivity, and thus are compatible with the presence of a number of functional groups in the substrates. The utility of these catalysts are evidenced by the variety of products prepared using them[29-34] (Fig. 11). The high turnover rates and extremely mild conditions used for many of the reactions make them very attractive for the preparation of a variety of pharmaceuticals. Fortunately, the diphenylacetylene moiety is ideally suited for preparation through homogeneous palladium catalysis.[29,30] The high yields of coupled products obtained in these reactions are as good, or better than those obtained through many other methods used for the preparation of this carbon framework.[22-27]

Initial work at McNeil on the palladium-catalyzed coupling reaction (Fig. 12) used conditions very similar to those reported in the literature[29]

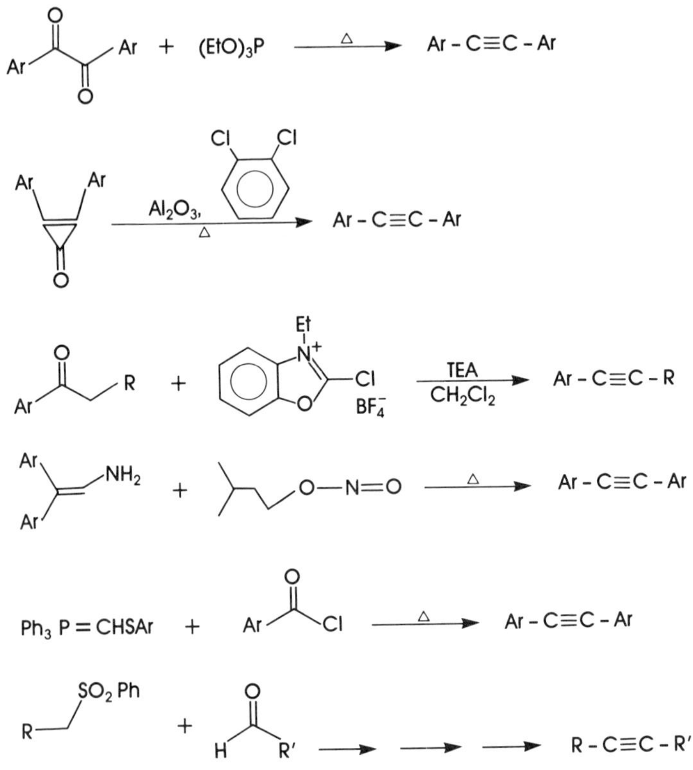

Fig. 10. Acetylene preparations taken from the literature.

but substituted triethylamine as solvent. By blowing argon through the reaction mixture to improve reproducibility, the coupling reaction was complete after 18 hours at room temperature. As the reaction proceeded, a solid precipitated from the reaction mixture. The reaction was reproduced on a 0.2 mole scale on numerous occasions, however, attempts to scale this reaction to 3.5 moles resulted in the formation of a rubber-like polymeric material. Further investigation of the reaction showed that replacement of triethylamine with diethylamine as solvent resulted in a homogeneous reaction, more rapid product formation, and in purer product.

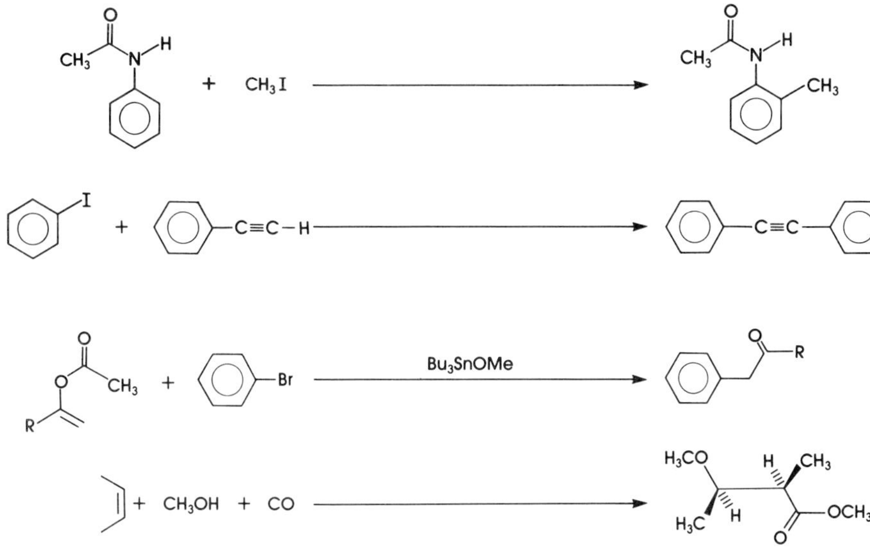

Fig. 11. Representative palladium catalyzed reactions taken from the literature.

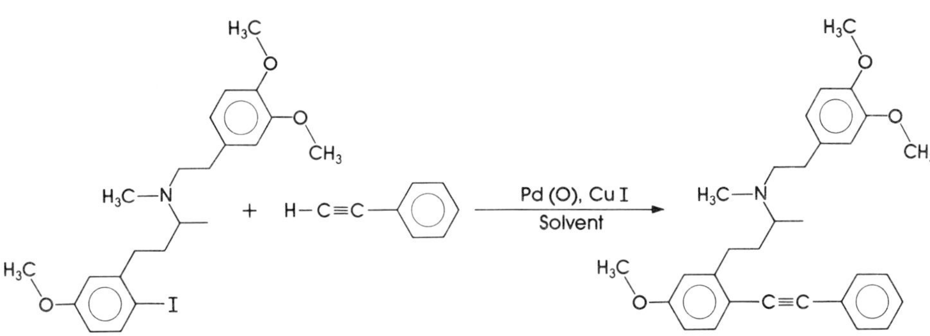

Fig. 12. Palladium catalyzed coupling reaction.

3. Removal of Palladium from the Product

Once the solvent was optimized, it was determined that the palladium could not be removed completely from the product. The palladium content in the coupling reaction was about 900-1000 ppm. Palladium in crude product varied from about 900 ppm to 100 ppm depending upon the specific work-up. Reduction of the palladium content to less than 20 ppm was attempted through a variety of procedures (Table III). Included in these methods were attempts to remove the palladium through reductive methods, extraction with chelating agents, and elution through silica gel. Although no experimental evidence has been obtained, it can be hypothesized that some of the palladium used in the reaction forms an N-sigma acetylene-pi bidentate complex (Fig. 13). This hypothesis helps explain why the palladium could not be completely removed through any of the methods noted above.

The hypothesis presented above, and the fact that some reduction in palladium content was observed upon treatment with sodium borohydride suggested that replacement of the basic nitrogen atom with a weaker palladium coordinator in the substrate could result in a satisfactory coupling reaction, with palladium levels of less than 20 ppm in the final product. The

TABLE III

Methods Used to Reduce the Palladium Content in McN-5691

WORKUP PROCEDURE	ppm Pd
Extraction with $NaBH_4$, filtration, treatment with $NaBH_4$ in MeOH	100 - 200
Hydrogenation in ethanol in the presence of carbon	930 - 950
Treatment of methylene chloride solution of product with sodium borohydride on alumina or silica gel	770 - 870
Chelation with dimethylglyoxime	259
Treatment of methylene chloride solution of product with Amborane® resin	112

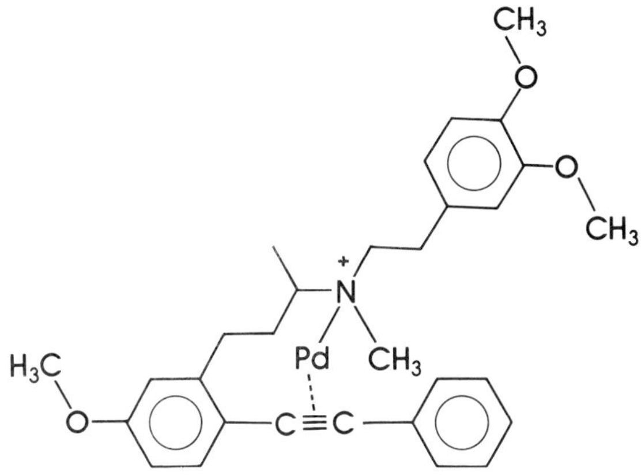

Fig. 13. Possible palladium complex.

reaction scheme was modified as shown in Fig. 14. Palladium catalyzed coupling of phenylacetylene to the iodoketone yielded the desired acetylenic ketone in greater than 80% yield. This ketone was then reductively aminated using sodium cyanoborohydride. Palladium levels in the resulting amine have been found to be less than 5 ppm. Thus, if the product in a reaction employing homogeneous catalysis is capable of multisite attachment to the metal ion, then high concentrations of the metal may be trapped in the product. One possible solution to this problem is to modify the synthetic sequence in such a way that the catalytic step generates an intermediate which is a poor ligand for the catalyst. Purifications in subsequent steps will then help to minimize the metal contamination in the final product.

III. CONCLUSION

Among the difficulties encountered by organic chemists in the use of catalytic processes are catalyst poisoning, incomplete catalyst recovery, and dramatic differences in the course of catalytic reactions caused by changes in reaction variables. Catalyst poisoning can sometimes be avoided by a purification step prior to the catalytic step. Since heterogeneous

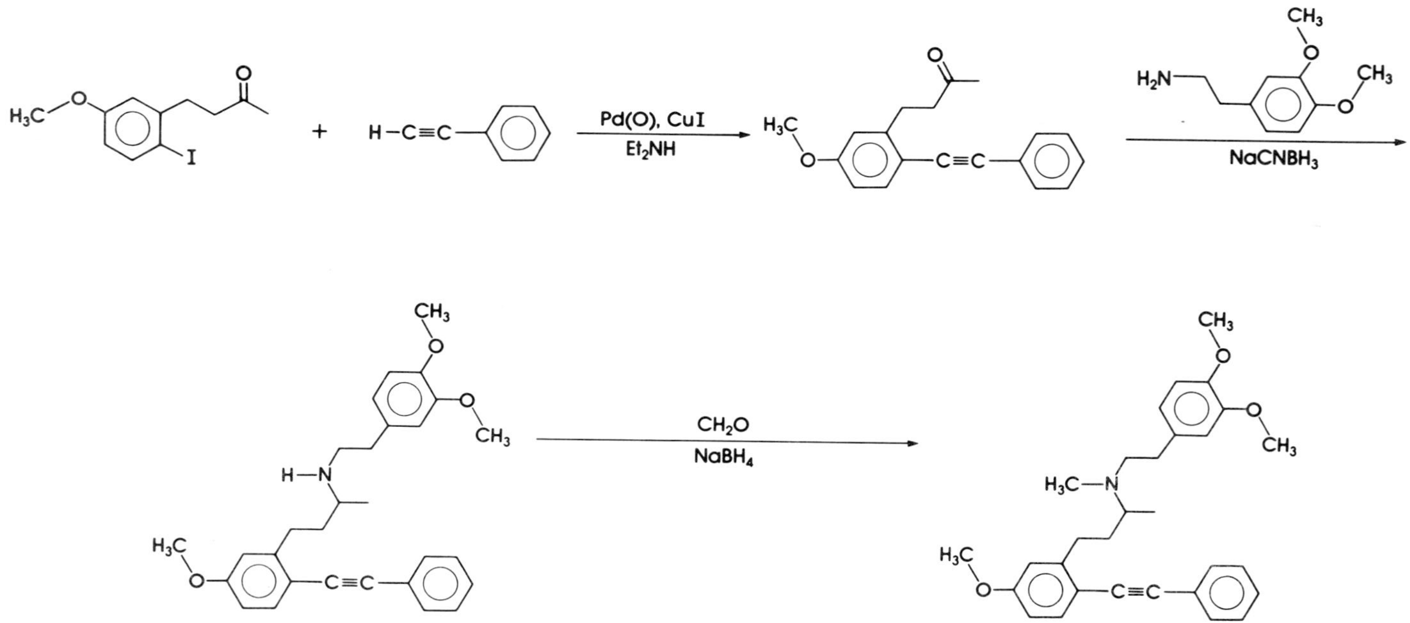

Fig. 14. Revised synthesis of McN-5691.

catalysts are susceptible to poisoning by trace impurities, filtration of a solution of the reaction substrate through five to ten percent by weight of a sorbant such as silica gel or alumina may remove enough of the poison to make the catalytic process viable.

Unfortunately, this approach is not applicable if the catalyst poison is formed as a product in the catalytic step. In this case, modification of the substrate or reaction sequence may result in an improvement in the catalytic step. Other alternatives include changing the metal used as the catalyst and/or modification of the catalyst support.

Catalyst recovery is especially important to improve the economics of a catalytic process when noble metal catalysts are used. However, the problem of environmental or product contamination by the catalyst should not be overlooked. Where product contamination by the catalyst has been demonstrated, rearrangement of the reaction sequence so that the product of the catalytic step is a poorer ligand for the catalyst may be the strategy of choice. Purification of the product may be easier and catalyst recovery may be simplified.

At McNeil Pharmaceutical, once a catalytic process has been identified in the process laboratory, it is evaluated in each working environment against non-catalytic processes. Economic and equipment constraints as well as novel organic chemistry can make non-catalytic processes competitive with those which are catalytic.

Catalytic processes are important tools for organic chemists, but many view catalysis as an "art". A better understanding, on the molecular level, of why minor changes in reaction conditions can produce major differences in catalytic processes should separate the "art" from the science.

REFERENCES

1. M.K. Scott, H.I. Jacoby, J.E. Mills, A.C. Bonfilio and C.R. Rasmussen, J. Med. Chem., **26**, 535 (1983).
2. S. Siegel in Advances in Catalysis and Related Subjects; W.G. Frankenburg, Ed.; Academic Press: New York, 1966; Vol. 16 p. 132.

3. H. Bredereck, R. Gompper, K. Klemm and H. Rempfer, Chem. Ber., **92**, 837 (1959).
4. H.G. Mandel and A.J. Hill, J. Am. Chem. Soc., **76**, 3978 (1954).
5. J. Besan, L. Kulcsar and M. Kovacs, Synthesis, 883 (1980).
6. H. Bredereck, F. Effenberger and G. Simchen, Chem. Ber., **96**, 1350 (1963).
7. P. Oxley, D.A. Peak and W.F. Short, J. Chem. Soc., 1618 (1948).
8. P. Reynaud, J.-D. Brion and G. Menard, Bull. Soc. Chim. Fr., (9-10):pt. II, 449 (1978).
9. J.L. Neumeyer, J. Pharm. Sci., **53**, 1539 (1964).
10. E. Haug and W. Kantlehner, Synthesis, 35 (1983).
11. A.-ur-Rahman, A. Basha, N. Waheed, and S. Ahmed, Tetrahedron Letters, 219 (1976).
12. D.J. Brown, J. Appl. Chem., **2**, 202 (1952).
13. A. Larizza, G. Brancaccio, and G. Lettieri, J. Org. Chem., **29**, 3697 (1964).
14. T. Saegusa, Y. Ito, S. Kobayashi, K. Hirota and H. Yoshioka, Tetrahedron Letters, 6121 (1966).
15. Y. Ito, Y. Inubushi and T. Saegusa, Tetrahedron Letters, 1283 (1974).
16. W.P. Weber and G.W. Gokel, Tetrahedron Letters, 1637 (1972).
17. G. Feth and J.E. Mills, US Patent 4499274, 1985.
18. M.D. Brannan, R.J. Carmosin, J.R. Carson, S.F. Flaim, A. Gill, M.M. Gleason, S.L. Keely, D.W. Ludovici, P.M. Pitis, F.J. Villani, Jr. and M.C. Rebarchak, Abstracts of Papers, 191st National Meeting of the American Chemical Society, New York, NY; Americn Chemical Society: Washington, DC, 1986; MEDI #13.
19. T. Nishio and Y. Omote, Chem. Lett., 1223 (1979).
20. R.L. Augustine, "Catalytic Hydrogenation", Marcel Dekker, Inc., New York 1965, p. 60.
21. P.N. Rylander and N. Himmelstein, Engelhard Ind., Tech. Bull. **4**, 131 (1964).
22. T. Mukaiyama, H. Nambu and T. Kumamoto, J. Org. Chem., **29**, 2243 (1964).
23. D.H. Wadsworth and B.A. Donatelli, Synthesis, 285 (1981).
24. T. Tsuji, Y. Watanabe and T. Mukaiyama, Chem. Lett., 481 (1979).

25. D.Y. Curtin, J.A. Kampmeier and B.R. O'Connor, J. Am. Chem. Soc., **87**, 863 (1965).

26. B. Lythgoe and I. Waterhouse, Tetrahedron Letters, 2625 (1978).

27. A.L. Braga, J.V. Comasseto and N. Petragnani, Tetrahedron Letters, 1111 (1984).

28. R.D. Stephens and C.E. Castro, J. Org. Chem., **28**, 3313 (1963).

29. K. Sonogashira, Y. Tohda and N. Hagihara, Tetrahedron Letters, 4467 (1975).

30. A.O. King, E.-i. Negishi, F.J. Villani, Jr. and A. Silveira, Jr., J. Org. Chem., **43**, 358 (1978).

31. S.J. Tremont and H.U. Rahman, J. Am. Chem. Soc., **106**, 5759 (1984).

32. H.A. Dieck and F.R. Heck, J. Organometallic Chem., **93**, 259 (1975).

33. M. Kosugi, I. Hagiwara, T. Sumiya, and T. Migita, J. Chem. Soc., Chem. Comm., 344 (1983).

34. J.K. Stille, D.E. James and L.F. Hines, J. Am. Chem. Soc., **95**, 5062 (1973).

19

Aspects of Cross-Coupling Reactions Catalyzed by Palladium and Nickel Complexes

Ei-ichi Negishi, Tamotsu Takahashi, and Kazunari Akiyoshi

Department of Chemistry
Purdue University
West Lafayette, Indiana

ABSTRACT

The Pd- or Ni-catalyzed cross-coupling reactions of organometals with organic halides or related electrophiles have proved to be one of the most versatile methods for carbon-carbon bond formation. This paper briefly describes the current synthetic scope of the Pd- or Ni-catalyzed cross-coupling reaction. A mechanistic scheme involving oxidative addition, transmetallation, and reductive elimination is presented, and its plausibility is discussed. Various factors affecting reductive elimination are discussed. They include the ligands, such as PPh_3, PPh_2Me, $PPhMe_2$ and PEt_3, the countercations (M), such as Li, Mg, Zn, Al, B, Sn, Zr, and Hg, and the carbon groups (R^1 and R^2) to be coupled. A special attention is paid to our recent findings that, in many favorable Pd-catalyzed cross-coupling reactions, transmetallation is the rate-determining step and that some highly electropositive metals, such as Li, inhibit reductive elimination through formation of palladate complexes of the Li_2PdR_4 type.

I. INTRODUCTION

The cross-coupling reaction of organometallic reagents with organic halides and related electrophiles provides one of the most straightforward methods for the formation of carbon-carbon single bonds (Eq. 1). Although traditional methods involving alkali metals, e.g., Li and Na, and Grignard reagents[1] are of limited synthetic utility, the Cu-promoted cross-coupling reactions[2] developed since the mid-1960's have made this methodology a respectable tool for organic synthesis. Even so, however, many problems had remained unsolved until the development of the Ni-or Pd-catalyzed cross-coupling reaction.[3]

$$R^1M + R^2X \rightarrow R^1R^2 + MX \qquad (1)$$

Although scattered examples of transition-metal catalyzed coupling reactions known as the Kharasch-type coupling reactions[4] have been known for many decades, they are generally of limited utility in cross coupling due to the generally low cross/homo ratio and other selectivity problems. The use of nickel-phosphine catalysts in the reaction of Grignard reagents with organic halides reported by Kumada[5] and by Corriu[6] in 1972 paved the way to the discovery and development of the d-catalyzed cross-coupling method by us and others in the mid-1970's.[7-11] Subsequent studies have established that the Ni- and Pd-catalyzed cross-coupling reactions are not only mutually complementary but also collectively complementary with the Cu-promoted methodology.[3]

II. SCOPE AND LIMITATIONS

The current available data indicate that the reactions of organometals containing various metals, such as K, Na, Li, Mg, Zn, Cd, Hg, B, Al, Sn, Zr, and Cu, with organic halides and related electrophiles can be effectively catalyzed by phosphine complexes of Ni and Pd.[3] Although not fully established, Pt complexes appear to be unsuitable as catalysts. This is presumably because diorganoplatinum complexes of the $R^1R^2Pt(PR_3)_2$ type are too stable to undergo reductive elimination fast enough to sustain catalytic cycles. Although there can be many exceptions, the following reaction conditions appear to be generally satisfactory, and some aspects of these conditions will be further discussed later:

A. Recommended reaction conditions

Precatalyst:	$Ni(0)(PPh_3)_4$, $Cl_2Ni(II)(PPh_3)_2$, $Ni(II)(acac)_2$+nPPh_3, $Pd(0)(PPh_3)_4$, $Cl_2Pd(II))(PPh_3)_2$, Li_2PdCl_4+nPPh_3, etc.
Solvent:	Tetrahydrofuran (THF), ether, etc.
Leaving group(X):	I, Br, OSO_2CF_3, $OPO(OPh)_2$, etc.
Countercation (M):	Zn, Mg, Al, B, Zr, Cu, etc.

A wide variety of Ni- or Pd-phosphine complexes containing Ni(0), Ni(II), Pd(0), or Pd(II) can be used as precatalysts. In some cases, it is desirable to preform Ni(0)- or Pd(0)-phosphine precatalysts. Some such complexes, e.g., $Pd(0)(PPh_3)_4$, are commercially available. Generally speaking, however, Ni(0)- and Pd(0)-phosphine complexes are considerably less stable than Ni(II) and Pd(II) complexes. For this reason, in situ generation of Ni(0)- and Pd(0)-phosphine complexes from the corresponding Ni(II) and Pd(II) complexes, respectively, is often desirable. Various reducing agents, such as i-Bu_2AlH,[9] have been used for this purpose. Our recent results show that the use of n-BuLi is particularly convenient.[12] Triphenylphosphine is not only one of the most economical phosphines but also one of the most effective in promoting the desired catalysis. As discussed later, its relatively low basicity and large cone angle (or steric requirement) presumably are responsible for its effectiveness.

No systematic study of the effects of various solvents used in the Pd- or Ni-catalyzed cross coupling has been made. However, THF appears to be very well suited for the reaction, and it has been most widely used in the past. The leaving groups (X) in R^2X have been mostly halogens. Their relative reactivity in oxidative addition is: I > Br > Cl.[13] In cases where alkenyl or aryl halides are used, only iodides and highly reactive bromides containing some electron-withdrawing substituents readily participate in the Pd-catalyzed cross coupling, while the corresponding Ni-catalyzed reactions proceed readily with a wider range of bromides and even with chlorides. With more reactive allyl, benzyl, and acyl derivatives, however, chlorides are usually reactive enough, and some even less reactive leaving groups, such as OAc, have also been successfully used in some cases.[3] In recent years, other oxygenated leaving groups, such as $OPO(OPh)_2$[14] and $OSO_2CF_3(OT_f)$,[15] have been successfully used. Since these derivatives can be derived from the corresponding ketones and aldehydes, they seem to provide some additional synthetic opportunities.

The reactivity of the organometals (RM) with respect to M shows an interesting trend. Highly electropositive metals, such as K, Na, Li, and even Mg, are often far less effective[16] than metals of intermediate electronegativity, such as Zn, Cd, Al, Zr, and Cu. This point will be discussed later in some detail. Highly electronegative metals are expectedly of lower reactivity than those mentioned above. In general, organosilanes appear to be too unreactive to be of practical utility, while some heavy metals, such as Hg and Tl, are prone to undesirable redox processes with Pd or Ni complexes, which may limit the scope of catalytic processes involving these heavy metals. In many cases, organometals containing Zn appear to exhibit the highest reactivity.[3b] Furthermore, the reactions of organometals containing other metals, such as Al,[17] Zr,[17] and Cu,[18] can be additionally catalyzed by Zn salts, such as $ZnCl_2$ (double metal catalysis).

From the viewpoint of cross coupling the organic groups may be classified into approximately ten or a dozen types leading to 100 to 150 binary combinations to be considered here. Although the current scope of the Cu-promoted method[2] and that of the Pd- or Ni-catalyzed method are both reasonably broad, they are nevertheless far from being general. The current scope of the Pd- or Ni-catalyzed cross coupling is indicated in Table I. The following brief comparison of the two methods helps delineate the scopes and the complementary nature of the two methods.

B. Cross coupling between two unsaturated carbon groups

The unsaturated carbon groups in this paper loosely refer to the alkenyl, aryl, and alkynyl groups. The Pd-catalyzed method is highly general and applicable to all nine binary combination cases except for alkynyl-alkynyl coupling. When Ni-phosphine catalysts are used, alkynes and dienes may not readily be obtained selectively and in high yields presumably because these compounds readily react further with Ni complexes. At present, the Pd-

TABLE I

Scope of Pd- or Ni-Catalyzed Cross Coupling

R^1M \ R^2M	alkenyl–M	ArX	≡–X	allyl–X	ArCH$_2$–X	propargyl–X	RX (R=alkyl)	RCOX
alkenyl–M	E	E	E	E	E	?		E
ArM	E	E	E	E	E	?		E
≡–M	E	E	*a*	?	?	?		E
allyl–M	E	E	?	*a*	E	?		
ArCH$_2$–M	E	E	?					E
propargyl–M			?					
RM (R=alkyl)	G to E	G to E						E
O=C–C–M	F to G	F to G	?	E	?	?		

E = excellent. G = good. F = fair. ? = Attempted but its scope not well delineated. [a]Cross-homo scrambling.

catalyzed alkenyl-alkenyl and alkenyl-alkynyl coupling reactions involving Al,[9] B,[16e,19] Cu,[18,20] Mg,[21] Zn,[16a,16b,17] and Zr[22] represent by far the most satisfactory cross-coupling method for the synthesis of conjugated dienes and enynes, respectively (Eq. 2). A large number of pheromones and

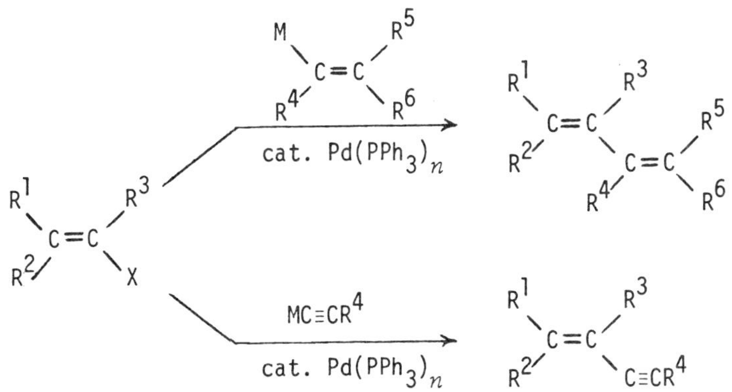

(2)

carotenoids have been synthesized by this method. The Ni- or Pd-catalyzed aryl-aryl coupling reaction involving arylzinc derivatives appears to be particularly well-suited for preparing unsymmetrical biaryls, as exemplified by its application to the synthesis of an antitumor agent, steganone[24] (Eq. 3). In the above-described cases, Cu-promoted method is generally unsatisfactory.

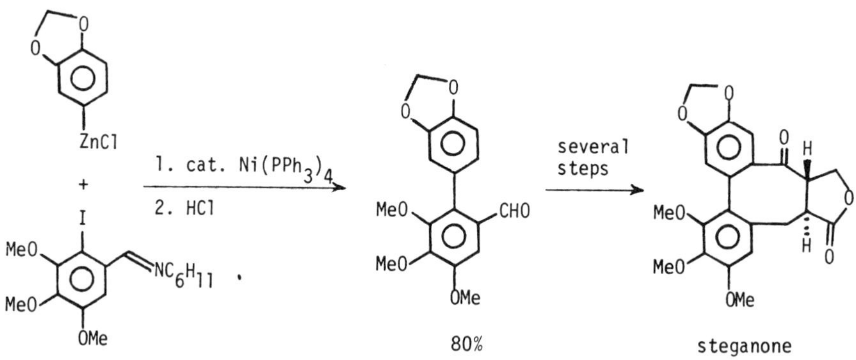

(3)

The Pd-catalyzed alkynyl-alkynyl coupling reaction proceeds readily and in high yields. Unfortunately, the products usually consist of a nearly statistical mixture of the cross- and homo-coupled products irrespective of M (Mg, Al, Zn, Sn, etc.) and X (I, Br, or Cl). Clearly, the Cu-catalyzed method of Cadiot and Chodkiewicz[25] is still the method of choice. Since enynes can be readily obtained by the Pd-catalyzed cross coupling, however, conjugated diynes can also be prepared via enynes,[26] as in Eq. 4.

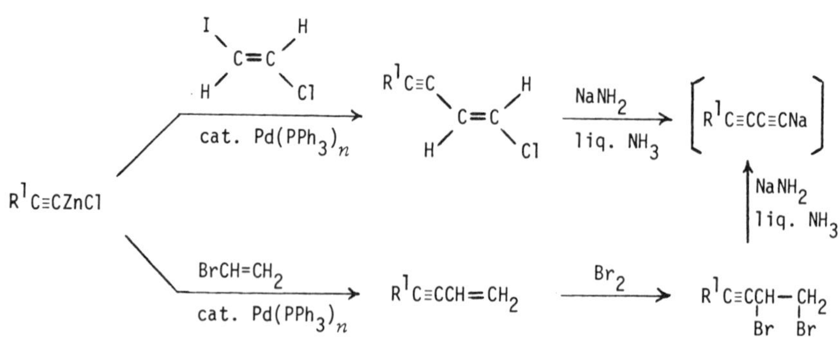

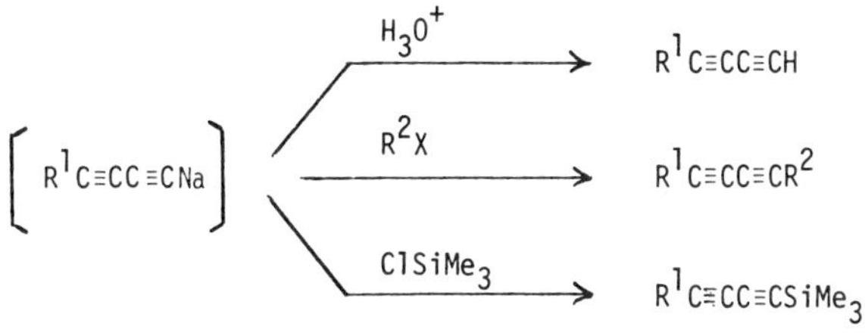

(4)

C. Cross coupling involving allyl, benzyl, and propargyl groups

Allylation, benzylation, and propargylation may be carried out by the reaction of organic halides containing these groups with appropriate Grignard and organolithium reagents.[1] The results can often be significantly improved through the use of Cu complexes, such as CuI and Li_2CuCl_4.[2] These methods are still considered to be the most reliable and widely applicable. In some cases, however, the Pd- or Ni-promoted reactions have been shown to be excellent. They include allylation and benzylation of alkenyl and aryl groups.[3] The use of in situ generated stereo- and regio-defined alkenylmetals readily obtainable via hydrometallation or carbometallation of alkynes with reagents containing Al,[27] B,[28] and Zr,[27,29] as exemplified by the transformation shown in Eq. 5, is particularly attractive.

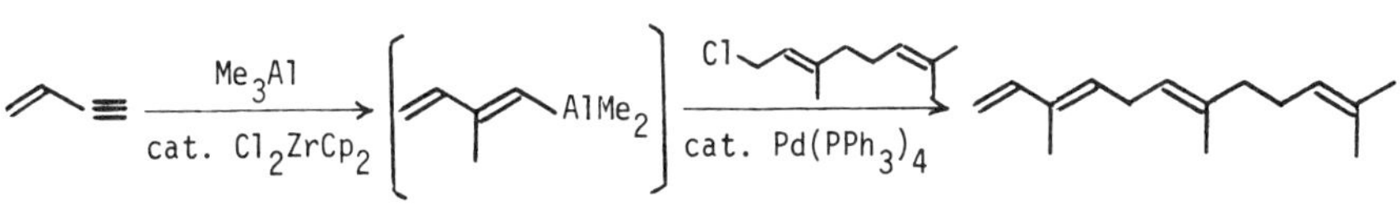

(5)

Although stoichimetric in Ni, the reaction of π-allylnickel derivatives with various organic halides[30] provides a general method for allylation. The reaction is neither stereospecific nor regiospecific, however.

D. Alkylation

In sharp contrast with the high reactivity of unsaturated organic halides and various electrophiles containing allyl, benzyl, and propargyl groups, simple alkyl halides are relatively unreactive toward Ni(O)- and Pd(O)-phosphine complexes, indicating that their interaction with a π system, i.e., alkenyl, aryl, or alkynyl, proximal to a carbon-halogen or related carbon-hetero atom bond may be responsible for the observed facile oxidative addition processes. In fact, there does not appear to be any report on successful Pd- or Ni-catalyzed cross coupling with simple alkyl halides. This represents a major limitation of this methodology. For this purpose, the Cu-promoted method together with the more conventional method involving Li and Mg should be considered.

It should be emphasized here, however, that the imcompatibility of the Pd- or Ni-catalyzed method with alkyl halides is definitely not due to the tendency of alkylmetals containing Pd or Ni to undergo facile β elimination. In fact, this potential difficulty can be circumvented by introducing an alkyl group as part of an alkylmetal rather than an alkyl halide. Thus, various primary, secondary, and even tertiary alkylmetals containing Mg and Zn undergo cross coupling with organic halides, such as aryl and alkenyl iodides and bromides, in the presence of a Pd- or Ni-phosphine catalyst.[31,32] Clearly, reductive elimination of a diorganometal intermediate containing Ni or Pd (1) is considerably faster than its β elimination (Eq. 6).

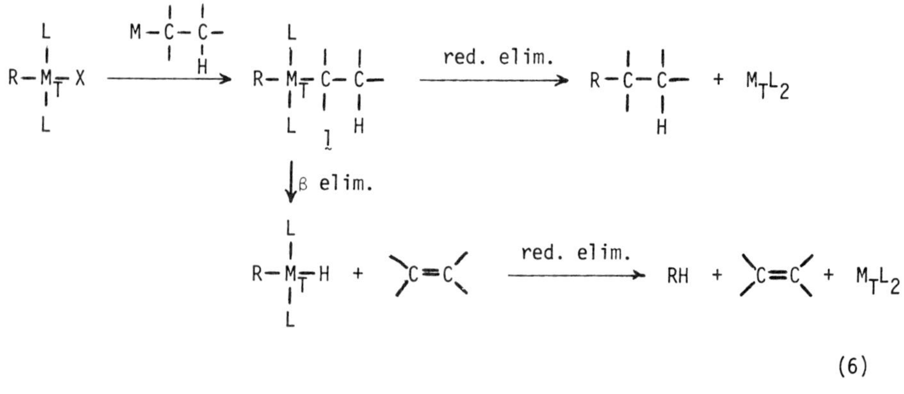

(6)

It is noteworthy that even homoallyl, homobenzyl, and homoproparygyl derivatives of Zn or Mg readily participate in cross coupling without the complication of β elimination[31] (Eq. 7). Also noteworthy is that the use of

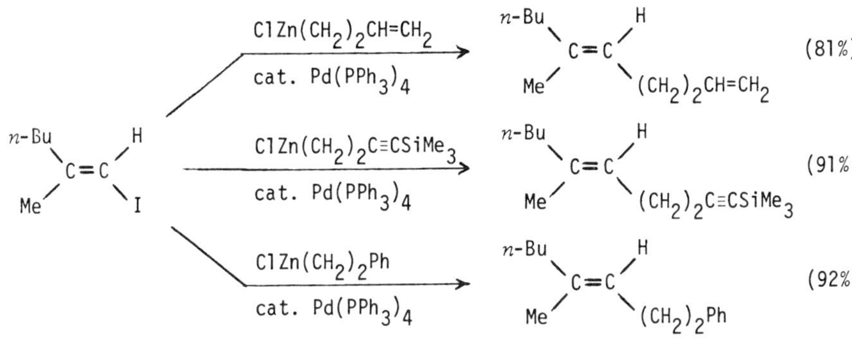

(7)

chiral phosphine-containing Ni or Pd catalysts permits asymmetric cross coupling, as in Eq. 8, although the extent of asymmetric induction is still somewhat modest.[32]

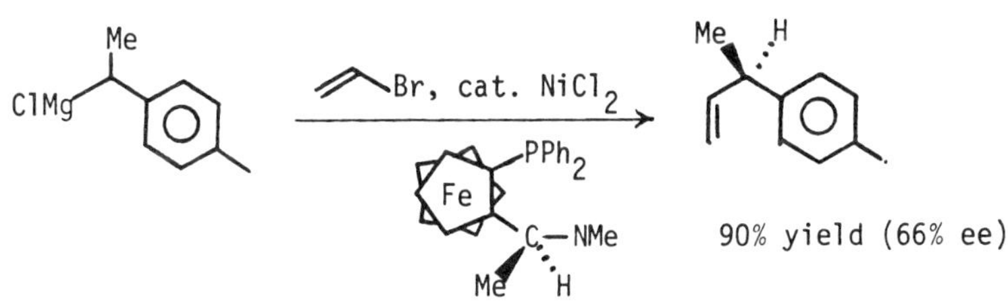

(8)

E. Acylation

In light of the facile oxidative addition of acyl halides with Pd complexes,[13] the reaction of organometals with acyl halides may be expected to be catalyzed by Pd complexes. Organometals containing Hg,[33] Sn,[34] and Zn[35] have indeed been shown to participate in such reactions. The Pd-catalyzed acylation of organozincs (Eq. 9) appears to be highly general, and organozincs containing alkyl, benzyl, alkenyl, aryl, and alkynyl groups have been successfully used.

$$R^1ZnCl + R^2COCl \xrightarrow{\text{cat. } Pd(PPh_3)_n} R^1COR^2 \qquad (9)$$

F. Cross coupling involving metal enolates

Metal enolates are synthetic equivalents to α-metallated ketones and aldehydes. Their carbon-carbon bond forming reactions with organic halides may therefore be viewed as special cases of cross coupling. Allylation of "stabilized" metal enolates, such as ethyl sodiomalonate, catalyzed by Pd complexes, i.e., the Tsuji-Trost reaction,[36] has been a well-developed methodology for more than a decade. On the other hand, the corresponding allylation of "ordinary" metal enolates, such as metal cyclohexenolates, had remained poorly developed until several years ago. Although alkali metal enolates themselves are poor reagents, their treatment with trialkylboranes, e.g., Et_3B, or $ZnCl_2$ produces satisfactory reagents for Pd-catalyzed allylation.[37] The high regio- and stereospecificity as well as the ready participation of allyl acetates and chlorides, as exemplified by Eq. 10,[37d]

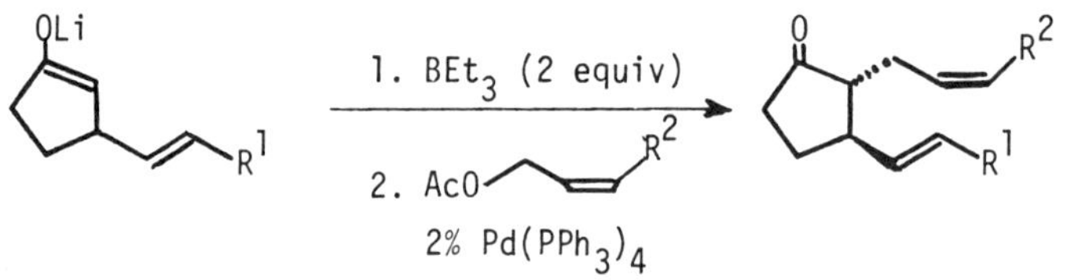

(10)

make this method unique among the known methods for allylating enolates. The use of $ClCH_2C(Cl)=CH_2$ and $ClCH_2CH=C(Cl)Me$ as allylating agents permits a convenient synthesis of 1,4- and 1,5-diketones, respectively[37b] (Eq. 11).

The Pd- or Ni-catalyzed reaction of metal enolates with alkenyl, aryl, or alkynyl halides can be a highly valuable synthetic operation. Despite some reports[38] on the reaction, it has not yet been developed into a dependable synthetic method. Its further development is highly desirable. As an alternate approach, the Ni-catalyzed reaction of β-bromovinyl ethers with Grignard reagents has been reported.[39]

In general, the Pd- or Ni-catalyzed cross-coupling reactions of proximally heterofunctional reagents have not been adequately investigated.[40] A large number of proximally heterofunctional organometals containing alkali metals and Mg and now known.[1] As such, they are

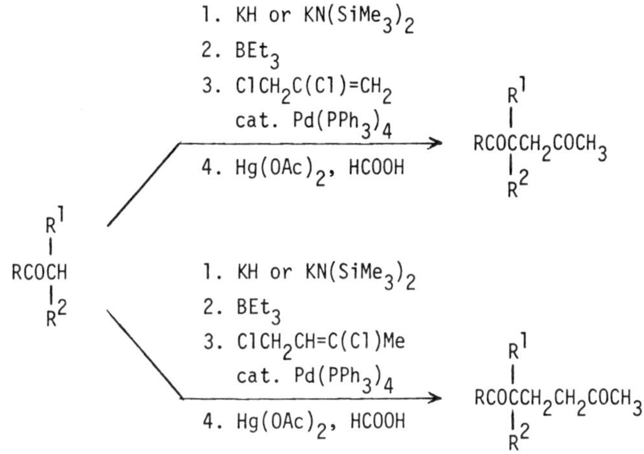

(11)

incapable of undergoing carbon-carbon bond formation with unsaturated organic halides. It is therefore highly desirable to catalyze such reactions with transition metal complexes. This represents one of the major remaining tasks in this area.

III. MECHANISM

A catalytic cycle consisting of oxidative addition of an organic halide, transmetallation, and reductive elimination for the Ni-catalyzed cross coupling (Scheme 1) was proposed in 1972,[5,6] but no mechanistic details were discussed. Similar mechanisms have also been repeatedly proposed for the Pd-catalyzed cross coupling.[3]

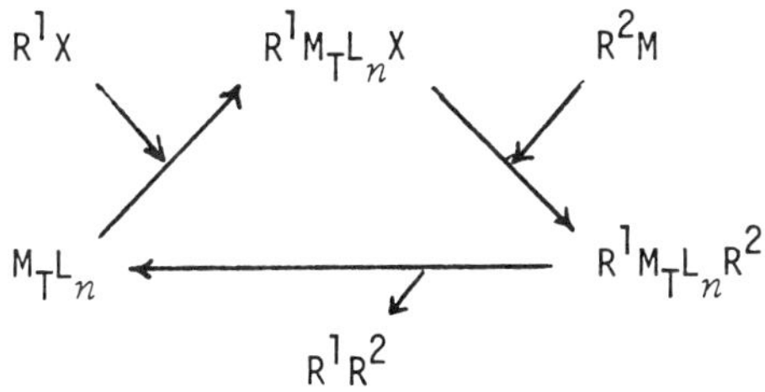

Scheme 1

An investigation of the stoichiometric reaction of preformed haloarylbis(phosphine)metals containing Ni, Pd, and Pt with Grignard reagents by NMR reported in 1974 by Parshall[41] appears to be the first mechanistic study in the field. This study showed the feasibility of Scheme 1. On the other hand, a later study by Kochi[42] indicates that although o-tolylmethylbis(triethylphosphine)nickel can undergo a thermal, intramolecular, first-order reaction to produce o-xylene in >90% yield in accordance with Scheme 1, its reaction in the presence of added aryl halide is a faster radical-chain process involving Ni(I) and Ni(III) species (Eq. 12).

$$
\begin{aligned}
&\mathrm{ArNi^{I}L_{\mathit{n}}} + \mathrm{ArX} \longrightarrow \mathrm{Ar_2Ni^{III}L_{\mathit{n}}X} \\
&\mathrm{Ar_2Ni^{III}L_{\mathit{n}}X} + \mathrm{ArNi^{II}L_{\mathit{n}}Me} \longrightarrow \mathrm{ArNi^{II}L_{\mathit{n}}X} + \mathrm{Ar_2Ni^{III}L_{\mathit{n}}Me} \qquad (12) \\
&\mathrm{Ar_2Ni^{III}L_{\mathit{n}}Me} \longrightarrow \mathrm{ArMe} + \mathrm{ArNi^{I}L_{\mathit{n}}}
\end{aligned}
$$

Another mechanism involving Pd(IV) species (Eq. 13) was proposed by

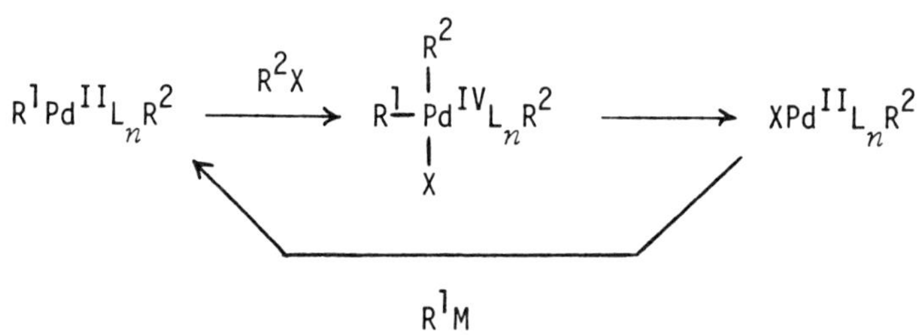

(13)

Stille[43] mainly to account for the accelerating effect of an added organic halide. While these mechanisms appear plausible for the reactions investigated, their generality remains to be established. In fact, reductive elimination of diorganopalladium complexes in many synthetically useful cases is very facile even in the absence of an organic halide (vide infra), indicating that the intermediacy of Pd(IV) species may not be required. Although of no direct relevance to the mechanism of catalytic processes, reductive elimination of dialkylpalladium species has been investigated.[44] A few theoretical studies of such reactions have also appeared.[45]

It is uniformly agreed that the Pd- or Ni-catalyzed cross coupling reaction using relatively electropositive metals, such as Li, Mg, or Zn, involves (a) oxidative addition of an organic halide or its equivalent, (b) transmetallation, and (c) reductive elimination of organometallic intermediates containing both carbon groups (R^1 and R^2) to be cross-coupled. Scheme 1 contains all and only these three steps. It therefore provides a reasonable scheme to be further scrutinized or modified as needed. Although not indicated in Scheme 1, it is also agreed that concerted reductive elimination requires the cis relationship between the R^1 and R^2 group.[43]

All of the previous mechanistic investigations have dealt with those cross-coupling reactions that involve reductive elimination of stable and characterizable diorganopalladium or -nickel intermediates in the rate-determining step. As such, they do not provide explanations for the marked countercation effects observed by us and others.[3] We therefore chose the Pd-catalyzed reaction of alkenylmetals with aryl halids as a representative class of rapid catalytic cross coupling and decided to probe its mechanistic details. Specifically, for example, the reaction of PhI with (E)-1-octenylzinc chloride catalyzed by 1-5 mole % of $Pd(PPh_3)_4$ at room temperature provides within a few hours (E)-1-octenylbenzene in excellent yield.[17]

A. Oxidative addition

Various factors affecting oxidative addition of organic halides with Pd or Ni complexes and the mechanism of oxidative addition are reasonably well delineated.[13,41,46] The oxidative-addition reaction of Ni, Pd, or Pt complexes with unsaturated organic halides appears to be a concerted process that can proceed with complete retention of the stereochemistry of alkenyl halides. The relative rates of oxidative addition are as follows:

Metal	Ni > Pd > Pt.
Halogen	I > Br > Cl >> F.
Phosphine	PEt_3 > PPh_3.

Although Pd(O)- or Ni(O)-phosphine complexes containing four or three phosphines can readily participate in oxidative addition, the active catalytic species under catalytic reaction conditions may well be "$Pd(PR_3)_2$" or "$Ni(PR_3)_2$", as discussed later.

B. Catalytic processes in which transmetallation is the rate-determining step

Treatment of PhI with $Pd(PPh_3)_4$ gives $PhPd(PPh_3)_2I$ (**2**)[46d] in nearly quantitative yield. The reaction of **2** with 1.2 equiv of (E)-1-octenylzinc chloride in a 2:1 mixture of CD_2Cl_2 and THF at 0°C smoothly produces (E)-1-octenylbenzene (**3**), the yields of **3** after 3, 7, and 12 min being 64, 84, and 95%, respectively. Significantly, the sum of **1** and **3** stays near the 100% level, and that of **2** and **3** in the 112 to 125% range throughout the course of the reaction, indicating that no buildup of any intermedites, such as **4**, occurs, even though **4** is believed to be actually formed (vide infra). At least in this case and probably in many other cross-coupling reactions where the rate of reaction is dependent on the countercation, the transmetallation step must be slower than that of reductive elimination. Although the exact nature of the transmetallation step is not clear, a four-center mechanism involving M--C--M three-center two-electron bonds appears to be plausible.

When the reaction of (E)-1-octenylzinc chloride with PhI is carried out in the presence of 0.5 equivalent of $Pd(PPh_3)_4$ in a 2:1 mixture of CD_2Cl_2 and THF at -10°C, the formation of **2** up to 18% yield (out of max. 50% yield) is observed after 20-25 min. along with the expected formation of **3** in 90% yield and disappearance of PhI within 1 h. The results of the two experiments presented above clearly indicate that the transmetallation step is the rate-determining and that, although the exact course of the formation of **3** from **2** and the nature of "$Pd(PPh_3)_2$" need to be further clarified, Scheme 2 provides a reasonable mechanism for these reactions.

C. Characterization of "$Pd(PPh_3)_2$"

The overall transformation in the above reaction demands that $Pd(PPh_3)_2$ or its derivative tentatively represented by "$Pd(PPh_3)_2$", such as THF-solvated or oligomerized species, be the other product. Addition of one equiv of PhI

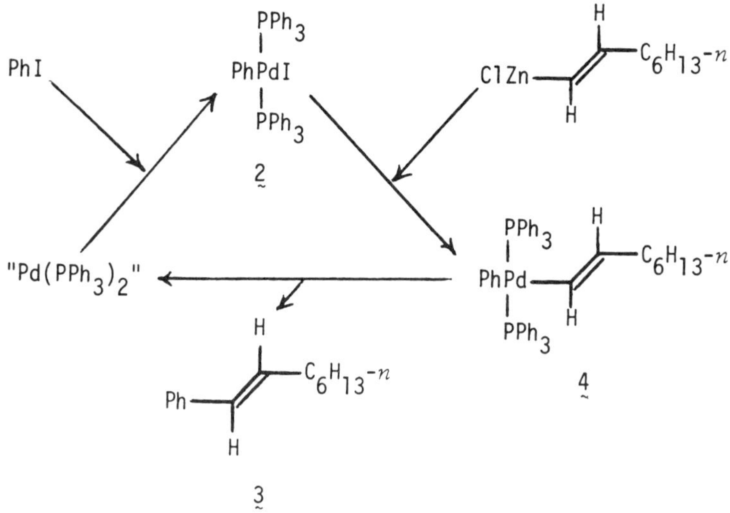

Scheme 2

to the mixture obtained by the reaction of **2** with 1.2 equiv of (E)-1-octenylzinc chloride gives **2** in 98% yield within several minutes at 0°C. Thus, "$Pd(PPh_3)_2$" appears to be the actual catalyst undergoing oxidative addition with PhI under the catalytic conditions, and the feasibility of the three steps that the presumed catalytic cycle consists of is clearly demonstrated (Scheme 2).

To further characterize "$Pd(PPh_3)_2$", the product mixture obtained above has been examined by ^{31}P NMR, which shows a singlet at 23.32 ppm relative to 85% H_3PO_4. Addition of one or two equivalents of PPh_3 cleanly converts the phosphorus-containing species into known $Pd(PPh_3)_3$[47] and $Pd(PPh_3)_4$,[48] which show ^{31}P NMR singlets at 20.54 and 15.79 ppm, respectively. Similarly, addition of one equivalent of maleic anhydride gives known 1:1 complex of $Pd(PPh_3)_2$ and maleic anhydride[49] showing a ^{31}P NMR signal at 27.75 ppm. Various bis(triorganophosphine)palladiums containing very bulky ligands, such as $P(Bu\text{-}t)_3$, are known and well-characterized,[50] and "$Pd(PPh_3)_2$" has been proposed by us and others.[3b] However, its characterization by NMR or other spectroscopic means does not appear to have been previously reported.

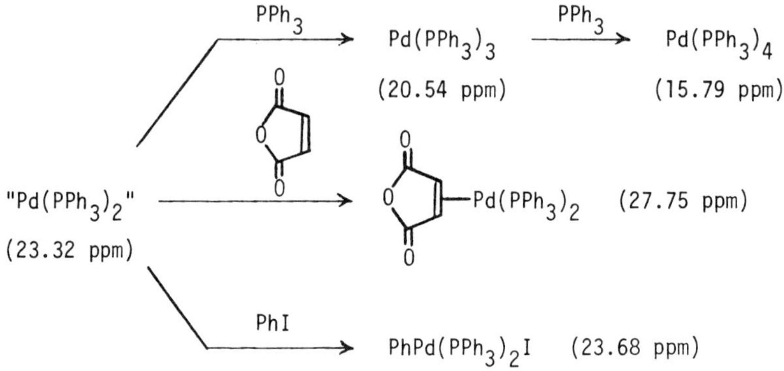

(14)

IV. FACTORS AFFECTING REDUCTIVE ELIMINATION

Although the plausibility of the mechanism represented by Scheme 1 for some representative -catalyzed cross-coupling reactions, e.g., Scheme 2, has now been experimentally supported, a number of fundamental questions still remain to be answered. From the synthetic viewpoint, the following are among some of the most important questions. "Can the Pd- or Ni-catalyzed or -promoted cross coupling be general?" Here, of course, we are not concerned about those cases that involve β elimination of alkylmetals. It is then not unreasonable to believe that essentially all diorganopalladium or -nickel derivatives represented by $R^1R^2Pd(PR_3)_2$ or $R^1R^2Ni(PR_3)_2$, respectively, can be prepared via transmetallation and/or oxidative addition. If so, all that is required is to induce their reductive elimination, leading to the formation of R^1R^2, preferably under catalytic conditions. "What factors affect reductive elimination?" "How?" From a more mechanistic viewpoint, it is of interest to learn further how reductive elimination processes under a wide variety of conditions actually proceed. With these questions and goals in mind, various factors involved in the Pd-or Ni-catalyzed cross coupling, such as phosphine ligands, organometals, and organic halides, have been examined.

A. Comparison of phosphines and carbon groups

To investigate the effects of phosphines and carbon groups on the relative ease of reductive elimination on Pd, we chose the following two types of reactions (Eqs. 15 and 16). As representative phosphines, PPh_3, PPh_2Me, $PPhMe_2$, and PEt_3 were chosen. As organolithiums (R^1Li), (E)-t-BuCH=CHLi, PhLi, and t-BuC≡CLi were chosen. Although of no direct relevance to catalytic processes, MeLi was also chosen for comparison. Reductive elimination of some diorganopalladium species,[41,43,44] especially $Me_2Pd(PR_3)_2$, has been previously studied. However, the currently available data do not permit direct comparisons of various phosphines and carbon groups.

$$Cl_2Pd(PR_3)_2 + 2\ R^1Li \longrightarrow (R^1)_2Pd(PR_3)_2 \longrightarrow R^1R^1 + \text{"}Pd(PR_3)_2\text{"} \qquad (15)$$

$$PhPd(PR_3)_2I + R^1Li \longrightarrow Ph(R^1)Pd(PR_3)_2 \longrightarrow PhR^1 + \text{"}Pd(PR_3)_2\text{"} \qquad (16)$$

The reaction of $Cl_2Pd(PR_3)_2$ with two equivalents of R^1Li or that of $PhPd(PR_3)_2I$ with one equivalent of R^1Li in THF is monitored by ^1H NMR and GLC. Although the reaction mixture is treated with 3N HCl, it is not clear at present if the diorganopalladium intermediates are fully quenched by this procedure. However, in cases where both ^1H NMR and GLC are used as analytical tools, the results obtained with them agree well within experimental errors. Some experimental results are summarized in Table II.

The results summarized in Table II indicate the following. First, the use of PPh_3 leads to the formation of the desired coupled products, R^1R^1 or R^1Ph, in $\geq$ 90% yields within 1 h. Second, the closely parallel results observed with $Cl_2Pd(PPh_3)_2$ on the one hand and $PhPd(PPh_3)_2I$ on the other indicate that the mode of Pd-C bond formation, i.e., whether the two carbon groups to be coupled are introduced both via transmetallation or via an oxidative addition-transmetallation sequence, has little or nothing to do with the thermal and "unassisted" reductive elimination step. Third, the use

TABLE II

Effects of Phosphines and Carbon Groups of R^1Li on the Ease of Reductive Elimination of $R^1_2Pd(PR_3)_2$ and $PhR^1Pd(PR_3)_2$

$Cl_2Pd(PR_3)_2$ or $PhPd(PR_3)_2I$	Time (h)	Yield of product obtained with R^1Li (%)			
		(*E*)-*t*-BuCH=CHLi	PhLi	*t*-BuC≡CLi	MeLi
$Cl_2Pd(PPh_3)_2$	1	95	85-95	>95	>95
$Cl_2Pd(PPh_2Me)_2$	1	90-95	63	<5	15
	3	—	66	5	40 (6h)
	24	—	—	10	90
$Cl_2Pd(PPhMe_2)_2$	1		32		
	3		37		
	24		—		
$Cl_2Pd(PEt_3)_2$	1	<2	8	<2	
	3	<2	9	<2	
	24	—	—	<2	
$PhPd(PPh_3)_2I$	1		97	>95	96[a]
$PhPd(PEt_3)_2I$	1			<2	<5
	3			<2	<5
	24			<2	—

[a]A mixture of toluene (52% yield) and biphenyl (44% of Ph accounted for).

of PEt_3 does not readily induce thermal and "unassisted" reductive elimination. Fourth, the rate of carbon-carbon bond formation via reductive elimination decreases in the order: $PPh_3 > PPh_2Me > PPhMe_2 > PEt_3$. The major factor appears to be their basicity rather than cone angles. The trend is also consistent with a reductive elimination mechanism[43-45] involving dissociation of a phosphine from $R^1R^2Pd(PR_3)_2$ to form $R^1R^2PdPR_3$. Fifth, the order of reactivity of the four carbon groups toward reductive elimination appears to be: (E)-t-BuCH=CH >Ph >Me >t-BuC ≡C. In a

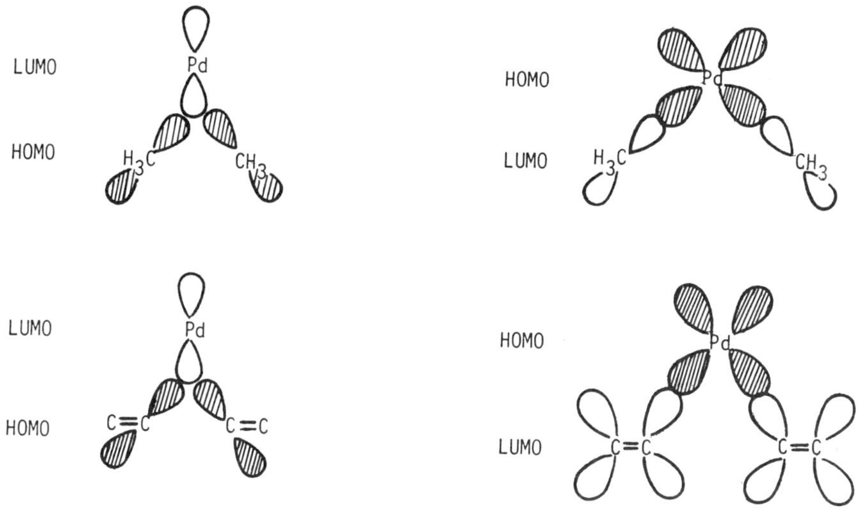

Scheme 3

concerted reductive elimination process, the HOMO-LUMO interactions shown in Scheme 3 may be important. If so, the reactivity difference between the $C_{sp}2$ groups, i.e., alkenyl and aryl, and the Me group, a $C_{sp}3$ group, may well be attributable to the relative LUMO levels of these carbon groups. However, this alone cannot explain the apparently lower reactivity of the t-BuC ≡C group relative to Me. Examination of (t-BuC ≡C)$_2$Pd(PPh$_2$Me)$_2$ and Me$_2$Pd(PPh$_2$Me)$_2$ by ^{1}H NMR indicates that the former is trans, whereas the latter is cis. It is therefore possible that the slow reductive elimination of the former is due to its slow trans-to-cis isomerization, whereas that of the latter is due to its slow coupling process. Clarification of this point, however, requires additional investigations.

B. Acceleration of reductive elimination by added reagents

So far, reductive elimination of $R^1R^2Pd(PR_3)_2$ (**5**) in the absence of any reagents other than LiCl and THF-pentane has been discussed. Under catalytic conditions, however, reductive elimination of **5** takes place in the presence of some other reagents, such as R^1M, R^2X, and free phosphines. It is therefore necessary to study the effects of these reagents on reductive elimination of **5**. Prior to our study, Kochi[42,51] reported accelerating effects of MeLi and aryl halides on reductive elimination of p-Tol(Me)Ni(PEt_3)$_2$ (**6**). The accelerating effect of MeLi is thought to involve nickelate formation, which may facilitate the trans-to-cis isomerization of **6**. On the other hand, oxidation of **5** and **6** by organic halides to give Pd(IV)[43] and Ni(III)[42] species, respectively, appears to cause acceleration of reductive elimination of **5** and **6**.

We have found that an excess of an organolithium can either accelerate or retard reductive elimination of **5**. In the reaction of PhLi with $Cl_2Pd(PR_3)_2$, where PR_3 is PPh_2Me, $PPhMe_2$, or PEt_3, but not PPh_3, the rate of biphenyl formation increases by increasing the ratio of PhLi to $Cl_2Pd(PR_3)_2$ from 2 to 4, as indicated by the results shown in Eq. 17. An

$$Cl_2Pd(PR_3)_2 \xrightarrow{n\ \text{PhLi}} \text{PhPh} + \text{PdL}_n \qquad (17)$$

PR_3	n	Yield (%) 1 h	3 h
PPh_2Me	2	63	66
"	4	97	100
$PPhMe_2$	2	32	37
"	4	77	84
PEt_3	2	8	9
"	4	41	42
"	8	70	72

even more dramatic effect was observed with (E)-t-BuCH=CHLi (**7**). As shown in Table II, the reaction of $Cl_2Pd(PEt_3)_2$ with two equivalents of **7** merely produces [(E)-t-BuCH=CH]$_2$Pd(PEt_3)$_2$ in essentially quantitative yield, the amount of the desired coupling product, i.e., (E,E)-t-BuCH=CHCH=CHBu-t (**8**), being no more than a trace, if any, even after 24 h at room temperature. On the other hand, addition of two equivalents of **7** to the above-mentioned reaction mixture induces formation of **8** in 87 and

97% yields in 0.5 and 1 h, respectively. Essentially the same results are obtained by treating $Cl_2Pd(PEt_3)_2$ with four equivalents of **7** (Eq. 18).

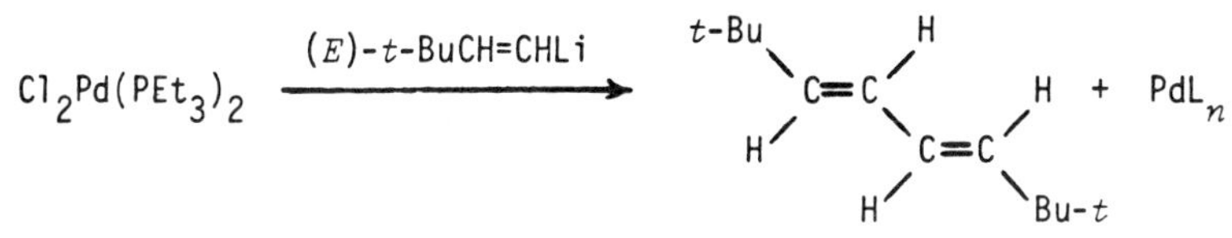

(18)

Equivalent	Yield (%) 0.5 h	1 h	24 h
2	<2	<2	<2
4 (or 2+2)	87	97	–

Although acceleration of reductive elimination by R^1M or R^2X does provide an attractive means of promoting transition metal-assisted carbon-carbon bond formation, its usefulness in Pd- or Ni-catalyzed cross coupling remains to be established. One potential difficulty is cross-home scrambling. If reductive elimination of $R^1R^2Pd(PR_3)_2$ or $R^1R^2Ni(PR_3)_2$ promoted by R^1M or R^2X were to proceed via 5- or 6-coordinate intermediates, such as **9** and **10**, a mixture consisting of the desired R^1R^2 and either R^1R^1 or R^2R^2 will be obtained in most cases. The high cross/homo ratios that we have observed in a large number of cases[3b] cast a serious doubt about the plausibility of any mechanism involving such intermediates for such cases. This matter is under investigation in our laboratories.

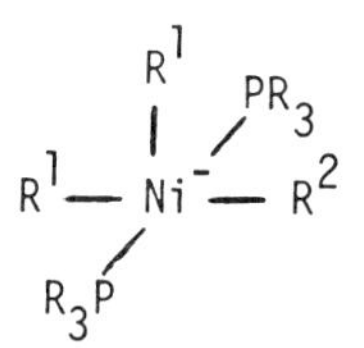

9

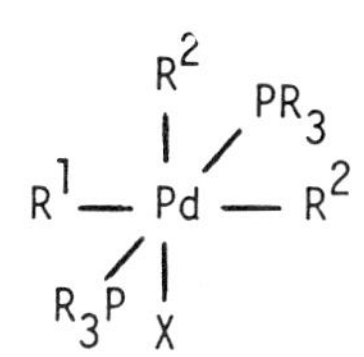

10

C. Inhibition or retardation of reductive elimination via "ate" complexation

One of the synthetically important findings in our investigation is that the cross-coupling reaction catalyzed by a Pd-PPh_3 complex is often very sluggish and/or low yielding, when organometals containing highly electropositive metals, e.g., Li, are used. Typically, for example, the reaction of t-BuC≡CLi with o-tolyl iodide in THF-pentane in the presence of 5 mole % of $Pd(PPh_3)_4$ at 22°C gives only a 5-10% yield of t-BuC≡CTol-o (**11**) even after 24 h, while the corresponding reaction of t-BuC≡ZnCl under the same conditions provides **11** in 85-95% yield within 3 h. Although less pronounced, the corresponding reactions of PhI also shows a similar countercation effect, i.e., 20 and 90% yield of t-BuC≡CPh (**12**) for Li and ZnCl, respectively.

As described earlier, the reaction of $PhPd(PPh_3)_2I$ (**2**) with one equivalent of either t-BuC≡CLi or t-BuC≡CZnCl gives **12** in >95% yield within 1 h, indicating that the intrinsic reactivity of t-BuC≡CLi toward **2** might not be lower than that of t-BuC≡CZnCl. On the other hand, the reaction of **2** was eight equivalents of t-BuC≡CLi first at -78°C and then at 22°C gives only a 20% yield of **12** after 1 h. Interestingly, essentially all PPh_3 was dissociated from Pd during this period. Although formation of $Li_2[PhPd(C{\equiv}CBu\text{-}t)_3]$ was suggested by ^1H NMR, further clarification of its structure was hampered by the complexity of NMR spectra as well as their gradual change with time. In sharp contrast to the Li reaction, the reaction of **2** with eight equivalents of t-BuC≡CZnCl does not show any sign of retardation. Nor does it show any NMR indication of an extensive dissociation of PPh_3.

Both inhibition of reductive elimination and formation of Pd(II) complexes of the Li_2PdR_4 type are clearly shown in the reaction of $Cl_2Pd(PPh_3)_2$ with t-BuC≡CLi. When two equivalents of t-BuC≡CLi are used, the reaction gives t-BuC≡CC≡CBu-t (**13**) in >95% yield within 1 h, the Pd-containing byproduct being "$Pd(PPh_3)_2$". On the other hand, the use of eight equivalents of t-BuC≡CLi produces $Li_2Pd(C{\equiv}Bu\text{-}t)_4$ in >95% yield along with 2 equivalents of PPh_3 and 4 equivalents of unreacted t-BuC≡CLi. Only a trace of **13** is detectable (Eq. 19). An analytically homogeneous

$$Cl_2Pd(PPh_3)_2 \xrightarrow{2\ t\text{-}BuC{\equiv}CLi} t\text{-}BuC{\equiv}CC{\equiv}CBu\text{-}t\ (\mathbf{13},\ >95\%) + \text{"}Pd(PPh_3)_2\text{"} + 2\ LiCl$$

$$Cl_2Pd(PPh_3)_2 \xrightarrow{8\ t\text{-}BuC{\equiv}CLi} Li_2Pd(C{\equiv}CBu\text{-}t)_4\ (>95\%) + 4\ t\text{-}BuC{\equiv}CLi + 2\ LiCl + 2\ PPh_3 \quad (19)$$

sample of $Li_2Pd(C{\equiv}CBu\text{-}t)_4$ can be prepared by the reaction of Li_2PdCl_4 with four equivalents of t-BuC≡CLi: ^1H NMR (THF) δ 1.04 (s) ppm; ^{13}C NMR (THF) δ 29.81, 33.42, 95.92, 115.36 ppm; IR (THF) 2076 cm^{-1}.

The dual nature of the mechanism of the Pd-catalyzed cross coupling reaction may be summarized as in Table III. We tentatively conclude that reductive elimination of $R_2Pd(PPh_3)_2$ involves dissociation of PPh_3 to give a three-coordinate complex, R_2PdPPh_3. It is not clear whether its reductive elimination is assisted by any reagents present in the reaction mixture. However, an organometallic reagent containing Li can inhibit its reductive elimination through formation of Li_2PdR_4. On the other hand, reductive

TABLE III

Some Notable Features of Coupling Reactions Catalyzed by Palladium-Phosphines

Feature	Phosphine: PPh_3	Phosphine: PEt_3
Rate of the stoichiometric reaction	fast	slow
Effect of an excess of RLi	• inhibition or retardation of reductive elimination	• acceleration of reductive elimination
	• dissociation of PPh_3	
	• formation of $Li_2Pd(II)R_4$	• formation of $Li_2Pd(0)R_2(PR'_3)_2$ (?)

elimination of $R_2Pd(PEt_3)_2$ is slow, presumably because the higher basicity of PEt_3 relative to that of PPh_3 does not permit its facile dissociation. However, reductive elimination of $R_2Pd(PEt_3)_2$ can be induced by the action of RLi, which appears to convert $R_2Pd(PEt_3)_2$ into $Li[R_3Pd(PEt_3)_2]$. Mechanistic details of these ractions are being further investigated.

ACKNOWLEDGMENTS

Our investigations of the Pd- or Ni-catalyzed cross-coupling reaction have been mainly supported by the National Science Foundation, the National Institutes of Health, and the etroleum Research Fund, administered by the American Chemical Society. We also acknowledge Engelhard Industries for a loan of palladium compounds. We are deeply indebted to a large number of coworkers whose names appear in our papers cited as references.

REFERENCES

1. For a review, see E. Negishi, Organometallics in Organic Synthesis, **Vol. 1**, Wiley-Interscience, New York, 1980.
2. G.H. Posner, Org. React., **22** 253 (1975).
3. (a) M. Kumada, Pure Appl. Chem., **52**, 669 (1980). (b) E. Negishi, Acc. Chem. Res., **15**, 340 (1982).
4. M.S. Kharasch and O. Reinmuth, **Grignard Reactions of Nonmetallic Substances**, Prentice-Hall, New York, 1954, Chapters V and XVI.
5. K. Tamao, K. Sumitani, and M. Kumada, J. Am. Chem. Soc., **94**, 4374 (1972).
6. R.J.P. Corriu and J.P. Masse, J. Chem. Soc., Chem. Comm., 144 (1972).
7. M. Yamamura, I. Moritani, and S.I. Murahashi, J. Organomet. Chem., **91**, C39 (1975).
8. L. Cassar, J. Organomet. Chem., **93**, 253 (1975).
9. (a) E. Negishi and S. Baba, J. Chem. Soc., Chem. Comm., 596 (1976). (b) S. Baba and E. Negishi, J. Am. Chem. Soc., **98**, 6729 (1976).
10. J.F. Fauvarque and A. Jutand, Bull. Soc. Chim. Fr., 765 (1976).

11. A. Sekiya and N. Ishikawa, J. Organomet. Chem., **118** 349 (1976).

12. Unpublished results.

13. J.K. Stille and K.S.Y. Lau Acc. Chem. Res., **10**, 434 (1977).

14. M. Sato, K. Taki, K. Oshima, and H. Nozaki, Tetrahedron Lett., **22**, 1609 (1981).

15. W.J. Scott, G.T. Crisp, and J.K. Stille, J. Am. Chem. Soc., **106**, 4630 (1984).

16. (a) A.O. King. N. Okukado, and E. Negishi, J. Chem. Soc., Chem. Comm., 683 (1977). (b) A.O. King. E. Negishi, F.J. Villani, Jr., and A. Silveira, Jr., J. Org. Chem., 93, 358 (1978). (c) E. Negishi in **Aspects of Mechanism and Organometallic Chemistry**, J.H. Brewster, ed., Plenum, New York, 1978, p. 285. (d) E. Negishi, F.T. Luo, R. Frisbee, and H. Matsuishita, Heterocycles, **18**, 117 (1982). (e) E. Negishi and F.T. Luo, J. Org. Chem. **48** 1560, (1983).

17. E. Negishi, N. Okukado, A.O. King, D.E. Van Horn, and B.I. Spiegel, J. Am. Chem. Soc., **100**, 2254 (1978).

18. N. Jabri, A. Alexakis, and J.F. Normant, Tetrahedron Lett., **22**, 959, 3851 (1981).

19. A. Suzuki, Acc. Chem. Res., **15**, 178 (1982).

20. K. Sonogashira, Y. Tohda, and N. Hagihara, Tetrahedron Lett., 4467 (1975).

21. H.P. Dang and G. Linstrumelle, Tetrahedron Lett., 191 (1978).

22. N. Okukado, D.E. Van Horn, W.I. Klima, and E. Negishi, Tetrahedron Lett., 1027 (1978).

23. E. Negishi, A.O. King, and N. Okukado, J. Org. Chem., **42**, 1821 (1977).

24. E.R. Larson and R.A. Raphael, Tetrahedron Lett., 5041 (1979).

25. P. Cadiot and W. Chodkiewicz in **Chemistry of Acetylenes**, H.G. Viehe, ed., Marcel Dekker, New York, 1969, Chapter 9.

26. E. Negishi, N. Okukado, S.F. Lovich, and F.T. Luo, J. Org. Chem., **49** 2629 (1984).

27. (a) H. Matsushita and E. Negishi, J. Am. Chem. Soc., **103**, 2882 (1981). (b) E. Negishi, S. Chatterjee, and H. Matsushita, Tetrahedron Lett., **22**, 3737 (1981). (c) H. Matsushita and E. Negishi, J. Chem. Soc., Chem. Comm., 160 (1982).

28. N. Miyaura, T. Yano, and A. Suzuki, Tetrahedron Lett., **21**, 2865 (1980).

29. J.S. Temple and J. Schwartz, J. Am. Chem. Soc., **102**, 7381 (1980).

30. M.F. Semmelhack, Org. React., **19**, 115 (1975).

31. (a) E. Negishi, L.F. Valente, and M. Kobayashi, J. Am. Chem. Soc., **102**, 3298 (1980). (b) M. Kobayashi and E. Negishi, J. Org. Chem., **45**, 5223 (1980). (c) E. Negishi, F.T. Luo, and C.L. Rand, Tetrahedron Lett., **23**, 27 (1982). (d) E. Negishi, H. Matsushita, M. Kobayashi, and C.L. Rand, Tetrahedron Lett., **24**, 3823 (1983).

32. (a) T. Hayashi, M. Konishi, and M. Kumada, Tetrahedron Lett., 1871 (1979). (b) K. Tamao, T. Hayashi, H. Matsumoto, H. Yamamoto, and M. Kumada, Tetrahedron Lett., 2155 (1979). (c) T. Hayashi, M. Konishi, K. Yokota, and M. Kumada, Chem. Lett., 767 (1980). (d) T. Hayashi, N. Nagashima, and M. Kumada, Tetrahedron Lett., **21**, 4623 (1980).

33. K. Takagi, T. Okamoto, Y. Sakakibara, A. Ohno, S. Oka, N. Hayama, Chem. Lett., **951**, (1975).

34. (a) D. Milstein and J.K. Stille, J. Am. Chem. Soc., **100**, 3636 (1978); J. Org. Chem., **44** 1613 (1979). (b) M.W. Logue and K. Teng, J. Org. Chem., **47**, 2549 (1982).

35. E. Negishi, V. Bagheri, S. Chatterjee, F.T. Luo, J.A. Miller, and A.T. Stoll, Tetrahedron Lett., **24**, 5181 (1983).

36. B.M. Trost, Acc. Chem. Res., **13**, 385 (1980).

37. (a) E. Negishi, H. Matsushita, S. Chatterjee, and R.A. John, J. Org. Chem., **47**, 3188 (1982). (b) E. Negishi, F.T. Luo, A.J. Pecora, and A. Silveira, Jr. J. Org. Chem., **48** 2427 (1983). (c) E. Negishi and R.A. John, J. Org. Chem., **48**, 4098 (1983). (d) F.T. Luo and E. Negishi, Tetrahedron Lett., **26**, 2177 (1985). (e) F.T. Luo and E. Negishi, J. Org. Chem., **50**, 4762 (1985).

38. (a) M.F. Semmelhack, B.P. Chong, R.P. Stauffer, T.D. Rogerson, A. Chong, and L.D. Jones, J. Am. Chem. Soc., **97**, 2507 (1975). (b) J.F. Fauvarque and A. Jutand, J. Organometal. Chem., **132**, C17 (1977). (c) I. Kuwajima and H. Urabe, J. Am. Chem. Soc., **104**, 6831 (1982).

39. K. Tamao, M. Zembayashi, and M. Kumada, Chem. Lett., 1237, 1239 (1976).

40. E. Negishi in **Current Trends in Organic Synthesis**, H. Nozaki, ed., Pergamon, Oxford, 1983, p. 269.
41. G.W. Parshall, J. Am. Chem. Soc., **96**, 2360 (1974).
42. G. Smith and J.K. Kochi, J. Organomet. Chem., **198**, 199 (1980).
43. (a) D. Milstein and J.K. Stille, J. Am. Chem. Soc., **101**, 4981, 4992 (1979). (b) A. Gillie and J.K. Stille, J. Am. Chem. Soc., **102**, 4933, (1980). (c) M.K. Loar and J.K. Stille, J. Am. Chem. Soc. **103**, 4174 (1981). (d) A. Moraviskiy and J.K. Stille, J. Am. Chem. Soc., **103**, 4182 (1981).
44. F. Ozawa, T. Ito, Y. Nakamura, and A. Yamamoto, Bull. Chem. Soc. Jpn., **54** 1868 (1981).
45. (a) K. Tatsumi, R. Hoffmann, A. Yamamoto, and J.K. Stille, Bull. Chem. Soc., Jpn., **54** 1857 (1981). (b) K. Tatsumi, A. Nakamura, S. Komiya, A. Yamamoto, and T. Yamamoto, J. Am. Chem. Soc., **106**, 8181 (1984).
46. (a) D.R. Fahey, J. Am. Chem. Soc., **92**, 402 (1970). (b) D.R. Fahey and J.E. Mahan, J. Am. Chem. Soc., **99**, 2501 (1977). (c) M. Hidai, T. Kashiwagi, T. Ikeuchi, and Y. Uchida, J. Organomet. Chem., **30**, 279 (1971). (d) P. Fitton, M.P. Johnson, and J.E. McKeon, J. Chem. Soc., 6 (1968). (e) P. Fitton and E.A. Rick, J. Organomet. Chem., **28**, 287 (1971).
47. B.E. Mawn and A. Musco, J. Chem. Soc. Dalton, 1673 (1975).
48. L. Malatesta and M. Angoletta, J. Chem. Soc., 1186 (1957).
49. S. Takahashi and N. Hagihara, Nippon Kagaku Zasshi (Japanese), **88**, 1306 (1967).
50. (a) R. van der Linde and R.O. der Jongh, J. Chem. Soc. D, 563 (1971). (b) K. Kudo, M. Hidai, and Y. Uchida, J. Organomet. Chem., **56**, 413 (1973). (c) A. Kusco, W. Kuran, A. Silvani, and M.W. Anker, J. Chem. Soc., Chem. Comm., 938 (1973). (d) S. Otsuka, T. Yoshida, M. Matsumoto, and K. Nakatsu, J. Am. Chem. Soc., **98**, 5850 (1976).
51. D.G. Morrell and J.K. Kochi, J. Sm. Chem. Soc. **97**, 7262 (1975)

20
Hydrogenation and Isomerization of (+)-Apopinene on Pd Catalysts

G.V. Smith*,≠ D. Ostgard*,
M. Bartok+, and F. Notheisz+

*Department of Chemistry and ≠Molecular Science Program
Southern Illinois University at Carbondale
Carbondale, IL
+Department of Organic Chemistry
Jozsef Attila University
Szeged, Hungary

ABSTRACT

A series of fourteen supported palladium catalysts ranging in percent exposed palladium atoms (percent dispersion, %D) from 16.7% to 100% have been examined with the hydrogenation reaction of the complex molecular probe (+)-apopinene, (6,6-dimethyl-lR,5R-bicyclo[3.1.1]hept-2-ene). The ratio of isomerization to addition, k_i/k_a, goes through a maximum at approximately 60%D, a maximum which is found also for supported Pt catalysts. This maximum occurs approximately at the maximum in the number of edge sites for fcc cubo-octahedral crystallites as a function of %D. Based on the data a scale of percent edge character is derived.

I. INTRODUCTION

We have been studying the surface structure of a series of Pd catalysts using the hydrogenation and isomerization of (+)-apopinene as a diagnostic chemical reaction.[1,2] This unique molecular probe is extremely sensitive to different preparations of supported and unsupported catalysts. Recently we characterized a series of platinum catalysts with this molecular probe.[3] In that study we measured the effect or percent dispersion (%D) on the rates of hydrogenation with deuterium and on the ratios of isomerization to addition, k_i/k_a. This ratio goes through a maximum at about 65%D and we have suggested that it is due to a maximum in edge sites, which are mainly responsible for isomerization. In this study we examine a series of palladium catalysts prepared with different dispersions to determine the effect of dispersion on the ratio of k_i/k_a.

II. EXPERIMENTAL

The apparatus and preparation of (+)-apopinene have been previously described.[1-3] Essentially, the hydrogenations were carried out in the liquid phase without solvent at ambient temperatures and at one atmosphere of deuterium pressure. The shaking rate, >1800/min, was believed to be adequate to essentially eliminate bulk phase diffusion.[2]

The Pd/SiO_2, Pd/Al_2O_3, and Pd/C catalysts were prepared using 80-140 mesh of the following supports: Davison Grade 56 silica gel, Katalco Alumina (82-6931), and Engelhard Carbon (22612). The support was washed five times in triple distilled water and dried at 573 K for 22 hours. Three different methods of preparation were used and all of the catalysts were reduced in the same way. They were reduced under flowing hydrogen (10 mL/min.) at 353 K for one hour, followed by temperatures rising to 391°K with a hold for one hour, and, finally, at 478 K with a hold for 2 hours. The three different methods of catalyst preparation used for this study were the Impregnation method (imp), the Ion Exchange method (ion ex), and the deposition of Palladium Acetylacetonate on the support (AcAc).

For the impregnation method the support was stirred in an aqueous 0.01 N HCl solution containing the requisite amount of $PdCl_2$. The mixture was then stirred for 24 hours, dried under an infrared lamp, heated for 24 hours at 413 K in air, and reduced.

In the ion exchange method the support was stirred in a 1.0 N NH_4OH solution for 24 hours, then, after adding a requisite amount of $PdCl_2$, stirred for another 24 hours. The resulting mixture was filtered, washed repeatedly with triple distilled water, dried at 413 K in air for 24 hours, and reduced.

The filtrate and all the wash solutions were kept for residual Pd analysis. This was done by forming the Pd-EDTA 1:1 complex at a pH of 1.8 and comparing its UV adsorption at a wave length of 377 nm to a set of eight standard solutions.

The palladium acetylacetonate method involved stirring the support in a benzene solution of the required amount of $Pd(AcAc)_2$ for 48 hours, after which the benzene was filtered and kept for residual Pd analysis. The treated support was dried in air at 413 K for 24 hours, dried in air at 573 K (calcined) for 24 hours, and reduced.

The percentage exposed Pd atoms (%D) were calculated for all the catalysts based on their chemisorption of hydrogen and carbon monoxide. The techniques and equipment used in these measurements have been discussed earlier.[3] Additionally, particle size measurements were made by electron microscopy on a Hitachi H500H transmission electron microscope. The preparation and characterization data for the Pd catalysts are shown in Table I.

TABLE I

Preparation and Characterization Data for Catalysts

Catalyst	Pd (wt%)	D %	Method
Pd/SiO_2	1.17	40.4	impregnation
Pd/SiO_2	0.46	60.4	ion exchange
Pd/SiO_2	0.12	100	AcAc
Pd/SiO_2	0.18	77.3	AcAc
Pd/Al_2O_3	1.17	16.7	impregnation
Pd/Al_2O_3	0.49	69.3	ion exchange
Pd/Al_2O_3	0.15	78.6	AcAc
Pd/Al_2O_3	0.29	57.7	AcAc
Pd/Al_2O_3	0.40	77.5	ion exchange
Pd/C	0.15	79.2	impregnation
Pd/C	0.49	53.7	impregnation
Pd/C	1.17	38.4	impregnation
Pd/C	0.39	29.6	impregnation

TABLE II

Results of Hydrogenation of (+)-Apopinene Over Pd/SiO_2, Pd/Al_2O_3, and Pd/C

Catalysts	%Add	%Or	k_i/k_a	Corr. Coef.[a]
Pd/SiO_2	2.8	75.7		
1.17%	4.9	57.6	5.6	0.999
%D=40.4	7.1	44.5		
Pd/SiO_2	2.8	38.4		
0.46%	4.3	18.7	17.5	0.997
%D-60.4	7.1	7.9		
Pd/SiO_2	1.8	93.7		
0.12%	2.2	89.2	2.5	0.973
%D=100	3.9	80.1		
	4.2	82.8		
Pd/SiO_2	3.1	74.2		
0.184%	6.5	57.2	4.2	0.999
%D=77.25	9.7	42.6		
Pd/Al_2O_3	5.3	63.5		
1.17%				
%D=16.7	7.5	43.0	5.2	0.984
Pd/Al_2O_3	4.1	61.6		
0.49%				
%D=69.3	5.6	43.3	6.8	0.975
Pd/Al_2O_3	4.7	65.2		
0.15%	4.9	66.8		
%D=78.6	5.3	58.0	4.7	0.994
	10.0	37.6		
Pd/Al_2O_3	3.3	54.8		
0.29%	7.2	20.8	10.5	0.998
%D=57.7	8.9	14.7		
Pd/Al_2O_3	2.2	68.1		
0.40%	2.3	68.1		
%D=77.5	2.4	66.0	6.8	0.982
	2.6	67.1		
	10.1	18.2		
	10.7	24.7		
Pd/C	9.48	21.95	1.38	0.990
0.15%	20.40	53.06		
%D=79.2				

TABLE II Continued

Catalysts	%Add	%Or	k_i/k_a	Corr. Coef.[a]
Pd/C	2.13	98.09		
0.49%	9.26	30.16	1.34	0.999
%D=53.7	19.47	56.99		
	28.66	10.54		
Pd/C	2.08	82.38		
1.17%	3.34	89.72	1.16	0.960
%D=38.4	9.52	77.13		
	18.25	58.85		
Pd/C	5.06	87.89	1.07	0.999
0.39%	15.86	68.51		
%D=29.6				

[a]correlation coefficient

III. RESULTS

Treatment of the measured data was the same as previously described.[2] The ratio k_i/k_a was calculated by the linear regression method from the equation

$$\ln(1-2\mathrm{Isom}) = [2k_i/k_aP_H]\ln(1-\mathrm{Add})$$

in which Isom is mole fraction apopinene isomerized and Add is mole function apopinene hydrogenated. The calculated data are shown in Table II.

IV. DISCUSSION

When the ratio of k_i/k_a is plotted against %D, the Pd/Al_2O_3, Pd/SiO_2, and Pd/C catalysts fall on different lines but all three pass through maxima at about 50-60 %D. We suggested earlier[3] that this maximum in k_i/k_a is due to a maximum in edge sites (Siegel ^2M sites[4]), which are responsible for isomerization, while both addition and isomerization occur on vertices or adatoms (Siegel ^3M sites[4]). Consequently,

$$k_i/k_a = R = (k_{i1}{}^2M + k_{i2}{}^3M)/k_a{}^3M \quad (1)$$

where k_{i1} = rate constant for isomerization on ^2M sites, k_{i2} = rate constant for isomerization on ^3M sites, and k_a = rate constant for addition.

To establish the minimum in R at 0%D we use the smallest value of approximately 0.5 which was found for a Pd-splat.[2] The 100% dispersed catalyst (0.12 Pd/SiO_2), should contain only ^3M sites, and because R on the Pd-splat is actually lower than R on the 100% dispersed catalyst, we assume that the active sites on the Pd-splat are also ^3M sites. However, the number of vertex sites decrease faster than the number of the edge sites with decreasing dispersion.[5] So, it is unlikely that the ^3M sites on the Pd-splat are vertices. We will assume they are adatoms or vertex-like protuberances.[1]

Therefore, ^3M = a+v, where a = surface concentration of adatom sites and v = surface concentration of vertex sites. On the nondispersed catalysts v = 0, and the smallest value for R is approximately 0.5 (Pd-splat).[2] If we further assume that ^2M = 0 on the splat then it follows that

$$R_{splat} = k_{i2}{}^3M/k_a{}^3M = k_{i2}/k_a = 0.5 \quad (2)$$

and $k_{i2} = 0.5k_a$. Substituting k_a for k_{i2} in equation (1) gives

$$R = (k_{i1}{}^2M + 0.5k_a{}^3M)/k_a{}^3M = (k_{i1}{}^2M/k_a{}^3M) + 0.5 \quad (3)$$

and

$$R-0.5 = k_{i1}{}^2M/k_a{}^3M \quad (4)$$

From Ref. 5 we can calculate the theoretical number of edge sites. The maximum concentration of edge sites as a function of dispersion depends on the crystal structure. The expected maximum values are approximately D=95%, D=71%, and D=50-65% for octahedral, cubic, and cubo-octahedral palladium crystallites respectively. The measured maximum value suggests that the cubo-octahedral structure is probable. For this structure in the vicinity of the maximum the number of vertices approximately equals the number of edge sites; that is $^2M = {}^3M$. Therefore, Eq. (4) becomes

$$R_{max} - 0.5 = k_{i1}/k_a \qquad (5)$$

The largest value for R measured in this work is 17.5 on the 0.46% Pd/SiO_2 catalyst, and we assume that the distribution of active sites on this catalyst is similar to the calculated values for the cubo-octahedral structure. On the basis of these assumptions Eq. (5) reduces to

$$k_{i1}/k_a = 17 \qquad (6)$$

Because the isomerization reaction on the edge sites (^2M) is much faster than both addition and isomerization on the vertex sites (^3M), the basic character of the correlation between the ratio, R, and %D is determined by the surface concentration of ^2M sites. Substituting (6) back into Eq. (4) gives

$$R - 0.5 = 17\,{}^2M/{}^3M$$

and

$$(R - 0.5)/17 = {}^2M/{}^3M \qquad (7)$$

Basically, the adatoms determine the catalytic properties of the Pd-splat, so this catalyst can be regarded as a ^3M-type catalyst. The same is true for the highly dispersed catalysts, for example, 0.12% Pd/SiO_2. On the other hand, the properties of the 0.46% Pd/SiO_2 catalyst are determined mainly by the edge sites, so this one is an edge-type or a ^2M-type catalyst. For comparison purposes we define percent edge character or percent ^2M character as

$$\%\ \text{edge character} = 100(R - 0.5)/17 \qquad (8)$$

Table III shows the calculated values for not only the catalysts used in this study but also those used in the earlier study.[2]

The data reveal a support effect. That is, R values are higher on Pd/SiO_2 than on Pd/C catalysts. The Pd/Al_2O_3 catalysts occupy a middle position. One possible explanation for this is that the real crystal structure

TABLE III

Percent Edge Character of a Series of Pd Catalysts

Catalyst	Pd (wt%)	D%	% Edge Character
Pd/SiO_2	0.46	60.4	100.0
Pd/Al_2O_3	1.0[2]	49.0	60.0
Pd/Al_2O_3	0.29	57.7	58.8
Pd-black	100[2]	1.7	56.5
Pd/Al_2O_3	1.04	22.0	48.2
Pd/Al_2O_3	0.40[2]	77.5	37.1(30.6)[2]
Pd/Al_2O_3	0.49	69.3	37.1
Pd/Al_2O_3	1.00[2]	41.0	33.5
Pd/SiO_2	1.17	40.4	30.0
Pd/Al_2O_3	1.17	16.7	27.6
Pd/$BaSO_4$	5[2]	73.1	25.9
Pd/Al_2O_3	0.15	78.6	24.7
Pd/SiO_2	0.18[4]	77.2	21.8
Pd/Al_2O_3	0.18[2]	82.5	20.6
18-PdO_2	100[2]	—	20.6
12-PdO_2	100[2]	—	17.1
Pd-powder	100[2]	—	17.1
Pd/Al_2O_3	1.11[2]	5.5	16.5
Pd/C	5[2]	12.5	16.5
Pd/Al_2O_3	0.05[2]	86.6	12.4
Pd/Al_2O_3	0.01[2]	103.0	11.8
Pd/SiO_2	0.12	100.0	11.8
6-PdO_2	100[2]	—	11.8
O-PdO_2	100[2]	—	8.2
Pd/C	0.15	79.2	5.2
Pd/C	0.49	53.7	4.9
Pd/C	1.17	38.4	3.9
Pd/C	0.39	29.6	3.4
Pd/foil	100[2]	—	2.7
$Pd_{80}Si_{20}$	80[2]	—	1.1
$Pd_{77}Ge_{23}$	77[2]	—	0.5
Pd-splat	100[2]	—	0.4

of the Pd/SiO_2 catalysts is similar to the theoretical, while the structure of the Pd/Al_2O_3 and, especially, of the Pd/C catalysts differ significantly from theoretically perfect crystallites. Because the Pd/SiO_2 and the Pd/C are the two extreme examples, it does not seem probable that this phenomenon is caused by electronic interactions.

We reported earlier[2] that the ratio measured on the Pd-black was much higher than on the PdO_2 catalyst. As shown in Table III the percent edge character of the Pd-black is also higher than that of the PdO_2 catalyst. It is possible, that the freshly reduced PdO_2 catalyst contains mainly ^3M sites while the Pd-black contains mainly edge sites (^2M). During the surface reconstruction treatment the edge character of the PdO_2 catalyst increases, in good agreement with the expected decrease of the concentration of ^3M sites. A similar phenomenon is observed for the splat cooled Pd-Si alloys.[1]

V. CONCLUSIONS

The hydrogenation reaction of (+)-Apopinene is able to differentiate between different preparations of supported and unsupported palladium catalysts. The selectivity determining structural feature on the palladium surfaces appears to be the edges, ^2M sites.

ACKNOWLEDGEMENTS

We gratefully acknowledge National Science Foundation support through Grant INT-8403357. Also, we are grateful for support from the Department of Energy Division of Materials Sciences through Grant DE-AC02-81ER10897. We thank the Materials Technology Center at SIUC for financial aid as well as critical support services. We thank Engelhard Industries for furnishing the carbon and Katalco for furnishing the alumina.

REFERENCES

1. G.V. Smith, A. Molnar, M.M. Khan, B. Richter and W.E. Brower, J. Catal., **83**, 238 (1983).

2. G.V. Smith, A. Molnar, M.M. Khan, D. Ostgard and N. Yoshida, J. Catal., **98**, 502 (1986).

3. F. Notheisz, M. Bartok, D. Ostgard and G.V. Smith, J. Catal. (in press, 1986).

4. S. Siegel, J. Outlaw, Jr., and N. Garti, J. Catal. **52**, 102 (1978); see also R.L. Augustine, and R.W. Warner, J. Catal. **80**, 358 (1983) for further discussion.

5. R. Hardeveld and F. Hartog, Adv. Catal., **22**, 75 (1972).

21
Activation of Commercially Available Nickel on Alumina Catalyst

Mike G. Scaros, Hugh L. Dryden, Jr., John P. Westrich*,
Owen J. Goodmonson and James R. Pilney

G.D. Searle & Company
Skokie, Illinois

ABSTRACT

Searle's hydrogenation group has developed a process for the activation of a commercially available nickel-on-alumina catalyst. The resulting activated catalyst has catalytic properties similar to Raney nickel but has the distinct advantage of being non-pyrophoric. In addition, the activated catalyst can be employed at pressures as low as 50 psig and temperatures as low as 50°C. The events leading to the catalyst activation and comparisons with non-activated catalyst and Raney nickel will be discussed.

I. INTRODUCTION

When it became evident that a process involving a Raney nickel-catalyzed hydrogenation was likely to be transferred to manufacturing, our group was prompted to find an alternate catalyst to replace Raney nickel due to concern about its pyrophoric properties.

Raney nickel had been selected as the catalyst of choice based on selectivity which could not be equalled by noble metal catalysts.

*Current affiliation: S.C. Johnson and Son, Inc.

A program was initiated to find a catalyst that would be selective, relatively non-pyrophoric, and active at 60 psig of hydrogen and 60°C in order to fit our existing manufacturing hydrogenation vessels. After screening a large number of candidates, one of which was non-pyrophoric nickel boride,[1] we concentrated on nickel-on-alumina E-230P* from Calsicat.

II. DISCUSSION

The hydrogenation of interest was the conversion of an alkyne to an alkane.

$$R\text{-}C\equiv C\text{-}\overset{\displaystyle O}{\overset{\|}{C}}\text{-}OCH_3 \rightarrow R\text{-}CH_2CH_2\overset{\displaystyle O}{\overset{\|}{C}}\text{-}OCH_3$$

The initial experiments with Calsicat E-230P nickel on alumina catalyst were carried out at 50 psig of hydrogen and 50°C. The results were encouraging as the major impurity was alkene (incomplete reduction) with complete retention of the unsaturation contained within the R group. The results obtained, although encouraging, were unacceptable due to the relatively high alkene impurity which exceeded the no more than 0.5% specification (Table I).

The results were encouraging enough for us to concentrate on trying to increase the activity of the catalyst.

We knew that this catalyst was active at higher pressure and temperature and felt that depassivation might enhance the catalytic activity sufficiently in order to obtain complete hydrogenation at 50 psig and 50°C. The following is a description of our attempts to depassivate the catalyst.

The catalyst was pre-reduced under 50 psig of hydrogen at various temperatures for various lengths of time. The alkyne was added and the reduction carried out at 50°C and 50 psig, with the following results Table II).

*Code E-230P is a powdered nickel catalyst containing 58-65% nickel supported on alumina. The catalyst is reduced and passivated.

TABLE I

Hydrogenation of the Alkyne Ester Ni/Al_2O_3

Run	Catalyst[a]	Catalyst Loading	Temp. °C	Press. psig	Induction[b] Time	Reduction[c] Time	% Alkene Impurity[d]
1	Ni/Al_2O_3	15%	50	50	150 min.	180 min.	11.0
2	Ni/Al_2O_3	15%	50	50	155 min.	170 min.	13.0
3	Ni/Al_2O_3	15%	50	50	180 min.	20 hrs.	2.6

[a]Calsicat type E-230P.

[b]Induction Time = Time elapsed after the agitator is turned on to when hydrogen uptake starts.

[c]Reduction Time = Time elapsed after hydrogen uptake starts to when hydrogen uptake ceases.

[d]Determined by Proton Magnetic Resonance

TABLE II

Reaction Conditions for the Hydrogenation of the Alkyne Ester Over Ni/Al_2O_3

Run	Pre-Reduction[a] Conditions	Hydrogenation Conditions Temp.	Hydrogenation Conditions Press.	Induction[b] Time	Reduction[c] Time	% Alkene Impurity
1	3 hrs. 50 psig. 50°C	50°C	50 psig.	165 min.	17.5 hrs.	0.12 (HPLC)
2	20 hrs. 50 psig. 50°C	50°C	50 psig.	60 min.	3.5 hrs.	17.5 (NMR)
3	3 hrs. 50 psig. 90°C	50°C	50 psig.	60 min.	21.0 hrs.	None (NMR)

[a] A 15% catalyst loading was employed.

[b] Amount of time that elapsed after the agitator is turned on to when hydrogen uptake starts.

[c] Amount of time elapsed after hydrogen uptake starts to when hydrogen uptake ceases.

The experiments indicated that pre-reduction under anticipated reaction conditions could activate the catalyst sufficiently to produce alkane of acceptable quality, but the hydrogenation time was too long. Product isolated after 3.5 hours contained an unacceptably high percentage of the alkene impurity. We were not interested in prereducing the catalyst under higher pressure or temperatures, so we turned to chemical reducing agents. Being aware that hydrazine had been used with Raney nickel to effect hydrogenation,[2] we tried to depassivate the catalyst with hydrazine. The nickel-on-alumina catalyst was suspended in a hydrazine/methanol solution, warmed to 50°C, and was used after bubbling ceased. The hydrazine pretreated catalyst produced acceptable quality alkane in 4 hours at 50 psig of hydrogen and 50°C. However, the hydrazine activation was not without problems since it took as long as 60 minutes. In addition, it had a variable onset of severe bubbling and foaming and required venting to release the ammonia, nitrogen, and hydrogen.

In an attempt to get around these problems, other reducing agents were tried. Activation treatments with borohydrides were not as effective as the hydrazine activation. We rationalized that the hydrazine activation contained ammonia from decomposed hydrazine, so we added ammonium hydroxide to a borohydride activation. The catalyst resulting from this potassium borohydride/ammonium hydroxide/methanol pretreatment produced acceptable alkane in 2 hrs at 50 psig and 50°C. Subsequent work focused on optimization of this potassium borohydride/ammonium hydroxide/methanol activation process.

A. Alkyne → Alkane

$$R\text{-}C\equiv C\text{-}\overset{\overset{\displaystyle O}{\|}}{C}\text{-}OCH_3 \rightarrow R\text{-}CH_2CH_2\overset{\overset{\displaystyle O}{\|}}{C}\text{-}OCH_3$$

The optimization resulted in a procedure in which the nickel on alumina catalyst is activated by adding a slurry of potassium borohydride in ammonium hydroxide and methanol to a stirred THF solution of the substrate and suspended Ni/Al_2O_3 catalyst. After 1-2 minutes, bubbling occurs as the potassium borohydride undergoes decomposition by the nickel catalyst. After another 3-4 minutes, the bubbling subsides and the reaction

TABLE III

Run	Catalyst[a]	Catalyst Loading	Time hrs.	% Alkane[b] (HPLC)	% Alkene Impurity	Induction[c] Period	Activation Reagents
1	E-230P	15%	2.0	89.0	0.2	None	$KBH_4/NH_4OH/MeOH$
2	E-230P	15%	4.0	91	0.2	75 min.	MeOH
3	E-230P	15%	4.0	88.3	0.2	80 min.	$KBH_4/MeOH$
4	E-230P	15%	4.0	84.0	3.6	None	$NH_4OH/MeOH$
5	E-230P	15%	3.25	--	0.2	40 min.	Externally Activated[e] $KBH_4/NH_4OH/MeOH$
6	Raney Ni	15%	2.5	91.4	0.2	None	—
7	E-230P	14%	2.0	85.8	0.2	None	Externally Activated[d] $KBH_4/NH_4OH/MeOH$ - Aged 1 Week
8	E-230P	14%	2.0	90.1	0.2	None	Externally Activated[d] $KBH_4/NH_4OH/MeOH$ - Aged 3 Weeks
9	E-230P	15%	2.0	86.5	0.2	None	Externally Activated[d] $KBH_4/NH_4OH/MeOH$ - Aged 13 Weeks
10	E-480P	10%	2.0	91.0	0.2	None	$KBH_4/NH_4OH/MeOH$
11	E-5132P	10%	2.0	91	0.2	None	$KBH_4/NH_4OH/MeOH$

vessel is purged with nitrogen followed by hydrogen. The hydrogenation is carried out at 50 psig and 50°C and takes about 3-4 hours (see experimental for details).

The laboratory work listed in Table III describes the results of the experiments performed in determining the optimal conditions for the Ni/Al_2O_3 activation and the comparison to Raney nickel.*

Run No. 1 which was activated with all three reagents started immediately and produced acceptable quality alkane. Runs No. 2, 3 and 4 where one or two reagents were used had 75-80 minute induction times or the percent alkene impurity exceeded the no more than 0.5% specification. However, when compared to Table I, Run No. 1, the induction time was shortened by either one or two of the activation reagents. Run No. 5 used freshly made externally activated catalyst which was washed to remove the activating reagents. The 40-minute induction time in Run No. 5 can be eliminated by adding methanol to the reaction mixture as can be seen in Runs No. 7-9 which used externally activated and aged catalysts. Runs #10 and 11 used newer, more active catalysts at a 10% loading.

A striking feature of this data is the profound effect of all three activating reagents and the importance of the presence of methanol in the reaction mixture. When methanol was omitted from hydrogenations with non-preactivated catalyst, Table I, Run No. 3, a three-hour induction time occurred and even after 20 hours the residual alkene was 2.6%. Other alcohols (i.e., 2-methoxyethanol, ethanol, propanol, and isopropanol) can be used but longer hydrogenation times or higher loadings are required.

The catalyst can be externally activated and stored for up to 12 weeks in water or 2-methoxyethanol. The aged catalyst is equivalent in activity to freshly activated catalyst. If activated catalyst is exposed to the air, a warming occurs, and the catalyst is rendered inactive. Both the externally activated Ni/Al_2O_3 and the in situ activated Ni/Al_2O_3 have been found to be non-pyrophoric. Searle has employed this catalyst for several years to

*After the process was optimized a thorough literature search uncovered a patent[3] where deactivated nonpyrophoric nickel, iron, and cobalt catalysts could be reactivated to pyrophoric catalysts.

produce thousands of tons of alkane. During that time no ignition or sparking has occurred.

B. Nitrile → Amine

The second system involves the reduction of 1,4-dicyanobutane to 1,6-hexanediamine. In this case three runs were made in order to compare activated Ni/Al_2O_3 and partially activated Ni/Al_2O_3 with W-2 Raney nickel (Table IV).

$$NC(CH_2)_4CN \rightarrow H_2N(CH_2)_6NH_2$$

Dicyanobutane 1,6-Hexanediamine

As can be seen from the above data, the yield of diamine with the activated catalyst is essentially the same as that with Raney nickel, 94% vs 97%. However, when the Ni/Al_2O_3 is treated only with methanol and ammonium hydroxide, the yield of diamine drops to about 51%. No further work was done on this particular system.

TABLE IV

Run	Catalyst	% Loading	Activation	Reduction Conditions: Press.	Reduction Conditions: Temp.	Reduction Conditions: Time	% Diamine
1	Raney Ni	22	—	50 psi	50°C	7 hrs	97.3
2	Ni/Al_2O_3	17.6	Reagents $KBH_4/NH_4OH/MeOH$	50 psi	50°C	10 hrs.	94.4
3	Ni/Al_2O_3[a]	17.6	Reagents $NH_4OH/MeOH$	50 psi	50°C	23 hrs.	51.3

III. EXPERIMENTAL

A. Hydrogenation of the ethynyl carboxylate employing the $KBH_4/NH_4OH/MeOH$ activated Ni/Al_2O_3.

$$R{-}C{=}C{-}\overset{\overset{O}{\|}}{C}{-}OCH_3 \rightarrow R{-}CH_2CH_2\overset{\overset{O}{\|}}{C}{-}OCH_3$$

Ethynyl carboxylate (0.25 moles) in tetrahydrofuran solution from the previous step is placed in a 1.0 L bomb which contains 15.5 g of the Ni/Al_2O_3 catalyst. The flask is then rinsed with 46 ml of tetrahydrofuran. A slurry is prepared by adding a mixture of 50 ml methanol and 9.8 ml of ammonium hydroxide to 0.70 g of potassium borohydride. This slurry is added to the contents of the bomb, while the contents are being stirred. Bubbling will occur in approximately 1 minute as the potassium borohydride begins to decompose. After 3-4 minutes the bubbling ceases. The bomb is sealed and flushed three times with nitrogen and two times with hydrogen. The bomb is then pressurized to 50 psig and the reduction is started allowing the temperature to increase to 50°C. The temperature is then maintained at 50-55°C during the course of the hydrogenation. Hydrogen uptake ceases after about 2 hours and the bomb is stirred for an additional 45 minutes. After venting and nitrogen purging, the bomb is opened and the catalyst removed by filtration at 45-50°C. The ethyl carboxylate is isolated and assayed by HPLC.

B. External activation of Ni/Al_2O_3 employing $KBH_4/NH_4OH/MeOH$

200 g of Ni/Al_2O_3 is stirred in 500 ml of solvent. A slurry of 9 g of KBH_4/126 ml of NH_4OH/400 ml of MeOH is added quickly. The slurry bubbles vigorously and the temperature rises to 40-45°C. Stirring is continued until bubbling subsides (10-15 mins). The activated Ni/Al_2O_3 (a pasty sludge which filters slowly) is filtered to remove the activating reagents and washed with 250 ml of solvent. Care must be taken during the filtration to keep the catalyst covered with solvent thus preventing exposure of the catalyst to oxygen which renders the catalyst inactive. The catalyst

is then stored under the desired solvent. Solvents which can be used include water, tetrahydrofuran, ethyl acetate and 2-methoxyethanol.

C. Hydrogenation of 2,4-dicyanobutane employing the $KBH_4/NH_4OH/MeOH$ activated Ni/Al_2O_3

$$N\equiv C(CH_2)_4C\equiv N \rightarrow H_2N(CH_2)_6NH_2$$

5.0 g of Ni/Al_2O_3 was placed in a reaction vessel containing a solution of 0.25 moles of 1,4-dicyanobutane in 125 ml tetrahydrofuran. A slurry of .22 g KBH_4, 7.5 ml methanol, and 100 ml NH_4OH was added to the stirring reaction vessel. The bubbling subsided after 3-4 minutes and the reaction vessel was sealed and flushed 3 times with nitrogen and twice with hydrogen. This vessel was pressurized to 50 psig and reduction began while warming vessel to 50°C. Reduction proceeded for 18 hours at which time the vessel was cooled to room temperature and vented. The catalyst was removed by filtration and solvent removed under vacuum. Quantitative gas chromatography indicated 94.4% of desired 1,6-hexanediamine.

IV. CONCLUSIONS

We feel that the data discussed here clearly indicates that the activated Ni/Al_2O_3, either when activated and used in situ, when externally activated and used freshly, or even when activated and stored, is superior to unactivated Ni/Al_2O_3 catalyst. In our hands, it has never produced any sparks or ignited any solvent, even when air has been drawn through the catalyst when it is wet with methanol. However, when the catalyst is exposed to air after filtration a warming occurs in which the temperature reaches a high of approximately 44°C for a couple of minutes, but again, no ignition has been noted during the years Searle has worked with the activated catalyst. For laboratory use the activated Ni/Al_2O_3 appears to have little advantage over Raney nickel since Raney nickel can be handled easily in the laboratory by one skilled in catalysis. However, for production scale, particularly in conjunction with organic solvents, the non-pyrophoricity and the mild conditions can be advantageous.

ACKNOWLEDGMENTS

The authors acknowledge Gatis Plume, Thomas Telinski and Kathy McLaughlin for supplying the various substrates. We also wish to extend our appreciation to John Damascus and staff for the spectral data, to Tom Davey and his staff for the HPLC work, and to Dr. John Witt and Dr. Chuck Ellefson for their encouragement and support. The authors are also indebted to Ann Worlatschek for her diligent and conscientious help in the preparation of this manuscript and to G.D. Searle and Company for permission to publish this work.

REFERENCES

1. R. Paul, P. Buisson, and N. Joseph, Ind. Eng. Chem., **44**, 1006 (1952)
2. D. Balcome and A. Furst, J. Am. Chem. Soc., **75**, 4334 (1953).
3. M. Jung and H.H. von Doehren, U.S. Patent 3,560,404 (1971).

Index

I